ELLIPSOIDAL FIGURES
OF EQUILIBRIUM

ELLIPSOIDAL FIGURES OF EQUILIBRIUM

S. Chandrasekhar

Morton D. Hull Distinguished Service Professor Emeritus
The University of Chicago

Dover Publications, Inc., New York

Published in Canada by General Publishing Company, Ltd., 30 Lesmill Road, Don Mills, Toronto, Ontario.
Published in the United Kingdom by Constable and Company, Ltd., 10 Orange Street, London WC2H 7EG.

This Dover edition, first published in 1987, is an unabridged, partly revised, and slightly corrected republication of the work first published by the Yale University Press, New Haven, Conn., 1969, as volume 42 in The Silliman Foundation Lectures series. The author has revised Chapter 8, Section 63, for this edition.

Manufactured in the United States of America
Dover Publications, Inc., 31 East 2nd Street, Mineola, N.Y. 11501

Library of Congress Cataloging-in-Publication Data

Chandrasekhar, S. (Subrahmanyan), 1910–
  Ellipsoidal figures of equilibrium.

  "An unabridged, partly revised, and slightly corrected republication of the work first published by the Yale University Press . . . 1969"—T.p. verso.
  Bibliography: p.
  Includes indexes.
  1. Rotating masses of fluid.   2. Ellipsoid.
I. Title.
QB410.C47   1987   523.01   86-13411
ISBN 0-486-65258-0

CONTENTS

There is a square; there is an oblong. The players take
the square and place it upon the oblong. They place it
very accurately; they make a perfect dwelling place.
Very little is left outside. The structure is now visible;
what was inchoate is here stated; we are not so various
or so mean; we have made oblongs and stood them upon
squares. This is our triumph; this is our consolation.

—Virginia Woolf

PREFACE

WHEN I was asked early in 1962 to give the Silliman Memorial Lectures for 1963, I was, together with Norman R. Lebovitz, in the midst of a re-evaluation of the classical work on rotating liquid masses in the framework of a new method and a new approach. And in the four lectures on the "Rotation of Astronomical Bodies" I gave in April 1963, I attempted a broad survey of the subject in the spirit of the following quotation from a magnificent lecture by E. T. Whittaker on "Spin in the Universe."

Rotation is a universal phenomenon; the earth and all the other members of the solar system rotate on their axes, the satellites revolve round the planets, the planets revolve round the Sun, and the Sun himself is a member of the galaxy or Milky Way system which revolves in a very remarkable way. How did all these rotatory motions come into being? What secures their permanence or brings about their modification? And what part do they play in the system of the world?

Only the last of the four lectures was devoted to what has now been expanded into this book. But in 1963 I did not know how incomplete the classical work was and how much remained to be done; and I was not aware even of the existence of the fundamental papers of Dirichlet, Dedekind, and Riemann. And so I have preferred to devote the entire book to an expansion of the last lecture; in this way the book may have a permanence that I cannot claim for my discussion of the larger topics that were the subjects of the other lectures.

The completion of the work, that is the substance of this book, owes much to Dr. Norman R. Lebovitz who has collaborated with me on most phases of it and contributed many essential ideas. I am also grateful to Miss Donna D. Elbert for her assistance in all stages; and to Dr. Edward P. Lee who read the entire manuscript and helped the elimination of oversights and errors.

The investigations in this book have in part been supported by the Office of Naval Research under contract Nonr–2121 (24) with the University of Chicago.

S. Chandrasekhar

April 16, 1968

1

HISTORICAL INTRODUCTION

1. Newton

THE study of the gravitational equilibrium of homogeneous uniformly
rotating masses began with Newton's investigation on the figure of the
earth (*Principia*, Book III, Propositions XVIII–XX). Newton showed
that the effect of a small rotation on the figure *must* be in the direction
of making it slightly oblate; and, further, that the equilibrium of the
body will demand a simple proportionality between the *effect* of rotation,
as measured by the ellipticity,

$$\epsilon = \frac{\text{equatorial radius} - \text{polar radius}}{\text{the mean radius } (R)}, \tag{1}$$

and its *cause*, as measured by

$$m = \frac{\text{centrifugal acceleration at the equator}}{\text{mean gravitational acceleration on the surface}}$$

$$= \frac{\Omega^2 R}{GM/R^2} = \frac{\Omega^2 R^3}{GM}, \tag{2}$$

where G denotes the constant of gravitation and M is the mass of the
body. More precisely, Newton established the relation

$$\epsilon = \tfrac{5}{4}m \tag{3}$$

in case the body is homogeneous. The arguments by which Newton
derived this relation are magisterial; and they are worth recalling.

Newton imagined a hole of unit cross-section drilled from a point on
the equator to the center of the earth and a similar hole drilled from the
pole to the center; and he further imagined that the "canals" so constructed were filled with a fluid (see Fig. 1, after Newton's original
illustration in the *Principia*). From the fact that the fluid in the canals
will be in equilibrium, Newton concludes that the "weights" of the
equatorial and the polar columns of the fluid must be equal. However,
along the equator the acceleration due to gravity is "diluted" by the
centrifugal acceleration; and since both these accelerations in a homogeneous body vary from the center proportionately with the distance,

the "dilution factor" remains constant and is given by its value at the boundary, namely m.

If a denotes the equatorial radius, the weight of the equatorial column is given by

$$\text{weight of equatorial column} = \tfrac{1}{2}ag_{\text{equator}}(1-m), \tag{4}$$

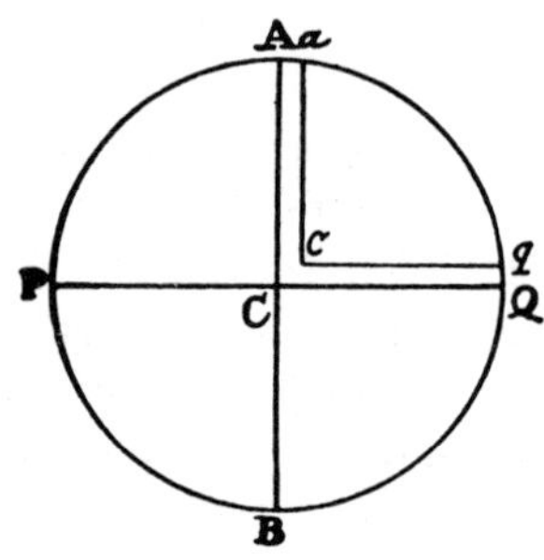

FIG. 1. Illustration from the *Principia* bearing on Newton's arguments for the rotational flattening of the earth.

where g_{equator} is the acceleration due to gravity at the equator. Similarly, if b denotes the polar radius,

$$\text{weight of polar column} = \tfrac{1}{2}bg_{\text{pole}}. \tag{5}$$

And since the two weights must be equal,

$$ag_{\text{equator}}(1-m) = bg_{\text{pole}}. \tag{6}$$

But for a slightly oblate body Newton knew that

$$\frac{g_{\text{pole}}}{g_{\text{equator}}} = 1+\tfrac{1}{5}\epsilon+O(\epsilon^2). \tag{7}$$

Equations (6) and (7) and the definition of ϵ $(= 1-b/a)$ now give

$$1-m = (1-\epsilon)(1+\tfrac{1}{5}\epsilon)+O(\epsilon^2) = 1-\tfrac{4}{5}\epsilon+O(\epsilon^2); \tag{8}$$

and Newton's relation (3) follows.

It was known already in Newton's time that

$$m = \tfrac{1}{290}. \tag{9}$$

Therefore, Newton concluded that if the earth were homogeneous, it should be oblate with an ellipticity

$$\epsilon = \tfrac{5}{4}\tfrac{1}{290} \simeq \tfrac{1}{230}. \tag{10}$$

This prediction of Newton was contrary to the astronomical evidence of the time and "two generations of the best astronomical observers formed in the school of the Cassinis struggled in vain against the authority and reasoning of Newton" (Todhunter's *History*, **1**, 100). The opposing ideas of Newton and Cassini are strikingly illustrated in the accompanying old caricature (Fig. 2). However, geodetic measurements made in Lapland by Maupertuis and Clairaut (1738) afforded data which conclusively showed the flattening of the earth at the poles. As Todhunter has written (**1**, 100), "The success of the arctic expedition may be ascribed in great measure to the skill and energy of Maupertuis; and his fame was widely celebrated. The engravings of the period represent him in the costume of a Lapland Hercules having a fur cap over his eyes; with one hand he holds a club and with the other he compresses the terrestrial globe." And Voltaire, then Maupertuis' friend, congratulated him warmly for having "aplati les poles et les Cassini." Later Maupertuis and Voltaire became involved in a heroic-comic controversy and Voltaire wrote

> Vouz avez confirmé dans les lieux pleins d'ennui
> Ce que Newton connut sans sortir de chez lui.

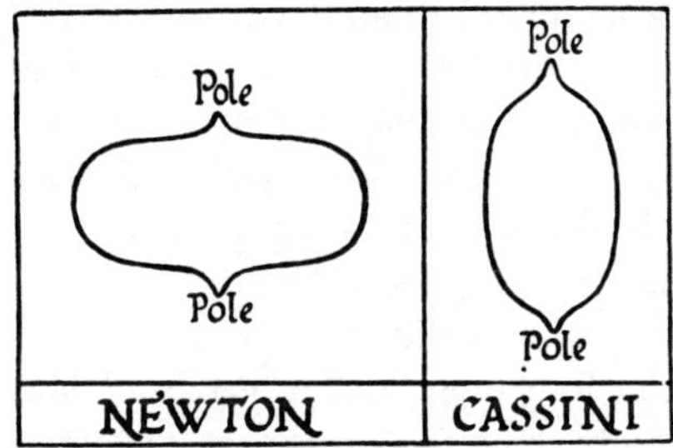

Fig. 2. An old-time caricature of the controversy between the opposing schools of Newton and Cassini with respect to the figure of the earth.

We know now that the actual ellipticity of the earth ($\sim 1/294$) is substantially smaller than Newton's predicted value ($\sim 1/230$); and this discrepancy is interpreted in terms of the inhomogeneity of the earth.

2. Maclaurin

The next advance (1742) in the theory was due to Maclaurin who generalized Newton's result to the case when the ellipticity caused by the rotation cannot be considered small.

Maclaurin had solved earlier the problem of the attraction of an oblate spheroid at an internal point; and he had shown in particular that the acceleration due to gravity at the equator and at the poles have the values

$$g_{\text{equator}} = 2\pi G \rho a \frac{(1-e^2)^{\frac{1}{2}}}{e^3}[\sin^{-1}e - e(1-e^2)^{\frac{1}{2}}]$$

and

$$g_{\text{pole}} = 4\pi G \rho a \frac{(1-e^2)^{\frac{1}{2}}}{e^3}[e - (1-e^2)^{\frac{1}{2}}\sin^{-1}e], \tag{11}$$

where ρ is the density of the spheroid, a its semi-major axis, and e its eccentricity. And since both the centrifugal acceleration in the equatorial plane and the acceleration due to gravity vary linearly with the coordinates, Newton's argument applies to this case equally well and we can write

$$g_{\text{equator}} - a\Omega^2 = g_{\text{pole}}(1-e^2)^{\frac{1}{2}},$$

or

$$\Omega^2 = \frac{1}{a}[g_{\text{equator}} - g_{\text{pole}}(1-e^2)^{\frac{1}{2}}]. \tag{12}$$

Inserting the expressions for g_{equator} and g_{pole} from equations (11), we obtain Maclaurin's formula

$$\frac{\Omega^2}{\pi G \rho} = \frac{(1-e^2)^{\frac{1}{2}}}{e^3} 2(3-2e^2)\sin^{-1}e - \frac{6}{e^2}(1-e^2). \tag{13}$$

Maclaurin realized that the foregoing derivation does not establish that a rapidly rotating mass will necessarily take the figure of an oblate spheroid. But he did show "(1) that the force which results from the attraction of the spheroid and those extraneous powers compounded together acts always in a right line perpendicular to the surface of the spheroid, (2) that the columns of the fluid sustain or balance each other at the center of the spheroid, and (3) that any particle in the spheroid is impelled equally in all directions."

To appreciate the foregoing qualifications of Maclaurin, one must remember that there was as yet no theory of hydrostatic equilibrium which provided *sufficient* conditions; so Maclaurin had to content himself with showing that all the conditions which had been recognized as *necessary* for equilibrium were satisfied. Considering the state of knowledge in his time, one can only admire Maclaurin's achievement in deriving the exact relation (13). And as Todhunter remarks (**1**, 175), "Maclaurin well deserves the association of his name with that of the great master in the inscription which records that he was appointed professor of mathematics at Edinburgh *ipso Newtono suadente*."

A remarkable feature of Maclaurin's relation was noticed by Thomas Simpson (1743): for any angular velocity less than a certain maximum value there are two and only two possible "oblata." This result is noteworthy in that we cannot deduce from the fact of a small equatorial angular velocity that the spheroid departs only slightly from a sphere; for as $\Omega^2 \to 0$, we have two solutions: a solution which, indeed, leads to a spheroid of small eccentricity and a second solution which leads to a highly flattened spheroid. It is generally believed that d'Alembert was the first to notice this feature of Maclaurin's solution; but as Todhunter has remarked (**1**, 181), "Although d'Alembert may have first explicitly published the statement, yet Simpson gives a table which distinctly implies the fact."

3. Jacobi

For nearly a century after Maclaurin's discovery of the spheroids (known after his name) it was believed that they represent the only admissible solution to the problem of the equilibrium of uniformly rotating homogeneous masses. The supposed generality of Maclaurin's solution was never questioned even though Lagrange in his *Mécanique Céleste* (1811) considered formally the possibility of ellipsoids with unequal axes satisfying the requirements of equilibrium. However, after obtaining two governing equations in which the two equatorial axes occur symmetrically (see equation (17) below), Lagrange infers that the two axes must be equal even though only the *sufficiency* (not the *necessity*) could be concluded. Jacobi (1834) recognized the inadequacy of Lagrange's demonstration† as he remarked, "One would make a grave mistake if one supposed that the spheroids of revolution are the only admissible figures of equilibrium even under the restrictive assumption of second degree surfaces." In making this last statement, Jacobi refers to the fact that while Maclaurin's solution provides, in the limit $\Omega^2 \to 0$, two solutions, one with $e \to 0$ and another with $e \to 1$, Legendre had shown that if one supposes that the figure is nearly spherical so that the attraction at a point on its surface can be expanded in powers of the departure from sphericity, then one obtains only the first of the two solutions "not in any approximation but with absolute geometrical rigor." According to Jacobi, the conclusion one must draw from Legendre's demonstration is that figures of equilibrium may exist that cannot be surmised from what one can establish in the limit of spherical figures. And Jacobi concludes, "In fact a simple consideration shows that ellipsoids with three unequal axes can very well be figures of equilibrium; and that one can assume an ellipse of arbitrary shape for the equatorial section and determine the third axis (which is also the least of the three axes) and the angular velocity of rotation such that the ellipsoid is a figure of equilibrium."

The existence of these ellipsoids of Jacobi can be established and the relations governing them can be determined by a simple extension of Newton's original argument.

At the time Jacobi made his discovery, it was known that the components of the attraction, g_i $(i = 1, 2, 3)$, along the directions of the principal axes of an ellipsoid can be expressed in the manner

$$g_i = 2\pi G \rho A_i x_i, \tag{14}$$

where

$$A_i = a_1 a_2 a_3 \int_0^\infty \frac{du}{(a_i^2 + u)\Delta} \tag{15}$$

and

$$\Delta^2 = (a_1^2 + u)(a_2^2 + u)(a_3^2 + u). \tag{16}$$

---

† Rather as Dirichlet states in his "Gedächtnissrede auf Carl Gustav Jacob Jacobi," Jacobi's suspicion was aroused by the qualification "necessary" in an account of Lagrange's considerations by the author of a "well-known textbook."

The formulas for the components of the attraction in the foregoing forms
were apparently first derived by Gauss (1813) and by Rodrigues (1815),
independently. However, in the less symmetrical forms in which one
generally writes them for purposes of reducing them to the standard
elliptic integrals of the two kinds, they were known much earlier: they
are (as Legendre has said) effectively included in Maclaurin's writings;
but explicitly, for ellipsoids with three unequal axes, they occur for the
first time† in Laplace's *Théorie du Mouvement et de la Figure Elliptique
des Planètes* (1784).

Returning to the extension of Newton's argument to the case of tri-
axial ellipsoids, we may imagine that three "canals" are drilled along
the directions of the three principal axes from the surface to the center
and further that they are all filled with a fluid. From the equilibrium
of the fluid in the three canals, we may infer the equality of the weights
of the three columns (per unit cross-section). We thus have

$$2A_1 a_1^2 - \frac{\Omega^2}{\pi G \rho} a_1^2 = 2A_2 a_2^2 - \frac{\Omega^2}{\pi G \rho} a_2^2 = 2A_3 a_3^2. \tag{17}$$

These relations require (if $a_1 \neq a_2 \neq a_3$)

$$\frac{\Omega^2}{\pi G \rho} = 2 \frac{A_1 a_1^2 - A_2 a_2^2}{a_1^2 - a_2^2} = 2a_1 a_2 a_3 \int_0^\infty \frac{u\, du}{(a_1^2 + u)(a_2^2 + u)\Delta}. \tag{18}$$

And we also have the purely geometrical condition

$$A_1 - \frac{a_3^2}{a_1^2} A_3 = A_2 - \frac{a_3^2}{a_2^2} A_3, \tag{19}$$

or

$$a_1^2 a_2^2 \frac{A_2 - A_1}{a_1^2 - a_2^2} = a_3^2 A_3. \tag{20}$$

This last relation explicitly has the form

$$a_1^2 a_2^2 \int_0^\infty \frac{du}{(a_1^2 + u)(a_2^2 + u)\Delta} = a_3^2 \int_0^\infty \frac{du}{(a_3^2 + u)\Delta}. \tag{21}$$

Equations (18) and (21), in *exactly* these forms, are given in Jacobi's
paper. And as Jacobi further states, for any assigned a_1 and a_2, equation

---

† As Todhunter has pointed out (**1**, 417), the formulas themselves appear in the
writing of d'Alembert though "he deliberately rejects them . . . this is perhaps the
strangest of all his [d'Alembert's] strange mistakes." And with regard to Laplace's
derivation, Todhunter says (**2**, 32), "Thus Laplace values and appropriates the treasure
which d'Alembert deliberately threw away."

(21) allows a solution for a_3 which satisfies the inequality

$$\frac{1}{a_3^2} > \frac{1}{a_1^2} + \frac{1}{a_2^2}, \tag{22}$$

and that when $a_1 = a_2$ equations (18) and (21) determine a configuration common to the spheroidal and the ellipsoidal sequences.

Referring to this discovery of Jacobi, Thomson and Tait in their *Natural Philosophy* (**2**, 530) say, "This curious theorem was discovered by Jacobi in 1834 and seems, simple as it is, to have been enunciated by him as a challenge to the French Mathematicians." In Todhunter's *History* there is no reference to Jacobi having issued a "challenge." But Todhunter (**2**, 381) does refer to a communication by Poisson to the French Academy on November 24, 1834 and states "Poisson begins by referring to a letter recently sent by Jacobi to the French Academy in which two results were enunciated. One was what we call Jacobi's theorem, namely, that an ellipsoid is a possible form of relative equilibrium for a rotating fluid; the other related to the attraction of a heterogeneous ellipsoid . . . Poisson's note related to the second result."

4. Meyer and Liouville

In his short and brief paper on the subject, Jacobi did not seriously examine the relationship of his ellipsoids to the Maclaurin spheroids. C. O. Meyer (1842) was the first to do so. Meyer's principal result was to show that the Jacobian sequence "bifurcates" (in the later terminology of Poincaré) from the Maclaurin sequence at the point where the eccentricity $e = 0 \cdot 81267$. This result can be readily deduced from Jacobi's equations (18) and (21). Thus, by setting $a_1 = a_2$ in these equations we obtain the relations

$$\frac{\Omega^2}{\pi G \rho} = 2a_1^2 a_3 \int_0^{\infty} \frac{u \, du}{(a_1^2 + u)^3 (a_3^2 + u)^{\frac{1}{2}}} \tag{23}$$

and

$$a_1^4 \int_0^{\infty} \frac{du}{(a_1^2 + u)^3 (a_3^2 + u)^{\frac{1}{2}}} = a_3^2 \int_0^{\infty} \frac{du}{(a_1^2 + u)(a_3^2 + u)^{\frac{3}{2}}}, \tag{24}$$

where $\Omega^2 / \pi G \rho$ on the left-hand side of equation (23) must now be identified with Maclaurin's function (13). It can be shown (see § 39) that both equations (23) and (24) are simultaneously satisfied when

$$e = 0 \cdot 81267 \quad \text{where} \quad \Omega^2 / \pi G \rho = 0 \cdot 37423. \tag{25}$$

Since it is known that the maximum value of $\Omega^2 / \pi G \rho$ along the Maclaurin sequence is $0 \cdot 4493$, it follows that for $\Omega^2 / \pi G \rho < 0 \cdot 37423$ there are three equilibrium figures possible: two Maclaurin spheroids and one Jacobi ellipsoid; for $0.4493 > \Omega^2/\pi G\rho > 0.3742$ only the Maclaurin figures are possible; and finally, for $\Omega^2/\pi G\rho > 0.4493$ no equilibrium figures are possible. This enumeration of the different possibilities is due to Meyer.

In 1846 Liouville restated Meyer's result using the angular momentum, instead of the angular velocity, as the variable; and he showed that while the angular momentum increases from zero to infinity along the Maclaurin sequence, the Jacobian figures are possible only for angular momenta exceeding a certain value (namely, that at the point of bifurcation along the Maclaurin sequence).

5. Dirichlet, Dedekind, and Riemann

The fact that no figures of equilibrium are possible for uniformly rotating bodies when the angular velocity exceeds a certain limit raises the question: What happens when the angular velocity exceeds this limit? Dirichlet addressed himself to this question during the winter of 1856–57; and though he included this topic in his lectures on partial differential equations in July 1857, he did not publish any detailed account of his investigations during his lifetime. Dirichlet's results were collated from some papers he left and were edited for publication by Dedekind. As Riemann wrote, "In his posthumous paper, edited for publication by Dedekind, Dirichlet has opened up, in a most remarkable way, an entirely new avenue for investigations on the motion of a self-gravitating homogeneous ellipsoid. The further development of his beautiful discovery has a particular interest to the mathematician even apart from its relevance to the forms of heavenly bodies which initially instigated these investigations."

The precise problem which Dirichlet considered in his paper is the following: Under what conditions can one have a configuration which, at every instant, has an ellipsoidal figure and in which the motion, in an inertial frame, is a linear function of the coordinates? Dirichlet formulated the general equations governing this problem (in a Lagrangian framework) and solved them in detail for the case when the bounding surface is a spheroid of revolution. Dirichlet did not seriously investigate the figures of equilibrium admissible under the general circumstances of his formulation. In the latter context, Dedekind (in an addendum to Dirichlet's paper) proved explicitly the following theorem (though, as Riemann remarks, it is already implicit in Dirichlet's equations): *Let a homogeneous ellipsoid with semi-axes a_1, a_2, and a_3 be in gravitational* *equilibrium with a prevalent motion whose components, resolved along the instantaneous directions of the principal axes of the ellipsoid and in an inertial frame, are given by*

$$\mathbf{u}^{(0)} = \begin{vmatrix} \alpha_{11} & \alpha_{12} & \alpha_{13} \\ \alpha_{21} & \alpha_{22} & \alpha_{23} \\ \alpha_{31} & \alpha_{32} & \alpha_{33} \end{vmatrix} \begin{vmatrix} x_1/a_1 \\ x_2/a_2 \\ x_3/a_3 \end{vmatrix} = \mathbf{A} \begin{vmatrix} x_1/a_1 \\ x_2/a_2 \\ x_3/a_3 \end{vmatrix} ; \tag{26}$$

then the same ellipsoid will also be a figure of equilibrium if the prevalent motion is that derived from the transposed matrix $\mathbf{A}^\dagger$, *i.e.* $\mathbf{u}^{(0)\dagger}$ *given by*

$$\mathbf{u}^{(0)\dagger} = \begin{vmatrix} \alpha_{11} & \alpha_{21} & \alpha_{31} \\ \alpha_{12} & \alpha_{22} & \alpha_{32} \\ \alpha_{13} & \alpha_{23} & \alpha_{33} \end{vmatrix} \begin{vmatrix} x_1/a_1 \\ x_2/a_2 \\ x_3/a_3 \end{vmatrix} = \mathbf{A}^\dagger \begin{vmatrix} x_1/a_1 \\ x_2/a_2 \\ x_3/a_3 \end{vmatrix} . \tag{26'}$$

We shall call the configuration with the motion derived from $\mathbf{A}^\dagger$ the *adjoint* of the configuration with the motion derived from $\mathbf{A}$.

Dedekind considered in particular the configurations which are congruent to the Jacobi ellipsoids and are their adjoints in the sense we have defined.

The motion of a Jacobi ellipsoid rotating uniformly with an angular velocity Ω about the x_3-axis can be represented in the manner

$$\mathbf{u}^{(0)} = \begin{vmatrix} 0 & -\Omega a_2 & 0 \\ \Omega a_1 & 0 & 0 \\ 0 & 0 & 0 \end{vmatrix} \begin{vmatrix} x_1/a_1 \\ x_2/a_2 \\ x_3/a_3 \end{vmatrix} . \tag{27}$$

The motion in the adjoint configuration will be given by

$$\mathbf{u}^{(0)\dagger} = \begin{vmatrix} 0 & \Omega a_1 & 0 \\ -\Omega a_2 & 0 & 0 \\ 0 & 0 & 0 \end{vmatrix} \begin{vmatrix} x_1/a_1 \\ x_2/a_2 \\ x_3/a_3 \end{vmatrix} ; \tag{28}$$

or, in terms of components,

$$u_1^{(0)\dagger} = \frac{\Omega a_1}{a_2} x_2, \qquad u_2^{(0)\dagger} = -\frac{\Omega a_2}{a_1} x_1, \qquad u_3^{(0)\dagger} = 0; \tag{29}$$

and this motion clearly satisfies the condition

$$\mathbf{u}^{(0)\dagger} \cdot \operatorname{grad}\left(\frac{x_1^2}{a_1^2} + \frac{x_2^2}{a_2^2} + \frac{x_3^2}{a_3^2} - 1\right) = 0 \tag{30}$$

required for the preservation of the ellipsoidal boundary. Also, the motion represented by (29) is one of uniform vorticity

$$\zeta = -\Omega\left(\frac{a_2}{a_1} + \frac{a_1}{a_2}\right) = -\frac{a_1^2 + a_2^2}{a_1 a_2}\Omega. \tag{31}$$

These *ellipsoids of Dedekind*, while they are congruent to the Jacobi ellipsoids, are stationary in an inertial frame and they maintain their ellipsoidal figures by the internal motions which prevail. (Lamb erroneously attributes to Love the discovery of this relation between the Jacobi and the Dedekind ellipsoids.) It is also clear that the ellipsoids of Dedekind bifurcate from the Maclaurin spheroids at the same point that the Jacobi ellipsoids do.

The complete solution to the problem of the stationary figures admissible under Dirichlet's general assumptions was given by Riemann in a paper of remarkable insight and power. Riemann first shows that under the restriction of motions which are linear in the coordinates, the most general type of motion compatible with an ellipsoidal figure of equilibrium consists of a superposition of a uniform rotation Ω and internal motions of a uniform vorticity ζ (in the rotating frame). More precisely he showed that ellipsoidal figures of equilibrium are possible only under the following three circumstances: (*a*) the case of uniform rotation with no internal motions, (*b*) the case when the directions of Ω and ζ coincide with a principal axis of the ellipsoid, and (*c*) the case when the directions of Ω and ζ lie in a principal plane of the ellipsoid. Case (*a*) leads to the sequences of Maclaurin and Jacobi. Case (*b*) leads to sequences of ellipsoids along which the ratio $f = \zeta/\Omega$ remains constant (the Jacobian and the Dedekind sequences are special cases of these general "Riemann sequences" for $f = 0$ and ∞, respectively). And finally, case (*c*) leads to three other classes of ellipsoids. Riemann wrote down the equations governing the equilibrium of these ellipsoids and specified their domain of occupancy in the (a_1, a_2, a_3)-space. (A more detailed description of the properties of these ellipsoids will be found in Chapter 7.) Riemann also sought to determine the stability of these ellipsoids by an energy criterion. But his criterion, as has recently been shown by Lebovitz, is erroneous and Riemann's conclusions, with the notable exception of those pertaining to the Maclaurin and the Riemann sequences for $f \geqslant -2$, are false.

While Riemann's paper made an impressive start towards the solution of Dirichlet's general problem, it left a large number of questions unanswered. Indeed, even the relation of Riemann's ellipsoids to the Maclaurin spheroids which they adjoin was left obscure. Nevertheless these questions were to remain unanswered for more than a hundred years. The reason for this total neglect must, in part, be attributed to a spectacular discovery by Poincaré (see § 6 below) which channeled all subsequent investigations along directions which appeared rich with possibilities; but the long quest it entailed turned out, in the end, to be after a chimera.

6. Poincaré and Cartan

The investigations relating to the equilibrium and the stability of ellipsoidal figures of equilibrium, for which Dirichlet and Riemann had laid such firm foundations, took an unexpected turn (from which it was not to be diverted for the next seventy-five years) when Poincaré discovered in 1885 that along the Jacobian sequence a point of bifurcation occurs similar to the one along the Maclaurin sequence and that even as the Jacobian sequence branches off from the Maclaurin sequence, a new sequence of pear-shaped configurations branches off from the Jacobian sequence. This result of Poincaré is equivalent to the statement (in current terminology) that along the Jacobian sequence there is a point where the ellipsoid allows a neutral mode of oscillation belonging to the third harmonics. A corollary which was also enunciated by Poincaré is that along the Jacobian sequence there must be further points of bifurcation where the Jacobian ellipsoid allows neutral modes of oscillation belonging to the fourth, fifth, and higher harmonics. And Poincaré conjectured "that the bifurcation of the pear-shaped body leads onward stably and continuously to a planet attended by a satellite, the bifurcation into the fourth zonal harmonic probably leads unstably to a planet with a satellite on each side, that into the fifth harmonics to a planet with two satellites on one and one on the other and so on" (Darwin). It was further conjectured by Darwin that one may look for the origin of the double stars in similar instabilities; the "fission theory" of the origin of double stars arose in this fashion. The grand mental panorama that was thus created was so intoxicating that those who followed Poincaré were not to recover from its pursuit. In any event, Darwin, Liapounoff, and Jeans spent years of effort towards the substantiation of these conjectures; and so single-minded was the pursuit† that one did not even linger to investigate the stability of the Maclaurin spheroids and the Jacobi ellipsoids from a direct analysis of normal modes. Finally, in 1924 Cartan established that the Jacobi ellipsoid becomes unstable at its first point of bifurcation and behaves in this respect differently from the Maclaurin spheroid which, in the absence

---

† For example, the question whether along the Dedekind sequence a neutral point occurs similar to the one along the congruent Jacobian sequence does not appear to have been considered or even raised.

of any dissipative mechanism, is stable on either side of the point of bifurcation where the Jacobian sequence branches off.

And at this point the subject quietly went into a coma.

7. Roche, Darwin, and Jeans

An important problem of secondary interest in the theory of ellipsoidal figures of equilibrium was formulated by Roche (1847–50). Roche considered the equilibrium of an infinitesimal satellite (of density ρ) rotating about a rigid spherical planet (of mass M') in a circular Keplerian orbit (of radius R); and he showed that no equilibrium figures are possible if the angular velocity (Ω) of orbital rotation exceeds the limit

$$\frac{\Omega^2}{\pi G\rho} = \frac{M'}{\pi\rho R^3} \leqslant 0 \cdot 090093. \tag{32}\dagger$$

The lower limit to R set by the foregoing inequality is called the *Roche limit*. Roche also considered the case when the mass of the satellite is finite; and showed that inequalities analogous to (32) exist. But in all of Roche's considerations, the assumption of a rigid spherical body for the distorting mass was retained. In 1906, Darwin, with a view toward application to double stars, attempted to allow for the mutual distortion of the components by an approximate procedure; but his efforts were only partially successful.

An interesting case, when in Roche's problem only the tidal forces are taken into account, was considered by Jeans (1916).

The equilibrium and the stability of the ellipsoids of Jeans, Roche, and Darwin form a separate chapter in the theory of ellipsoidal figures of homogeneous masses.

In the present book an attempt will be made to consolidate these classical researches on the ellipsoidal figures of equilibrium. The presentation will not be historical; it will be based, almost exclusively, on the series of papers which Professor N. R. Lebovitz and the author have published, separately and in collaboration, during the years 1961–69, mainly in the *Astrophysical Journal* (vols. 134–57). (These papers, numbered by roman numerals, are listed in the Appendix, pp. 245–7; and, when the occasion arises, they will be referred to in the text by their appropriate roman numerals.) In these papers a new method of approach was systematically exploited to investigate the entire class of problems whose historical background has been reviewed in this

---

† Roche originally gave the value $0 \cdot 092$ for the constant on the right-hand side of this inequality; the value quoted is that obtained from more recent investigations.

chapter. However, in contrast to the original papers, the principal effort in this book will be toward bringing to the subject order and coherence in a common outlook.

BIBLIOGRAPHICAL NOTES

An indispensable source book for all investigations bearing on the topic of this book from the time of Newton to that of Laplace is:

I. TODHUNTER, *History of the Mathematical Theories of Attraction and the Figure of the Earth* (London, Constable, 1873; reprint ed. New York, Dover Publications, 1962).

§§ 1 and 2. The account in these sections relating to the contributions of Newton and Maclaurin is largely based on Todhunter's *History*. The interested reader may, however, wish to consult:

F. CAJORI, *Newton's Principia* (Berkeley, University of California Press, 1946).
COLIN MACLAURIN, *A Treatise on Fluxions* (1742).

§ 3. The account is based on:

C. G. J. JACOBI, "Über die Figur des Gleichgewichts," *Poggendorff Annalen der Physik und Chemie*, **33** (1834), 229–38; reprinted in *Gesammelte Werke* **2** (Berlin, G. Reimer, 1882), 17–72.

G. LEJEUNE DIRICHLET, "Gedächtnissrede auf Carl Gustav Jacob Jacobi gehalten in der Akademie der Wissenschaften am 1 Juli 1852," *Gesammelte Werke*, **2** (Berlin, G. Reimer, 1897), 243.

See also:

W. THOMSON and P. G. TAIT, *Treatise on Natural Philosophy* (Cambridge, England, Cambridge University Press, 1883), pt. 2, pp. 324–35.

§§ 4 and 5. A useful historical account of the subject with special reference to the contributions made during 1830–80 is given by:

W. M. HICKS, "Recent progress in hydrodynamics," *Reports to the British Association* (1882), pp. 57–61.

The particular references to the papers of Dirichlet, Dedekind, and Riemann are:

G. LEJEUNE DIRICHLET, "Untersuchungen über ein Problem der Hydrodynamik," *J. Reine Angew. Math.*, **58** (1860), 181–216.

R. DEDEKIND, "Zusatz zu der vorstehenden Abhandlung," *J. Reine Angew. Math.*, **58** (1860), 217–28.

B. RIEMANN, "Untersuchungen über die Bewegung eines flüssigen gleichartigen Ellipsoides," *Abh. d. Königl. Gesell. der Wis. zu Göttingen*, **9** (1860), 3–36; see also *Gesammelte Mathematische Werke* (Leipzig, B. G. Teubner, 1892), p. 182.

A partial account of Riemann's work on the equilibrium figures will be found in:

A. BASSETT, *A Treatise on Hydrodynamics*, **2** (Cambridge, England, Deighton, Bell and Company, 1888; reprint ed. New York, Dover Publications, 1961).

See also:

SIR HORACE LAMB, *Hydrodynamics* (Cambridge, England, Cambridge University Press, 1932), pp. 722–23.

§ 6. The references to the basic papers of Poincaré and Cartan are:

H. POINCARÉ, "Sur l'équilibre d'une masse fluide animée d'un mouvement de rotation," *Acta Math.*, **7** (1885), 259–380.

H. CARTAN, "Sur les petites oscillations d'une masse fluide," *Bull. Sci. Math.*, **46** (1922), 317–52, 356–69.

H. CARTAN, "Sur la stabilité ordinaire des ellipsoides de Jacobi," *Proc. International Math. Congress, Toronto, 1924*, **2** (Toronto, University of Toronto Press, 1928), 2–17.

For an account of Cartan's methods and a derivation of his principal results see:

R. A. LYTTLETON, *The Stability of Rotating Liquid Masses* (Cambridge, England, Cambridge University Press, 1953), chap. 9.

§ 7. The references to the papers of Roche, Darwin, and Jeans are:

M. ÉD. ROCHE, "Mémoire sur la figure d'une masse fluide (soumise à l'attraction d'un point éloigné)," *Acad. des Sci. de Montpellier* **1** (1847–50), 243–62, 333–48.

G. H. DARWIN, "On the figure and stability of a liquid satellite," *Phil. Trans. R. Soc. (London)*, **206** (1906), 161–248; see also *Scientific Papers*, **3** (Cambridge, England, Cambridge University Press, 1910), 436.

J. H. JEANS, "The motion of tidally-distorted masses, with special reference to theories of cosmogony," *Mem. Roy. Astron. Soc. London*, **62** (1917), 1–48.

J. H. JEANS, *Problems of Cosmogony and Stellar Dynamics* (Cambridge, England, Cambridge University Press, 1919); also *Astronomy and Cosmogony* (Cambridge, England, Cambridge University Press, 1929), chaps. 7 and 8.

A perceptive account of Jeans's investigations is given by:

E. A. MILNE, *Sir James Jeans, a Biography* (Cambridge, England, Cambridge University Press, 1952), chap. 9.

The historical account given in this chapter (except for § 7) is a reproduction of Paper XXXIII (see Appendix, p. 246).

2

THE VIRIAL EQUATIONS OF THE VARIOUS ORDERS

8. Introduction

A STANDARD technique for treating the integro-differential equations of mathematical physics is to take the moments of the equations concerned and consider suitably truncated sets of the resulting equations. An advantage in considering such moment equations is that the equations of the lowest orders often have simple physical interpretations; and, moreover, in many instances, their solutions with suitable assumptions of "closure" suggest methods of obtaining approximate solutions of the exact equations in a systematic way. The *virial method,* as developed in this book, is essentially the method of the moments applied to the solution of hydrodynamical problems in which the gravitational field of the prevailing distribution of matter is taken into account. The *virial equations* of the various orders are, in fact, no more than the moments of the relevant hydrodynamical equations.

While the equations derived in this chapter are of general validity, their particular usefulness for the problems reviewed in Chapter 1 is that, in these instances, they enable their *exact* solutions to be obtained in an entirely elementary way.

9. The moments describing the distribution of density, pressure, and velocity

As we have stated, in the virial method we take moments of the equations of motion. These equations must naturally involve the moments of the distribution of density, pressure, velocity, and gravitational potential. In §§ 9 and 10 we consider these moments.

Consider a distribution of mass in a volume V. Choose a frame of reference whose origin (momentarily) is at the center of mass. Then the choice of the frame of reference requires

$$I_i = \int_V \rho(\mathbf{x})x_i \, d\mathbf{x} = 0 \quad (i = 1, 2, 3), \tag{1}$$

where $\rho(\mathbf{x})$ is the density† at the point $\mathbf{x}$. The mass,

$$M = \int_V \rho(\mathbf{x})\, d\mathbf{x}, \tag{2}$$

and its moment of inertia,

$$I = \int_V \rho(\mathbf{x})|\mathbf{x}|^2\, d\mathbf{x} = \tfrac{3}{5}Mk^2, \tag{3}$$

are gross integral properties that determine a *linear scale* for the distribution: the radius of gyration, k, is the radius of an *equivalent sphere*. The *moment of inertia tensor*,

$$I_{ij} = \int_V \rho x_i x_j\, d\mathbf{x}, \tag{4}$$

gives more information about the distribution: it defines an *equivalent oriented ellipsoid*. The tensor I_{ij} is clearly symmetric in its indices; and, moreover, its trace is the scalar moment of inertia:

$$I_{ii} = I. \tag{5}$$

For most problems of elementary mechanics, it is not necessary to consider moments of the density distribution higher than the second; however, in our work, we shall have occasions to consider also the third and the fourth moments defined by the higher-order tensors

$$I_{ijk} = \int_V \rho x_i x_j x_k\, d\mathbf{x} \quad\text{and}\quad I_{ijkl} = \int_V \rho x_i x_j x_k x_l\, d\mathbf{x}. \tag{6}$$

The prevailing distribution of pressure p (assumed isotropic, here) similarly leads us to a consideration of the moments

$$\Pi = \int_V p\, d\mathbf{x}, \qquad \Pi_i = \int_V p x_i\, d\mathbf{x}, \qquad \Pi_{ij} = \int_V p x_i x_j\, d\mathbf{x}, \quad\text{etc.} \tag{7}$$

Since we are primarily interested in the hydrodynamical motions that may occur, we shall require certain moments of the prevailing distribution of macroscopic motions to describe their gross features. The most important quantity in this connection is, of course, the total kinetic energy of the motions in the system:

$$\mathfrak{T} = \tfrac{1}{2}\int_V \rho|\mathbf{u}|^2\, d\mathbf{x}, \tag{8}$$

where $|\mathbf{u}|^2$ $(= u_1^2 + u_2^2 + u_3^2)$ is the square of the velocity of the fluid

† All such quantities introduced (including the volume V) will, in general, be functions of time, t, as well. But the considerations of §§ 9 and 10 are in no way affected by this dependence on time; and it is suppressed on this account.

element at $\mathbf{x}$. As in the case of the moment of inertia, the *kinetic-energy tensor*

$$\mathfrak{T}_{ij} = \tfrac{1}{2} \int_V \rho u_i u_j \, d\mathbf{x}, \tag{9}$$

whose trace is $\mathfrak{T}$, is a more informative quantity; and like I_{ij} it is also symmetric in its indices. Further moments which we shall find useful are

$$\mathfrak{T}_{ij;k} = \tfrac{1}{2} \int_V \rho u_i u_j x_k \, d\mathbf{x}$$

and

$$\mathfrak{T}_{ij;kl} = \tfrac{1}{2} \int_V \rho u_i u_j x_k x_l \, d\mathbf{x}; \tag{10}$$

we shall not have occasion to consider moments of higher order.

It should be noted that when considering the energy content of a system, the integral over the pressure, p, will represent the internal energy so that $\tfrac{1}{2}\Pi\delta_{ij}$ is comparable to $\mathfrak{T}_{ij}$.

10. The tensor potentials and the potential-energy tensors

The gravitational effect at a point $\mathbf{x}$, due to a distribution of matter with density $\rho(\mathbf{x})$, is determined by the Newtonian potential

$$\mathfrak{B}(\mathbf{x}) = G \int_V \frac{\rho(\mathbf{x}')}{|\mathbf{x}-\mathbf{x}'|} \, d\mathbf{x}', \tag{11}$$

where G denotes the constant of gravitation. Associated with this gravitational potential is the potential energy

$$\mathfrak{W} = -\tfrac{1}{2} \int_V \rho \mathfrak{B} \, d\mathbf{x}. \tag{12}$$

As in the case of the moment of inertia I and the kinetic energy $\mathfrak{T}$, we shall need tensor generalizations, $\mathfrak{B}_{ij}$ and $\mathfrak{W}_{ij}$, of $\mathfrak{B}$ and $\mathfrak{W}$ such that

$$\mathfrak{B} = \mathfrak{B}_{ii} \quad \text{and} \quad \mathfrak{W} = \mathfrak{W}_{ii}. \tag{13}$$

The required generalizations are provided by

$$\mathfrak{B}_{ij}(\mathbf{x}) = G \int_V \rho(\mathbf{x}') \frac{(x_i-x_i')(x_j-x_j')}{|\mathbf{x}-\mathbf{x}'|^3} \, d\mathbf{x}' \tag{14}$$

and

$$\mathfrak{W}_{ij} = -\tfrac{1}{2} \int_V \rho \mathfrak{B}_{ij} \, d\mathbf{x}. \tag{15}$$

As defined, these tensors are manifestly symmetric; and the trace conditions (13) are clearly satisfied.

An alternative form of $\mathfrak{W}_{ij}$, which represents a tensor generalization of the known result

$$\mathfrak{W} = \int_V \rho x_i \frac{\partial \mathfrak{B}}{\partial x_i} \, d\mathbf{x}, \tag{16}$$

follows from the definition of $\mathfrak{B}_{ij}$. Thus

$$\mathfrak{W}_{ij} = -\tfrac{1}{2} \int_V \rho \mathfrak{B}_{ij} \, d\mathbf{x}$$

$$= -\tfrac{1}{2} G \int_V \int_V \rho(\mathbf{x})\rho(\mathbf{x}') \frac{(x_i - x_i')(x_j - x_j')}{|\mathbf{x} - \mathbf{x}'|^3} \, d\mathbf{x} d\mathbf{x}'$$

$$= -G \int_V \int_V \rho(\mathbf{x})\rho(\mathbf{x}') \frac{x_i(x_j - x_j')}{|\mathbf{x} - \mathbf{x}'|^3} \, d\mathbf{x} d\mathbf{x}'$$

$$= G \int_V \int_V d\mathbf{x} \, \rho(\mathbf{x}) x_i \frac{\partial}{\partial x_j} \int d\mathbf{x}' \frac{\rho(\mathbf{x}')}{|\mathbf{x} - \mathbf{x}'|}, \tag{17}$$

or

$$\mathfrak{W}_{ij} = \int_V \rho x_i \frac{\partial \mathfrak{B}}{\partial x_j} \, d\mathbf{x}; \tag{18}$$

and the contraction of this last result gives equation (16). Incidentally, equation (18) establishes the symmetry (not manifest) of the integral on the right-hand side with respect to the indices i and j.

Moments of $\mathfrak{B}_{ij}$, analogous to $\mathfrak{T}_{ij;k}$ and $\mathfrak{T}_{ij;kl}$, are

$$\mathfrak{W}_{ij;k} = -\tfrac{1}{2} \int_V \rho \mathfrak{B}_{ij} x_k \, d\mathbf{x} \tag{19}$$

and

$$\mathfrak{W}_{ij;kl} = -\tfrac{1}{2} \int_V \rho \mathfrak{B}_{ij} x_k x_l \, d\mathbf{x}. \tag{20}$$

However, it will appear that whenever $\mathfrak{W}_{ij;kl}$ occurs, an associated quantity which also occurs is

$$\mathfrak{W}_{ij;k;l} = -\tfrac{1}{2} \int_V \rho \mathfrak{D}_{ij;k} x_l \, d\mathbf{x}, \tag{21}$$

where

$$\mathfrak{D}_{ij;k}(\mathbf{x}) = G \int_V \rho(\mathbf{x}') x_k' \frac{(x_i - x_i')(x_j - x_j')}{|\mathbf{x} - \mathbf{x}'|^3} \, d\mathbf{x}' \tag{22}$$

is the " $\mathfrak{B}_{ij}$ " induced by the distribution ρx_k. Inserting this last expression for $\mathfrak{D}_{ij;k}$ in equation (21) and inverting the order of the integrations, we readily show that

$$\mathfrak{W}_{ij;k;l} = -\tfrac{1}{2} \int_V \rho \mathfrak{D}_{ij;l} x_k \, d\mathbf{x}, \tag{23}$$

so that $\mathfrak{W}_{ij;k;l}$, like $\mathfrak{W}_{ij;kl}$ and $\mathfrak{T}_{ij;kl}$, is symmetric in k and l (in addition to being manifestly symmetric in i and j).

A question of some practical importance, in connection with the various tensor potentials we have introduced, concerns the most convenient way in which they can be determined. It appears that for this purpose, they are best expressed in terms of the hierarchy of Newtonian potentials

$$\mathfrak{D}_i, \quad \mathfrak{D}_{ij}, \quad \mathfrak{D}_{ijk}, \quad \text{etc.,} \tag{24}$$

derived from the distributions

$$\rho x_i, \quad \rho x_i x_j, \quad \rho x_i x_j x_k, \quad \text{etc.,} \tag{25}$$

respectively. These Newtonian potentials can be expressed as integrals in the forms

$$\mathfrak{D}_i(\mathbf{x}) = G \int_V \frac{\rho(\mathbf{x}')x_i'}{|\mathbf{x}-\mathbf{x}'|} \, d\mathbf{x}', \qquad \mathfrak{D}_{ij}(\mathbf{x}) = G \int_V \frac{\rho(\mathbf{x}')x_i' x_j'}{|\mathbf{x}-\mathbf{x}'|} \, d\mathbf{x}', \quad \text{etc.} \tag{26}$$

We shall now show how $\mathfrak{B}_{ij}$ can, for example, be expressed in terms of $\mathfrak{D}_i$. We have

$$\frac{\partial \mathfrak{D}_i}{\partial x_j} = -G \int_V \rho(\mathbf{x}') \frac{x_i'(x_j-x_j')}{|\mathbf{x}-\mathbf{x}'|^3} \, d\mathbf{x}'$$

and

$$\frac{\partial \mathfrak{B}}{\partial x_j} = -G \int_V \rho(\mathbf{x}') \frac{(x_j-x_j')}{|\mathbf{x}-\mathbf{x}'|^3} \, d\mathbf{x}'. \tag{27}$$

From these equations it follows that

$$\mathfrak{B}_{ij} = -x_i \frac{\partial \mathfrak{B}}{\partial x_j} + \frac{\partial \mathfrak{D}_i}{\partial x_j}. \tag{28}$$

This last equation is easily generalized; thus

$$\mathfrak{D}_{ij;k} = -x_i \frac{\partial \mathfrak{D}_k}{\partial x_j} + \frac{\partial \mathfrak{D}_{ik}}{\partial x_j}, \tag{29}$$

since this is the "$\mathfrak{B}_{ij}$" induced by the distribution ρx_k so that $\mathfrak{D}_k$ plays the role of "$\mathfrak{B}$" in equation (28). Similarly, $\mathfrak{D}_{ij;kl}$, which represents the "$\mathfrak{B}_{ij}$" induced by the distribution $\rho x_k x_l$, is given by

$$\mathfrak{D}_{ij;kl} = -x_i \frac{\partial \mathfrak{D}_{kl}}{\partial x_j} + \frac{\partial \mathfrak{D}_{ikl}}{\partial x_j}. \tag{30}$$

In many ways it is natural that when describing a distribution of matter, grossly, in terms of moments, we need integrals associated not only with the gravitational potential $\mathfrak{B}$ but also with the potentials $\mathfrak{D}_i, \mathfrak{D}_{ij}$, etc., derived from the kernels of the different moment of inertia tensors.

An alternative expression for $\mathfrak{B}_{ij}$ in terms of the scalar potential

$$\chi(\mathbf{x}) = -G \int_V \rho(\mathbf{x}') |\mathbf{x}-\mathbf{x}'| \, d\mathbf{x}', \tag{31}$$

instead of $\mathfrak{D}_i$, is sometimes useful. Thus, since

$$\frac{\partial \chi}{\partial x_i} = -G \int_V \rho(\mathbf{x}') \frac{x_i - x_i'}{|\mathbf{x}-\mathbf{x}'|} \, d\mathbf{x}' \tag{32}$$

and

$$\frac{\partial^2 \chi}{\partial x_i \, \partial x_j} = -\delta_{ij} G \int_V \frac{\rho(\mathbf{x}')}{|\mathbf{x}-\mathbf{x}'|} \, d\mathbf{x}' + G \int_V \rho(\mathbf{x}') \frac{(x_i - x_i')(x_j - x_j')}{|\mathbf{x}-\mathbf{x}'|^3} \, d\mathbf{x}', \tag{33}$$

we have

$$\mathfrak{B}_{ij} = \delta_{ij} \mathfrak{B} + \frac{\partial^2 \chi}{\partial x_i \, \partial x_j}. \tag{34}$$

Contracting this last equation, we obtain

$$\nabla^2 \chi = -2\mathfrak{B}; \tag{35}$$

and applying the operator ∇^2 to this equation, we find

$$\nabla^4 \chi = 8\pi G \rho. \tag{36}$$

Thus, instead of considering the hierarchy of the Newtonian potentials $\mathfrak{D}_i$, $\mathfrak{D}_{ij}$, etc., we might consider the hierarchy of scalar potentials $\mathfrak{B}$, χ, Φ, etc., determined by the equations

$$\nabla^2 \mathfrak{B} = -4\pi G \rho, \qquad \nabla^4 \chi = 8\pi G \rho, \qquad \nabla^6 \Phi = -32\pi G \rho, \quad \text{etc.} \tag{37}$$

As we shall see in detail in Chapter 3, not only the Newtonian potential $\mathfrak{B}$, but also the potentials $\mathfrak{D}_i$, $\mathfrak{D}_{ij}$, $\mathfrak{D}_{ijk}$, etc., can be explicitly given for homogeneous ellipsoids.

11. The virial equations of the various orders

We consider an ideal fluid described in terms of a density $\rho(\mathbf{x}, t)$ and an isotropic pressure $p(\mathbf{x}, t)$. We suppose further that the only force to which the fluid motions are subject, apart from the pressure gradients that may exist, is that derived from its own gravitation. Under these circumstances, the hydrodynamic equation governing the motions, referred to an inertial frame of reference, is given by

$$\rho \frac{du_i}{dt} = -\frac{\partial p}{\partial x_i} + \rho \frac{\partial \mathfrak{B}}{\partial x_i}, \tag{38}$$

where

$$\frac{d}{dt} = \frac{\partial}{\partial t} + u_j \frac{\partial}{\partial x_j} \tag{39}$$

is the *total* time derivative: it defines the variation with time of any quantity associated with a fluid element as we follow the element in its motion.

The virial equations of the various orders are now obtained by simply multiplying equation (38), successively, by 1, x_j, $x_j x_k$, $x_j x_k x_l$, etc., and integrating over the entire volume V instantaneously occupied by the fluid. But first we shall explicitly state a Lemma which we shall constantly be using in the derivations.

LEMMA. *If $Q(\mathbf{x}, t)$ is any attribute of a fluid element, then*

$$\frac{d}{dt} \int_V \rho(\mathbf{x},t) Q(\mathbf{x},t)\, d\mathbf{x} = \int_V \rho(\mathbf{x},t) \frac{dQ}{dt}\, d\mathbf{x}. \qquad (40)$$

This Lemma is manifest from the constancy of the mass of each fluid element during its motion:

$$\frac{d}{dt} \int_V \rho(\mathbf{x},t)\, d\mathbf{x} = \frac{dM}{dt} = 0. \qquad (41)$$

(a) The equations of the first order

The equations of the first order are obtained by simply integrating equation (38) over the instantaneous volume V occupied by the fluid. We obtain

$$\frac{d}{dt} \int_V \rho u_i\, d\mathbf{x} = - \int_V \frac{\partial p}{\partial x_i}\, d\mathbf{x} + \int_V \rho(\mathbf{x}) \frac{\partial \mathfrak{B}}{\partial x_i}\, d\mathbf{x}$$

$$= - \int_S p\, dS_i - G \int_V \int_V \rho(\mathbf{x})\rho(\mathbf{x}') \frac{x_i - x_i'}{|\mathbf{x} - \mathbf{x}'|^3}\, d\mathbf{x}d\mathbf{x}', \qquad (42)$$

where we have used the Lemma in writing the expression on the left-hand side. Also, in the second line on the right-hand side, S denotes the *free surface* bounding the volume V and dS_i is an element of this surface. The two integrals in the second line of equation (42) vanish: the first on account of the condition that the pressure must vanish on a free boundary and the second on account of the manifest antisymmetry of the integrand in $\mathbf{x}$ and $\mathbf{x}'$. We are thus left with (cf. equation (1))

$$\frac{d}{dt} \int_V \rho u_i\, d\mathbf{x} = \frac{d^2}{dt^2} \int_V \rho x_i\, d\mathbf{x} = \frac{d^2 I_i}{dt^2} = 0, \qquad (43)$$

an equation which simply expresses the uniform motion of the center of mass; it provides no essentially new information.

(b) The equations of the second order

These equations are obtained by multiplying equation (38) by x_j and integrating over the volume V. By making use of the Lemma and the definition of the tensor $\mathfrak{T}_{ij}$ (equation (9)), we can reduce the term arising from the left-hand side of the equation to the form

$$\int_V \rho \frac{du_i}{dt} x_j\, d\mathbf{x} = \int_V \rho\left[\frac{d}{dt}(u_i x_j) - u_i u_j\right] d\mathbf{x}$$

$$= \frac{d}{dt}\int_V \rho u_i x_j\, d\mathbf{x} - 2\mathfrak{T}_{ij}. \tag{44}$$

And the first term on the right-hand side of equation (38), after an integration by parts, gives (cf. equation (7))

$$-\int_V x_j \frac{\partial p}{\partial x_i}\, d\mathbf{x} = \delta_{ij}\int_V p\, d\mathbf{x} = \delta_{ij}\,\Pi, \tag{45}$$

where the vanishing of p on the boundary of V has been used. And finally, making use of equation (18), we have

$$\int_V \rho x_j \frac{\partial \mathfrak{B}}{\partial x_i}\, d\mathbf{x} = \mathfrak{W}_{ij}. \tag{46}$$

By combining the foregoing results, we obtain the basic equation

$$\frac{d}{dt}\int_V \rho u_i x_j\, d\mathbf{x} = 2\mathfrak{T}_{ij} + \mathfrak{W}_{ij} + \delta_{ij}\,\Pi. \tag{47}$$

Equation (47) provides a set of nine moment-equations.

Since all the tensors on the right-hand side of equation (47) are symmetric in i and j, the antisymmetric part of the tensor on the left-hand side must vanish, i.e.

$$\frac{d}{dt}\int_V \rho(u_i x_j - u_j x_i)\, d\mathbf{x} = 0. \tag{48}$$

This equation expresses simply the conservation of the angular momentum of the system.

Under conditions of a stationary state, equation (47) gives

$$2\mathfrak{T}_{ij} + \mathfrak{W}_{ij} = -\delta_{ij}\,\Pi. \tag{49}$$

Equation (49) provides six integral relations which must obtain whenever the conditions are stationary.

It may also be noted that by considering the symmetric part of the tensor on the left-hand side of equation (47), we obtain

$$\frac{1}{2}\frac{d}{dt}\int_V \rho(u_i x_j + u_j x_i)\,d\mathbf{x} = \frac{1}{2}\frac{d^2}{dt^2}\int_V \rho x_i x_j\,d\mathbf{x} = \frac{1}{2}\frac{d^2 I_{ij}}{dt^2}. \tag{50}$$

Hence we must have

$$\frac{1}{2}\frac{d^2 I_{ij}}{dt^2} = 2\mathfrak{T}_{ij} + \mathfrak{W}_{ij} + \delta_{ij}\,\Pi. \tag{51}$$

(c) *The equations of the third order*

These equations are obtained by multiplying equation (38) by $x_j x_k$ and integrating over the volume V.

Making use of the definitions (7) and (10), we can now write

$$\int_V \rho\,\frac{du_i}{dt}\,x_j x_k\,d\mathbf{x} = \int_V \rho\left[\frac{d}{dt}(u_i x_j x_k) - (u_i u_j x_k + u_i u_k x_j)\right]d\mathbf{x}$$

$$= \frac{d}{dt}\int_V \rho u_i x_j x_k\,d\mathbf{x} - 2(\mathfrak{T}_{ij;k} + \mathfrak{T}_{ik;j}) \tag{52}$$

and

$$-\int_V \frac{\partial p}{\partial x_i}\,x_j x_k\,d\mathbf{x} = \delta_{ij}\,\Pi_k + \delta_{ik}\,\Pi_j. \tag{53}$$

Similarly, by making use of the definitions (14) and (19), we have

$$\int_V \rho\,\frac{\partial\mathfrak{W}}{\partial x_i}\,x_j x_k\,d\mathbf{x}$$

$$= -G\int_V\int_V \rho(\mathbf{x})\rho(\mathbf{x}')\frac{x_j x_k(x_i - x_i')}{|\mathbf{x} - \mathbf{x}'|^3}\,d\mathbf{x}d\mathbf{x}'$$

$$= -\tfrac{1}{2}G\int_V\int_V \rho(\mathbf{x})\rho(\mathbf{x}')\frac{(x_j x_k - x_j' x_k')(x_i - x_i')}{|\mathbf{x} - \mathbf{x}'|^3}\,d\mathbf{x}d\mathbf{x}'$$

$$= -\tfrac{1}{2}G\int_V\int_V \rho(\mathbf{x})\rho(\mathbf{x}')\frac{[x_j(x_k - x_k') + x_k'(x_j - x_j')](x_i - x_i')}{|\mathbf{x} - \mathbf{x}'|^3}\,d\mathbf{x}d\mathbf{x}'$$

$$= -\tfrac{1}{2}\int_V \rho(\mathbf{x}')x_k'\mathfrak{B}_{ij}(\mathbf{x}')\,d\mathbf{x}' - \tfrac{1}{2}\int_V \rho(\mathbf{x})x_j\mathfrak{B}_{ik}(\mathbf{x})\,d\mathbf{x}$$

$$= \mathfrak{W}_{ij;k} + \mathfrak{W}_{ik;j}. \tag{54}$$

Combining the foregoing results, we obtain

$$\frac{d}{dt}\int_V \rho u_i x_j x_k\,d\mathbf{x} = 2(\mathfrak{T}_{ij;k} + \mathfrak{T}_{ik;j}) + \mathfrak{W}_{ij;k} + \mathfrak{W}_{ik;j} + \delta_{ij}\,\Pi_k + \delta_{ik}\,\Pi_j. \tag{55}$$

Equation (55) provides a set of eighteen moment-equations.

Under conditions of a stationary state, equation (55) gives

$$2(\mathfrak{T}_{ij;k}+\mathfrak{T}_{ik;j})+\mathfrak{W}_{ij;k}+\mathfrak{W}_{ik;j} = -\delta_{ij}\,\Pi_k-\delta_{ik}\,\Pi_j. \tag{56}$$

And finally, we may note that by symmetrizing equation (55) with respect to all three indices i, j, and k, we obtain

$$\frac{1}{6}\frac{d^2 I_{ijk}}{dt^2} = 2(\mathfrak{T}_{ij;k}+\mathfrak{T}_{jk;i}+\mathfrak{T}_{ki;j})+\mathfrak{W}_{ij;k}+\mathfrak{W}_{jk;i}+\mathfrak{W}_{ki;j}+$$
$$+\delta_{ij}\,\Pi_k+\delta_{jk}\,\Pi_i+\delta_{ki}\,\Pi_j. \tag{57}$$

(d) The equations of the fourth order

The equations of the fourth order are obtained by multiplying equation (38) by $x_j\,x_k\,x_l$ and integrating over the volume V. The reduction of the integrals involving du_i/dt and $\partial p/\partial x_i$ can be carried out as in the two earlier instances. But the reduction of the integral involving $\partial\mathfrak{W}/\partial x_i$ requires some explanation. The integral in question is

$$\int_V \rho\,\frac{\partial\mathfrak{W}}{\partial x_i}\,x_j\,x_k\,x_l\,d\mathbf{x} = -\tfrac{1}{2}G\int_V\int_V \rho(\mathbf{x})\rho(\mathbf{x}')\frac{(x_j\,x_k\,x_l-x_j'\,x_k'\,x_l')(x_i-x_i')}{|\mathbf{x}-\mathbf{x}'|^3}\,d\mathbf{x}d\mathbf{x}'. \tag{58}$$

Now replacing $(x_j\,x_k\,x_l-x_j'\,x_k'\,x_l')$ by

$$\tfrac{1}{3}[(x_j-x_j')x_k\,x_l+x_j'(x_k-x_k')x_l+x_j'\,x_k'(x_l-x_l')+$$
$$+(x_k-x_k')x_l\,x_j+x_k'(x_l-x_l')x_j+x_k'\,x_l'(x_j-x_j')+$$
$$+(x_l-x_l')x_j\,x_k+x_l'(x_j-x_j')x_k+x_l'\,x_j'(x_k-x_k')], \tag{59}$$

and making use of the definitions (14), (19), (20), and (21), we observe that the integral in question is

$$\tfrac{1}{3}(2\mathfrak{W}_{ij;kl}+2\mathfrak{W}_{ik;lj}+2\mathfrak{W}_{il;jk}+\mathfrak{W}_{ij;k;l}+\mathfrak{W}_{ik;l;j}+\mathfrak{W}_{il;j;k}). \tag{60}$$

The virial equation of the fourth order is, then,

$$\frac{d}{dt}\int_V \rho u_i\,x_j\,x_k\,x_l\,d\mathbf{x} = 2(\mathfrak{T}_{ij;kl}+\mathfrak{T}_{ik;lj}+\mathfrak{T}_{il;jk})+$$
$$+\tfrac{1}{3}(2\mathfrak{W}_{ij;kl}+2\mathfrak{W}_{ik;lj}+2\mathfrak{W}_{il;jk}+\mathfrak{W}_{ij;k;l}+\mathfrak{W}_{ik;l;j}+\mathfrak{W}_{il;j;k})+$$
$$+\delta_{ij}\,\Pi_{kl}+\delta_{ik}\,\Pi_{lj}+\delta_{il}\,\Pi_{jk}. \tag{61}$$

Equation (61) provides an additional set of thirty moment-equations.

In view of their complexity and numbers, the fourth-order virial equations have not been greatly used. On this account, we shall restrict our consideration only to the second- and the third-order equations in the remainder of this chapter.

12. The virial equations in a rotating frame of reference

In the subsequent chapters, we shall be concerned, mainly, with configurations that are rotating uniformly with an angular velocity $\boldsymbol{\Omega}$. For problems relating to the equilibrium and the stability of these configurations, it is convenient to refer the equations of motion to a frame of reference rotating with this angular velocity. In such a rotating frame, equation (38) is replaced by

$$\rho \frac{du_i}{dt} = -\frac{\partial p}{\partial x_i} + \rho \frac{\partial \mathfrak{B}}{\partial x_i} + \tfrac{1}{2}\rho \frac{\partial}{\partial x_i} |\boldsymbol{\Omega} \times \mathbf{x}|^2 + 2\rho\epsilon_{ilm} u_l \Omega_m, \qquad (62)$$

where $\tfrac{1}{2}|\boldsymbol{\Omega} \times \mathbf{x}|^2$ and $2\mathbf{u} \times \boldsymbol{\Omega}$ represent the centrifugal potential and the Coriolis acceleration, respectively. The contribution of these two additional terms to the virial equations are readily written down.

(a) The equations of the second order

In place of equation (47) we now have

$$\frac{d}{dt} \int_V \rho u_i x_j \, d\mathbf{x} = 2\mathfrak{T}_{ij} + \mathfrak{B}_{ij} + \Omega^2 I_{ij} - \Omega_i \Omega_k I_{kj} + \delta_{ij} \Pi +$$

$$+ 2\epsilon_{ilm} \Omega_m \int_V \rho u_l x_j \, d\mathbf{x}. \qquad (63)$$

If the state is steady, equation (63) gives

$$2\mathfrak{T}_{ij} + \mathfrak{B}_{ij} + \Omega^2 I_{ij} - \Omega_i \Omega_k I_{kj} + \delta_{ij} \Pi +$$

$$+ 2\epsilon_{ilm} \Omega_m \int_V \rho u_l x_j \, d\mathbf{x} = 0. \qquad (64)$$

The particular relations which equation (64) provides when there are no relative motions (in the rotating frame considered) are of interest. In that case, it is convenient to choose the x_3-axis to be along the direction of $\boldsymbol{\Omega}$; with this choice of the orientation of the coordinate axes, equation (64) under static conditions gives

$$\mathfrak{B}_{ij} + \Omega^2(I_{ij} - \delta_{i3} I_{3j}) = -\delta_{ij} \Pi. \qquad (65)$$

Writing out equation (65) explicitly for the different components, we have

$$\mathfrak{B}_{11} + \Omega^2 I_{11} = \mathfrak{B}_{22} + \Omega^2 I_{22} = \mathfrak{B}_{33} = -\Pi, \qquad (66)$$

$$\mathfrak{B}_{12} + \Omega^2 I_{12} = \mathfrak{B}_{21} + \Omega^2 I_{21} = 0,$$

$$\mathfrak{B}_{13} + \Omega^2 I_{13} = 0, \qquad \mathfrak{B}_{31} = 0,$$

$$\text{and} \qquad \mathfrak{B}_{23} + \Omega^2 I_{23} = 0, \qquad \mathfrak{B}_{32} = 0. \qquad (67)$$

In view of the symmetry of the tensors $\mathfrak{B}_{ij}$ and I_{ij}, it follows from equations (67) that

$$\mathfrak{B}_{13} = \mathfrak{B}_{23} = 0 \quad \text{and} \quad I_{13} = I_{23} = 0. \qquad (68)$$

But it is not required that $\mathfrak{B}_{12}$ and I_{12} vanish identically; it is only required that they are related by

$$\mathfrak{B}_{12} + \Omega^2 I_{12} = 0. \qquad (69)$$

It is often convenient to eliminate Π from the relations (66) and obtain two integral relations which are in no way dependent on the constitutive relations that may exist. Two such relations are

$$\mathfrak{W}_{11}-\mathfrak{W}_{22}+\Omega^2(I_{11}-I_{22}) = 0 \tag{70}$$

and

$$\mathfrak{W}_{11}+\mathfrak{W}_{22}-2\mathfrak{W}_{33}+\Omega^2(I_{11}+I_{22}) = 0. \tag{71}$$

It is to be particularly noted that *equations (69)–(71) do not require $I_{11} = I_{22}$ and $\mathfrak{W}_{11} = \mathfrak{W}_{22}$ as conditions for equilibrium; and neither is it required that $\mathfrak{W}_{12} = 0$ and $I_{12} = 0$.* The maximum that can be deduced about the tensors $\mathfrak{W}_{ij}$ and I_{ij} (apart from the symmetry required by definitions) is that they are reducible to the forms

$$\mathfrak{W}_{ij} = \begin{vmatrix} \mathfrak{W}_{11} & \mathfrak{W}_{12} & 0 \\ \mathfrak{W}_{21} & \mathfrak{W}_{22} & 0 \\ 0 & 0 & \mathfrak{W}_{33} \end{vmatrix} \quad \text{and} \quad I_{ij} = \begin{vmatrix} I_{11} & I_{12} & 0 \\ I_{21} & I_{22} & 0 \\ 0 & 0 & I_{33} \end{vmatrix}. \tag{72}$$

However, by a rotation about the x_3-axis, we can arrange that I_{12} vanishes; then $\mathfrak{W}_{12}$ must also vanish and the tensors $\mathfrak{W}_{ij}$ and I_{ij} become diagonal.

It is important to observe that the virial equations do not provide any substance to the common expectation that symmetry about the rotational axis should be associated with any form produced by an axisymmetric rotational field. As we have already noted in Chapter 1, Jacobi's discovery in 1834 that a rotating incompressible mass can assume the form of a genuine tri-axial ellipsoid (if its angular momentum exceeds a certain determinate value) was most unexpected.

(b) The equations of the third order

In place of equation (55) we now have

$$\frac{d}{dt}\int_V \rho u_i x_j x_k \, d\mathbf{x} = 2(\mathfrak{T}_{ij;k}+\mathfrak{T}_{ik;j})+\mathfrak{W}_{ij;k}+\mathfrak{W}_{ik;j}+$$

$$+\Omega^2 I_{ijk}-\Omega_i\Omega_l I_{ljk}+\delta_{ij}\,\Pi_k+\delta_{ik}\,\Pi_j+$$

$$+2\epsilon_{ilm}\Omega_m\int_V \rho u_l x_j x_k \, d\mathbf{x}. \tag{73}$$

If the state is steady, equation (73) gives

$$2(\mathfrak{T}_{ij;k}+\mathfrak{T}_{ik;j})+\mathfrak{W}_{ij;k}+\mathfrak{W}_{ik;j}+\Omega^2 I_{ijk}-\Omega_i\Omega_l I_{ljk}+$$

$$+2\epsilon_{ilm}\Omega_m\int_V \rho u_l x_j x_k \, d\mathbf{x} = -\delta_{ij}\,\Pi_k-\delta_{ik}\,\Pi_j. \tag{74}$$

The particular relations which equation (74) provides when there are no relative motions (in the rotating frame considered) are of interest.

In that case, it is convenient to choose the x_3-axis to be along the direction of Ω. Equation (74) then gives

$$\mathfrak{W}_{ij;k}+\mathfrak{W}_{ik;j}+\Omega^2(I_{ijk}-\delta_{i3}I_{3jk}) = -\delta_{ij}\Pi_k-\delta_{ik}\Pi_j. \tag{75}$$

Suspending the summation convention, and letting $i \neq j \neq k$ denote distinct indices, we can group the eighteen relations included in equation (75) as follows:

$$2\mathfrak{W}_{ii;i}+\Omega^2(I_{iii}-\delta_{i3}I_{3ii}) = -2\Pi_i, \tag{76}$$

$$\mathfrak{W}_{jj;i}+\mathfrak{W}_{ij;j}+\Omega^2(I_{ijj}-\delta_{j3}I_{3ij}) = -\Pi_i, \tag{77}$$

$$2\mathfrak{W}_{ij;j}+\Omega^2(I_{ijj}-\delta_{i3}I_{3jj}) = 0, \tag{78}$$

and
$$\mathfrak{W}_{ij;k}+\mathfrak{W}_{ik;j}+\Omega^2(I_{ijk}-\delta_{i3}I_{3jk}) = 0 \tag{79}$$

$$(i \neq j \neq k;\ \text{and no summation over repeated indices}).$$

There are three equations of type (76); six equations of each of the types (77) and (78); and (only) three equations of type (79) (since it is symmetrical in j and k).

Writing out explicitly the three relations provided by equation (79), we have

$$\mathfrak{W}_{12;3}+\mathfrak{W}_{13;2}+\Omega^2 I_{123} = 0,$$

$$\mathfrak{W}_{23;1}+\mathfrak{W}_{21;3}+\Omega^2 I_{123} = 0,$$

and
$$\mathfrak{W}_{31;2}+\mathfrak{W}_{32;1} = 0. \tag{80}$$

From these equations it follows that

$$\mathfrak{W}_{23;1} = \mathfrak{W}_{13;2} = 0 \quad \text{and} \quad \mathfrak{W}_{12;3}+\Omega^2 I_{123} = 0. \tag{81}$$

And we may also note in this connection that by letting $i = 3$ and $j = 1$ or 2 in equation (78), we find

$$\mathfrak{W}_{31;1} = \mathfrak{W}_{32;2} = 0. \tag{82}$$

As in the case of the second-order equations, it is useful to eliminate the Π_i's from the relations (76)–(79) and consider the fifteen relations which are in no way dependent on the constitutive relations that may exist. By combining equations (76), (77), and (78) with the factors 1, -2, and -1, respectively,† we find

$$S_{ijj}+\Omega^2[I_{iii}-3I_{ijj}-\delta_{i3}(I_{3ii}-I_{3jj})+2\delta_{j3}I_{3ij}] = 0, \tag{83}$$

where
$$S_{ijj} = -4\mathfrak{W}_{ij;j}-2\mathfrak{W}_{jj;i}+2\mathfrak{W}_{ii;i} \tag{84}$$

$$(i \neq j;\ \text{and no summation over repeated indices}).$$

Equation (83) together with equations (78) and (79) provide the total of fifteen equations required.

---

† The reason for the choice of this particular combination is for later convenience (see Chapter 6, § 42).

The equations of the fourth order can be similarly discussed. But since we shall not have occasion to use them to any substantial extent, we shall omit their consideration.

13. The variations resulting from small departures from equilibrium. The Lagrangian and the Eulerian changes

The virial equations are useful not only in providing integral relations which must obtain under conditions of equilibrium; they are also useful in investigations pertaining to departures from equilibrium. For these latter purposes, it is necessary to obtain the "linearized" versions of the virial equations which govern such departures. As a preliminary to the derivation of such equations, we shall consider in this section certain general principles that underlie the analytical treatment of "neighboring flows."

Let Q be any attribute (such as velocity, density, or gravitational potential) that can be associated with an element of the fluid. In general the value of the attribute will depend on the instantaneous location $\mathbf{x}$ of the element. We shall denote by $Q(\mathbf{x}, t)$ the value of the attribute which an element of fluid, located at $\mathbf{x}$ and at time t, has. In particular, let $Q_0(\mathbf{x}, t)$ be this function for some initial flow. (In most instances the initial flow considered will be stationary; but this restriction is not essential to our present considerations.) Let the initial flow be slightly perturbed. We shall describe the perturbed flow by specifying, at each instant, the *displacement* $\boldsymbol{\xi}(\mathbf{x}, t)$ which an element of the fluid, in the perturbed flow, has experienced *relative* to its location $\mathbf{x}$ at time t in the unperturbed flow. Defined in this manner, $\boldsymbol{\xi}(\mathbf{x}, t)$ is called *the Lagrangian displacement caused by the perturbation*. We shall suppose that $\boldsymbol{\xi}(\mathbf{x}, t)$ is small and that quantities of the second and higher orders in $\boldsymbol{\xi}$ can be ignored. In other words, we shall be concerned only with perturbations causing *infinitesimal Lagrangian displacements*.

Now consider the value of an attribute Q, in the perturbed flow, of an element of fluid located at $\mathbf{x} + \boldsymbol{\xi}(\mathbf{x}, t)$ at time t. This element, in the unperturbed flow, would have been at $\mathbf{x}$ (at the same time t). We call the difference

$$\Delta Q = Q(\mathbf{x} + \boldsymbol{\xi}(\mathbf{x}, t), t) - Q_0(\mathbf{x}, t) \tag{85}$$

the *Lagrangian change in Q caused by the perturbation*. This change is clearly not the same as that which will be noted by an external observer who compares the values of Q, in the perturbed and in the unperturbed flows, *at the same location and at the same time*. This latter change

$$\delta Q = Q(\mathbf{x}, t) - Q_0(\mathbf{x}, t) \tag{86}$$

is called the *Eulerian change* in Q. It is clear from the definitions (85) and (86) that

$$\Delta Q = \delta Q + \xi_j \frac{\partial Q}{\partial x_j}. \tag{87}$$

Since this relation is valid for any attribute, we may write that the *operations* Δ and δ are related by

$$\Delta = \delta + \xi_j \frac{\partial}{\partial x_j}. \tag{88}$$

The velocity $\mathbf{u}$ is clearly an attribute of an element of the fluid. The Lagrangian change $\Delta\mathbf{u}$ in $\mathbf{u}$ is, by definition, the velocity, with which the element (in the perturbed flow) finds itself at $\mathbf{x}+\boldsymbol{\xi}(\mathbf{x}, t)$ at time t, *relative* to the velocity of the same element (in the unperturbed flow) at $\mathbf{x}$ and at time t. From this definition it follows that (cf. equation (39))

$$\Delta\mathbf{u} = \frac{\partial\boldsymbol{\xi}}{\partial t} + u_j \frac{\partial\boldsymbol{\xi}}{\partial x_j} = \frac{d\boldsymbol{\xi}}{dt}, \tag{89}$$

where $\mathbf{u}$ defines the velocity field in the unperturbed flow.

Some elementary consequences of the foregoing definitions will now be noted.

If the attribute Q is not *intrinsic* to the element (like pressure or density) but something which it assumes simply by virtue of its location (as the value, for example, of the centrifugal potential $\frac{1}{2}|\boldsymbol{\Omega}\times\mathbf{x}|^2$ acting on it) then its Eulerian change vanishes; but the Lagrangian change does not. If we restrict the symbol $F(\mathbf{x})$ to such an *extrinsic attribute*, then

$$\delta F = 0 \quad \text{and} \quad \Delta F = \xi_j \frac{\partial F}{\partial x_j}. \tag{90}$$

On the other hand, for an *intrinsic attribute*, such as density or pressure, neither the Eulerian nor the Lagrangian change, generally, vanishes. In particular, from the conservation of the mass of any element of fluid during its motion, it follows that

$$\Delta\rho = -\rho\,\text{div}\,\boldsymbol{\xi} \tag{91}$$

and, in consequence,

$$\delta\rho = -\rho\,\text{div}\,\boldsymbol{\xi} - \boldsymbol{\xi}.\text{grad}\,\rho = -\text{div}(\rho\boldsymbol{\xi}). \tag{92}$$

If the changes in pressure of a fluid element, accompanying the changes in density, take place adiabatically, then

$$\frac{\Delta p}{p} = \gamma \frac{\Delta\rho}{\rho}, \tag{93}$$

where γ denotes the "ratio of the specific heats." Accordingly, by equations (88) and (91),

$$\Delta p = -\gamma p \operatorname{div} \boldsymbol{\xi}$$

and

$$\delta p = -\gamma p \operatorname{div} \boldsymbol{\xi} - \boldsymbol{\xi}.\operatorname{grad} p. \tag{94}$$

It is evident from the definition of Eulerian changes (namely, that they are changes in an attribute at a given location and at a given time) that the operations of δ and of partial differentiation commute:

$$\delta \frac{\partial Q}{\partial t} = \frac{\partial}{\partial t} \delta Q \quad \text{and} \quad \delta \frac{\partial Q}{\partial x_j} = \frac{\partial}{\partial x_j} \delta Q. \tag{95}$$

In contrast, the operations of Δ and of partial differentiation do not commute. Thus,

$$\Delta \left(\frac{\partial Q}{\partial x_j}\right) = \left(\delta + \xi_k \frac{\partial}{\partial x_k}\right)\frac{\partial Q}{\partial x_j} = \frac{\partial}{\partial x_j} \delta Q + \xi_k \frac{\partial^2 Q}{\partial x_k \partial x_j}$$

$$= \frac{\partial}{\partial x_j}\left(\Delta Q - \xi_k \frac{\partial Q}{\partial x_k}\right) + \xi_k \frac{\partial^2 Q}{\partial x_k \partial x_j}, \tag{96}$$

or

$$\Delta \left(\frac{\partial Q}{\partial x_j}\right) = \frac{\partial}{\partial x_j}(\Delta Q) - \frac{\partial \xi_k}{\partial x_j}\frac{\partial Q}{\partial x_k}. \tag{97}$$

Similarly,

$$\Delta \left(\frac{\partial Q}{\partial t}\right) = \frac{\partial}{\partial t}(\Delta Q) - \frac{\partial \xi_k}{\partial t}\frac{\partial Q}{\partial x_k}. \tag{98}$$

An important consequence of the foregoing relations is that *the operations*

$$\Delta \quad \text{and} \quad \frac{d}{dt} = \frac{\partial}{\partial t} + u_j \frac{\partial}{\partial x_j} \tag{99}$$

commute. Thus, by equations (97) and (98)

$$\Delta \left(\frac{dQ}{dt}\right) = \Delta \left(\frac{\partial Q}{\partial t} + u_j \frac{\partial Q}{\partial x_j}\right)$$

$$= \frac{\partial}{\partial t}(\Delta Q) - \frac{\partial \xi_k}{\partial t}\frac{\partial Q}{\partial x_k} + u_j \frac{\partial}{\partial x_j}\Delta Q - u_j \frac{\partial \xi_k}{\partial x_j}\frac{\partial Q}{\partial x_k} + \Delta u_k \frac{\partial Q}{\partial x_k}, \tag{100}$$

or

$$\Delta \left(\frac{dQ}{dt}\right) = \frac{d}{dt}(\Delta Q) + \left(\Delta u_k - \frac{d\xi_k}{dt}\right)\frac{\partial Q}{\partial x_k}. \tag{101}$$

By virtue of equation (89) the second term on the right-hand side of this last equation vanishes and the theorem stated follows directly. On the other hand

$$\delta \left(\frac{dQ}{dt}\right) = \frac{d}{dt}(\Delta Q) - \xi_j \frac{\partial}{\partial x_j}\left(\frac{dQ}{dt}\right) \tag{102}$$

14. The equations governing small departures from a given initial flow

The equation governing an initial flow, in a frame of reference rotating uniformly with an angular velocity Ω, can be written in the form

$$\frac{du_i}{dt} = -\frac{1}{\rho}\frac{\partial p}{\partial x_i} + \frac{\partial U}{\partial x_i} + 2\epsilon_{ilm}u_l\Omega_m, \tag{103}$$

where
$$U = \mathfrak{B} + \tfrac{1}{2}|\Omega \times \mathbf{x}|^2 \tag{104}$$

is the total potential. We shall consider the equations governing both the Eulerian and the Lagrangian changes caused by a small perturbation of the initial flow.

As we have noted in § 13, the operation δ does not permute with d/dt; but by virtue of equations (88), (89), (101), and (102) we can write

$$\delta\left(\frac{du_i}{dt}\right) = \Delta\left(\frac{du_i}{dt}\right) - \xi_j\frac{\partial}{\partial x_j}\left(\frac{du_i}{dt}\right) = \frac{d}{dt}(\Delta u_i) - \xi_j\frac{\partial}{\partial x_j}\left(\frac{du_i}{dt}\right)$$

$$= \frac{d^2\xi_i}{dt^2} - \xi_j\frac{\partial}{\partial x_j}\left(\frac{du_i}{dt}\right). \tag{105}$$

The effect of the operator δ on the right-hand side of equation (103) can be written down at once in view of its permutability with partial differentiations. Thus, again making use of equations (88) and (89), we obtain

$$\frac{d^2\xi_i}{dt^2} - \xi_j\frac{\partial}{\partial x_j}\left(\frac{du_i}{dt}\right) = \frac{\delta\rho}{\rho^2}\frac{\partial p}{\partial x_i} - \frac{1}{\rho}\frac{\partial\delta p}{\partial x_i} + \frac{\partial\delta U}{\partial x_i} +$$

$$+ 2\epsilon_{ilm}\Omega_m\left(\frac{d\xi_l}{dt} - \xi_j\frac{\partial u_l}{\partial x_j}\right). \tag{106}$$

Considering next the effect of operating on equation (103) by Δ and remembering that now Δ and d/dt do permute, we obtain (by making further use of the relations (89), (97), and (98))

$$\frac{d^2\xi_i}{dt^2} = \frac{\Delta\rho}{\rho^2}\frac{\partial p}{\partial x_i} - \frac{1}{\rho}\frac{\partial\Delta p}{\partial x_i} + \frac{\partial\Delta U}{\partial x_i} + 2\epsilon_{ilm}\frac{d\xi_l}{dt}\Omega_m +$$

$$+ \frac{\partial\xi_j}{\partial x_i}\left(\frac{1}{\rho}\frac{\partial p}{\partial x_j} - \frac{\partial U}{\partial x_j}\right). \tag{107}$$

It can be readily verified that by replacing the Lagrangian changes, which occur on the right-hand side of equation (107), in terms of their respective Eulerian changes in accordance with equation (88), we recover equation (106) (if use is made of the equation governing the initial flow). If the initial state considered is one of static equilibrium (in the rotating frame), then

$$u_i = 0 \quad \text{and} \quad \frac{1}{\rho}\frac{\partial p}{\partial x_i} = \frac{\partial U}{\partial x_i}, \tag{108}$$

and equations (106) and (107) become

$$\frac{\partial^2 \xi_i}{\partial t^2} = \frac{\delta\rho}{\rho^2}\frac{\partial p}{\partial x_i} - \frac{1}{\rho}\frac{\partial \delta p}{\partial x_i} + \frac{\partial \delta U}{\partial x_i} + 2\epsilon_{ilm}\frac{\partial \xi_l}{\partial t}\Omega_m \tag{109}$$

and

$$\frac{\partial^2 \xi_i}{\partial t^2} = \frac{\Delta\rho}{\rho^2}\frac{\partial p}{\partial x_i} - \frac{1}{\rho}\frac{\partial \Delta p}{\partial x_i} + \frac{\partial \Delta U}{\partial x_i} + 2\epsilon_{ilm}\frac{\partial \xi_l}{\partial t}\Omega_m. \tag{110}$$

We observe that *in this case the equations governing the Eulerian and the Lagrangian changes are of identical forms.*

15. The first variations of the various integral properties and the linearized form of the virial equations

Consider the integral

$$J = \int_V Q_0(\mathbf{x}, t)\, d\mathbf{x} \tag{111}$$

of an attribute Q (in an initial flow) over the volume V instantaneously occupied by the fluid. The same integral defined with respect to the perturbed flow is

$$\int_{V+\Delta V} Q(\mathbf{x}, t)\, d\mathbf{x}, \tag{112}$$

where $V+\Delta V$ denotes the volume derived from V by subjecting its boundary to the displacement $\boldsymbol{\xi}$. We call

$$\delta J = \int_{V+\Delta V} Q(\mathbf{x}, t)\, d\mathbf{x} - \int_V Q_0(\mathbf{x}, t)\, d\mathbf{x} \tag{113}$$

the first variation of the integral J caused by the perturbation. By the transformation

$$\mathbf{x} - \xi(\mathbf{x}, t) = \mathbf{x}', \tag{114}$$

in the first integral, we can make the volumes, over which the two integrals defining δJ are effected, the same. But the integrand of the first integral will then have the additional factor $(1+\text{div}\,\boldsymbol{\xi})$ representing the Jacobian of the transformation (114). Therefore, to the first order in the displacement,

$$\delta J = \int_V (1+\text{div}\,\boldsymbol{\xi})Q(\mathbf{x}'+\boldsymbol{\xi}, t)\, d\mathbf{x}' - \int_V Q_0(\mathbf{x}, t)\, d\mathbf{x}$$

$$= \int_V [Q(\mathbf{x}+\boldsymbol{\xi}, t) - Q_0(\mathbf{x}, t) + Q_0(\mathbf{x}, t)\text{div}\,\boldsymbol{\xi}]\, d\mathbf{x}, \tag{115}$$

or

$$\delta J = \int_V (\Delta Q + Q\,\text{div}\,\boldsymbol{\xi})\, d\mathbf{x}, \tag{116}$$

where ΔQ is the Lagrangian change in Q consequent to the displacement $\boldsymbol{\xi}$. Equation (116) is the basic formula of the theory.

Applying the formula (116) to the integral

$$\int_V \rho \, d\mathbf{x}, \tag{117}$$

we obtain

$$\delta \int_V \rho \, d\mathbf{x} = \int_V (\Delta\rho + \rho \operatorname{div} \boldsymbol{\xi}) \, d\mathbf{x} = 0, \tag{118}$$

by virtue of the relation (91). The result (118), required by the constancy of the mass included in an *arbitrary* volume V occupied by the *same* elements of the fluid, can be used, conversely, to *deduce* the relation (91).

An immediate consequence of the result (118) is that for any attribute

$$\delta \int_V \rho Q \, d\mathbf{x} = \int_V \rho \Delta Q \, d\mathbf{x}. \tag{119}$$

In particular, if $F(x)$ is any extrinsic attribute, then by equation (90),

$$\delta \int_V \rho F \, d\mathbf{x} = \int_V \rho \xi_j \frac{\partial F}{\partial x_j} \, d\mathbf{x}; \tag{120}$$

or more generally,

$$\delta \int_V \int_V \rho(\mathbf{x})\rho(\mathbf{x}')F(\mathbf{x}, \mathbf{x}') \, d\mathbf{x}d\mathbf{x}'$$

$$= \int_V \int_V \rho(\mathbf{x})\rho(\mathbf{x}')\left[\xi_j(\mathbf{x})\frac{\partial}{\partial x_j} + \xi_j(\mathbf{x}')\frac{\partial}{\partial x_j'}\right]F(\mathbf{x}, \mathbf{x}') \, d\mathbf{x}d\mathbf{x}'. \tag{121}$$

Equations (120) and (121) can be used to derive the first variations of many of the integrals appearing in the virial equations. Thus

$$\delta I_{ij} = \delta \int_V \rho x_i x_j \, d\mathbf{x} = \int_V \rho(\xi_i x_j + \xi_j x_i) \, d\mathbf{x}; \tag{122}$$

and similarly

$$\delta I_{ijk} = \delta \int_V \rho x_i x_j x_k \, d\mathbf{x} = \int_V \rho(\xi_i x_j x_k + \xi_j x_k x_i + \xi_k x_i x_j) \, d\mathbf{x}. \tag{123}$$

We shall have occasion to define

$$V_{ij} = \delta I_{ij} \quad \text{and} \quad V_{ijk} = \delta I_{ijk}; \tag{124}$$

the corresponding *unsymmetrized* quantities will be denoted by

$$V_{i;j} = \int_V \rho \xi_i x_j \, d\mathbf{x} \quad \text{and} \quad V_{i;jk} = \int_V \rho \xi_i x_j x_k \, d\mathbf{x}. \tag{125}$$

We shall now obtain the first variations of the potential-energy tensors

$\mathfrak{W}_{ij}$ and $\mathfrak{W}_{ij;k}$. Considering first $\delta\mathfrak{W}_{ij}$, we can apply the result (121) to the integral defining $\mathfrak{W}_{ij}$; thus (cf. equation (17))

$$-2\delta\mathfrak{W}_{ij} = \delta G \int_V \int_V \rho(\mathbf{x})\rho(\mathbf{x}')\frac{(x_i-x_i')(x_j-x_j')}{|\mathbf{x}-\mathbf{x}'|^3}\,d\mathbf{x}d\mathbf{x}'$$

$$= G\int_V d\mathbf{x}\,\rho(\mathbf{x})\xi_l(\mathbf{x})\frac{\partial}{\partial x_l}\int_V d\mathbf{x}'\rho(\mathbf{x}')\frac{(x_i-x_i')(x_j-x_j')}{|\mathbf{x}-\mathbf{x}'|^3}+$$

$$+G\int_V d\mathbf{x}'\rho(\mathbf{x}')\xi_l(\mathbf{x}')\frac{\partial}{\partial x_l'}\int_V d\mathbf{x}\,\rho(\mathbf{x})\frac{(x_i'-x_i)(x_j'-x_j)}{|\mathbf{x}-\mathbf{x}'|^3}$$

$$= 2\int_V \rho(\mathbf{x})\xi_l(\mathbf{x})\frac{\partial\mathfrak{B}_{ij}(\mathbf{x})}{\partial x_l}\,d\mathbf{x}, \tag{126}$$

or

$$\delta\mathfrak{W}_{ij} = -\int_V \rho\xi_l\frac{\partial\mathfrak{B}_{ij}}{\partial x_l}\,d\mathbf{x}. \tag{127}$$

Contracting this last equation, we obtain the known result

$$\delta\mathfrak{W} = -\int_V \rho\xi_l\frac{\partial\mathfrak{B}}{\partial x_l}\,d\mathbf{x}. \tag{128}$$

Considering next the first variation of $\mathfrak{W}_{ij;k}$, we have

$$-2\delta\mathfrak{W}_{ij;k} = \delta\int_V \rho x_k\,\mathfrak{B}_{ij}\,d\mathbf{x}$$

$$= \int_V \rho\xi_k\,\mathfrak{B}_{ij}\,d\mathbf{x}+$$

$$+G\int_V d\mathbf{x}\,\rho(\mathbf{x})x_k\,\xi_l(\mathbf{x})\frac{\partial}{\partial x_l}\int_V d\mathbf{x}'\rho(\mathbf{x}')\frac{(x_i-x_i')(x_j-x_j')}{|\mathbf{x}-\mathbf{x}'|^3}+$$

$$+G\int_V d\mathbf{x}'\rho(\mathbf{x}')\xi_l(\mathbf{x}')\frac{\partial}{\partial x_l'}\int_V d\mathbf{x}\rho(\mathbf{x})x_k\frac{(x_i'-x_i)(x_j'-x_j)}{|\mathbf{x}-\mathbf{x}'|^3}$$

$$= \int_V \rho\xi_k\,\mathfrak{B}_{ij}\,d\mathbf{x}+\int_V \rho x_k\xi_l\frac{\partial\mathfrak{B}_{ij}}{\partial x_l}\,d\mathbf{x}+\int_V \rho\xi_l\frac{\partial\mathfrak{D}_{ij;k}}{\partial x_l}\,d\mathbf{x}, \tag{129}$$

where we have made use of the definition of $\mathfrak{D}_{ij;k}$ given in equation (22). Alternatively, we can write

$$-2\delta\mathfrak{W}_{ij;k} = \int_V \rho\xi_l\frac{\partial}{\partial x_l}(x_k\,\mathfrak{B}_{ij}+\mathfrak{D}_{ij;k})\,d\mathbf{x}. \tag{130}$$

The first variations of $\mathfrak{T}_{ij}$ and $\mathfrak{T}_{ij;k}$ follow readily from the definitions of these quantities and equations (89) and (119). Thus,

$$2\delta\mathfrak{T}_{ij} = \delta \int_V \rho u_i u_j\, d\mathbf{x} = \int_V (\rho u_j\,\Delta u_i + \rho u_i\,\Delta u_j)\, d\mathbf{x}$$

$$= \int_V \rho\left(u_j\frac{d\xi_i}{dt} + u_i\frac{d\xi_j}{dt}\right) d\mathbf{x} \tag{131}$$

and

$$2\delta\mathfrak{T}_{ij;k} = \int_V \rho[x_k(u_j\,\Delta u_i + u_i\,\Delta u_j) + u_i u_j \xi_k]\, d\mathbf{x}. \tag{132}$$

Similarly,

$$\delta \int_V \rho u_l x_j\, d\mathbf{x} = \int_V \rho\Delta u_l x_j\, d\mathbf{x} + \int_V \rho u_l \xi_j\, d\mathbf{x} \tag{133}$$

and

$$\delta \int_V \rho u_l x_j x_k\, d\mathbf{x} = \int_V \rho\Delta u_l x_j x_k\, d\mathbf{x} + \int_V \rho u_l(\xi_j x_k + \xi_k x_j)\, d\mathbf{x}. \tag{134}$$

The evaluation of the first variation of the integral on the right-hand side of equation (63) is somewhat more complicated. Clearly the operations of δ and d/dt *outside* an integral sign are permutable. We can, therefore, write

$$\delta\frac{d}{dt} \int_V \rho u_i x_j\, d\mathbf{x} = \frac{d}{dt} \int_V \rho\Delta u_i x_j\, d\mathbf{x} + \frac{d}{dt} \int_V \rho\xi_j u_i\, d\mathbf{x}. \tag{135}$$

Eliminating the first integral on the right-hand side of equation (135) with the aid of the formula (cf. equation (125))

$$\frac{dV_{i;j}}{dt} = \frac{d}{dt} \int_V \rho\xi_i x_j\, d\mathbf{x} = \int_V \rho\Delta u_i x_j\, d\mathbf{x} + \int_V \rho\xi_i u_j\, d\mathbf{x}, \tag{136}$$

we obtain

$$\delta\frac{d}{dt} \int_V \rho u_i x_j\, d\mathbf{x} = \frac{d^2 V_{i;j}}{dt^2} + \frac{d}{dt} \int_V \rho(\xi_j u_i - \xi_i u_j)\, d\mathbf{x}. \tag{137}$$

In the same way, we find

$$\delta\frac{d}{dt} \int_V \rho u_i x_j x_k\, d\mathbf{x} = \frac{d}{dt} \int_V \rho\Delta u_i x_j x_k\, d\mathbf{x} +$$

$$+ \frac{d}{dt} \int_V \rho u_i(\xi_j x_k + \xi_k x_j)\, d\mathbf{x}; \tag{138}$$

and since

$$\frac{dV_{i;jk}}{dt} = \int_V \rho\Delta u_i x_j x_k\, d\mathbf{x} + \int_V \rho\xi_i(u_j x_k + u_k x_j)\, d\mathbf{x}, \tag{139}$$

we can write

$$\delta\frac{d}{dt} \int_V \rho u_i x_j x_k\, d\mathbf{x} = \frac{d^2 V_{i;jk}}{dt^2} + \frac{d}{dt} \int_V \rho[(\xi_j x_k + \xi_k x_j)u_i -$$

$$- (u_j x_k + u_k x_j)\xi_i]\, d\mathbf{x}. \tag{140}$$

In Chapter 7, devoted to the Riemann ellipsoids, we shall be interested in the special case when the velocity field in an initial steady state is a linear function of the coordinates and is of the form

$$u_i = Q_{ij}x_j, \tag{141}$$

where Q_{ij} is a certain constant matrix. In this special case, equations (131)–(140) take the following forms:

$$\int_V \rho \Delta u_i x_j \, d\mathbf{x} = \frac{dV_{i;j}}{dt} - Q_{jl}V_{i;l}, \tag{142}$$

$$\delta \frac{d}{dt}\int_V \rho u_i x_j \, d\mathbf{x} = \frac{d^2 V_{i;j}}{dt^2} + Q_{ik}\frac{dV_{j;k}}{dt} - Q_{jk}\frac{dV_{i;k}}{dt}, \tag{143}$$

$$2\delta \mathfrak{I}_{ij} = Q_{jk}\int_V \rho\Delta u_i x_k \, d\mathbf{x} + Q_{ik}\int_V \rho\Delta u_j x_k \, d\mathbf{x}$$

$$= Q_{jk}\frac{dV_{i;k}}{dt} + Q_{ik}\frac{dV_{j;k}}{dt} - (Q_{jl}^2 V_{i;l} + Q_{il}^2 V_{j;l}), \tag{144}$$

$$\delta \int_V \rho u_l x_j \, d\mathbf{x} = \frac{dV_{l;j}}{dt} + Q_{lk}V_{j;k} - Q_{jk}V_{l;k}, \tag{145}$$

$$\int_V \rho\Delta u_i x_j x_k \, d\mathbf{x} = \frac{dV_{i;jk}}{dt} - (Q_{jm}V_{i;km} + Q_{km}V_{i;jm}), \tag{146}$$

$$\delta \frac{d}{dt}\int_V \rho u_i x_j x_k \, d\mathbf{x} = \frac{d^2 V_{i;jk}}{dt^2} + Q_{il}\frac{d}{dt}(V_{j;kl} + V_{k;lj}) -$$

$$- Q_{jl}\frac{dV_{i;kl}}{dt} - Q_{kl}\frac{dV_{i;jl}}{dt}, \tag{147}$$

$$2\delta \mathfrak{I}_{ij;k} = Q_{jl}\int_V \rho\Delta u_i x_k x_l \, d\mathbf{x} + Q_{il}\int_V \rho\Delta u_j x_k x_l \, d\mathbf{x} +$$

$$+ Q_{il}Q_{jm}\int_V \rho\xi_k x_l x_m \, d\mathbf{x}$$

$$= Q_{jl}\left(\frac{dV_{i;kl}}{dt} - Q_{km}V_{i;lm} - Q_{lm}V_{i;km}\right) +$$

$$+ Q_{il}\left(\frac{dV_{j;kl}}{dt} - Q_{km}V_{j;lm} - Q_{lm}V_{j;km}\right) + Q_{il}Q_{jm}V_{k;lm}, \tag{148}$$

and

$$\delta \int_V \rho u_l x_j x_k \, d\mathbf{x} = \frac{dV_{l;jk}}{dt} - (Q_{jm}V_{l;km} + Q_{km}V_{l;jm}) +$$

$$+ Q_{lm}(V_{j;km} + V_{k;jm}). \tag{149}$$

It remains to consider the first variations of quantities such as Π and Π_i. If the changes in pressure, accompanying the changes in density, take place adiabatically, then the Lagrangian change, Δp, in p is given by equation (94); and we find by applications of the formula (116) that

$$\delta\Pi = \int (\Delta p + p\,\mathrm{div}\,\boldsymbol{\xi})\,d\mathbf{x} = -\int_V (\gamma-1)p\,\mathrm{div}\,\boldsymbol{\xi}\,d\mathbf{x} \qquad (150)$$

and

$$\delta\Pi_i = \int_V (p\xi_i + x_i\Delta p + x_i p\,\mathrm{div}\,\boldsymbol{\xi})\,d\mathbf{x}$$

$$= \int_V p[\xi_i - x_i(\gamma-1)\mathrm{div}\,\boldsymbol{\xi}]\,d\mathbf{x}. \qquad (151)$$

If the fluid should be incompressible, then $\delta\Pi$ and $\delta\Pi_i$ should be eliminated from the equations and the reduced number of equations that will remain must be supplemented by the conditions that follow from the requirement that in this case only displacements which are divergence free should be considered (see § 23 (c)).

Turning finally to the first variations of the virial equations (63) and (73), of the second and the third orders, we have

$$\delta\frac{d}{dt}\int_V \rho u_i x_j\,d\mathbf{x} = 2\delta\mathfrak{T}_{ij} + \delta\mathfrak{W}_{ij} + \Omega^2 V_{ij} - \Omega_i\Omega_k V_{kj} +$$

$$+\delta_{ij}\delta\Pi + 2\epsilon_{ilm}\Omega_m\delta\int_V \rho u_l x_j\,d\mathbf{x} \qquad (152)$$

and

$$\delta\frac{d}{dt}\int_V \rho u_i x_j x_k\,d\mathbf{x} = 2(\delta\mathfrak{T}_{ij;k} + \delta\mathfrak{T}_{ik;j}) + \delta\mathfrak{W}_{ij;k} + \delta\mathfrak{W}_{ik;j} +$$

$$+\Omega^2 V_{ijk} - \Omega_i\Omega_l V_{ljk} + \delta_{ij}\delta\Pi_k + \delta_{ik}\delta\Pi_j + 2\epsilon_{ilm}\Omega_m\delta\int_V \rho u_l x_j x_k\,d\mathbf{x}. \qquad (153)$$

In the foregoing equations we can now insert the expressions we have derived for the first variations of the various quantities that occur.

BIBLIOGRAPHICAL NOTES

Information, supplementary to this chapter, will be found in Papers I, II, IV, VII, IX, and XXXV in the list on pp. 245-7; and the general accounts in Papers VI, XXII (Chapters 1 and 2), and XXIII emphasize different aspects of the subject.

§ 13. The treatment of the small departures from an initial flow in this section is derived from:

S. CHANDRASEKHAR and NORMAN R. LEBOVITZ, "Non-radial oscillations of gaseous masses," *Astrophys. J.*, **140** (1964), 1517–28; see particularly the Appendix, pp. 1525–28.

See also:

D. LYNDEN-BELL and J. P. OSTRIKER, "On the stability of differentially rotating bodies," *M.N.R.A.S.*, **136** (1967), 293–310.

3

THE POTENTIALS OF HOMOGENEOUS AND HETEROGENEOUS ELLIPSOIDS

16. Introduction

THE determination of the gravitational potential of a homogeneous ellipsoid was a major problem in mathematics for almost a century after Newton. It attracted the flattering attention of some of the most eminent mathematicians of the time: Maclaurin, d'Alembert, Legendre, Laplace, Gauss, Jacobi, Poisson, and Dirichlet. An important theorem in the subject due to Maclaurin was described by Kelvin as "Maclaurin's splendid theorem." And while the formulas giving the potential, as we know them, are attributed to Gauss (1813) and Rodrigues (1815), independent (or alternative) derivations were given by Poisson, Cayley, and Dirichlet.

The subsequent investigations on the potentials of *heterogeneous ellipsoids* (by Green, by Poisson himself, Cayley, Ferrers, Dyson, Niven, Routh, and Hobson) attracted, in contrast, curiously little attention. Nevertheless, for the purposes of this book a paper by Ferrers (1877) in which expressions are given for the potentials of an ellipsoid in which the density varies as $x_1^l x_2^m x_3^n$ is crucial: they are no more than the potentials $\mathfrak{D}_i$, $\mathfrak{D}_{ij}$, $\mathfrak{D}_{ijk}$, etc., (introduced in Chapter 2, § 10) for homogeneous ellipsoids.† As we shall see, the systematic use of Ferrers' results eliminates, in toto, the need for ellipsoidal harmonics—a discipline introduced into the subject by Poincaré and which, in spite of Darwin's conspicuous efforts and Cartan's notable achievement with respect to the stability of the Jacobi ellipsoid, hardly succeeded in encompassing (or even clarifying) the subject.

In view, then, of the importance of the potentials of an ellipsoid for the subject of this book, this chapter will be devoted to them.

---

† It is relevant to note in this context that no reference to the potentials of heterogeneous ellipsoids (and none to Ferrers' results) is to be found in any of the standard books devoted to the theory of the potential. The only book in which the author was able to find an account of the topic is Routh's *Analytical Statics*, **2** (1892). But Routh was, of course, interested in the subject himself.

17. Newton's and related theorems on the potential interior to homoeoidal shells

We begin with a definition.

Definition. A homoeoid is a shell bounded by two similar concentric ellipsoids in which strata of equal density are also ellipsoids that are concentric with and similar to the bounding ellipsoids.

THEOREM 1 (*Newton's Theorem*). *The attraction at any internal point of a homogeneous homoeoid is zero.*

Proof. With any point O inside the homoeoid as vertex, draw an elementary cone of solid angle $d\omega$ (see Fig. 3). Let the generators of the cone intersect the homoeoid in frusta $PQQ'P'$ and $RSS'R'$. If ρ is the density, the mass of an elementary slice of the cone of thickness dr at a distance r from O is $\rho r^2 dr d\omega$; and its contribution to the attraction at O is $G\rho\,d\omega\,dr$. The attractions of the two intercepted frusta of the cone are, then, $G\rho\,d\omega\,PQ$ and $G\rho\,d\omega\,RS$, in opposite directions. But since the bounding ellipsoids are similar, they make equal intercepts on any chord; PQ is, therefore, equal to RS and the attractions of the two frusta are equal and opposite; and the theorem follows by summing over all solid angles.

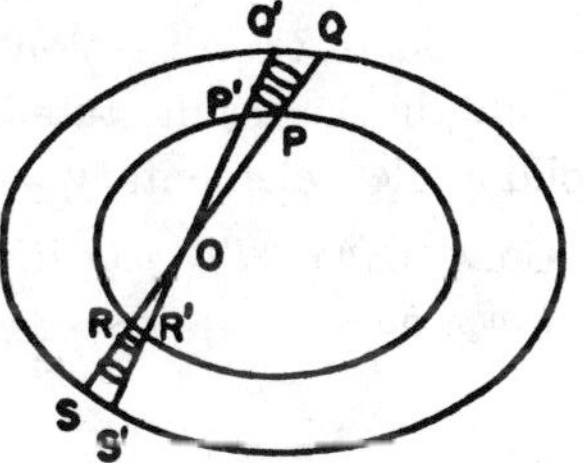

FIG. 3. Illustrating the proof of Newton's theorem on the vanishing of the attraction in the interior of homoeoidal shells.

COROLLARY. *The theorem is true for any heterogeneous homoeoid in which the strata of equal density are ellipsoids concentric with and similar to the bounding ellipsoids.*

Proof. The corollary is an immediate consequence of the theorem since the potential due to the heterogeneous homoeoid can be considered as the sum of infinitesimal homoeoidal shells of constant density.

THEOREM 2. *The constant potential inside a homogeneous homoeoidal shell enclosed between the ellipsoids with the semi-axes a_i and ma_i ($a_1 \leqslant a_2 \leqslant a_3$ and $m < 1$) is given by*

$$\mathfrak{B} = \tfrac{1}{2}G\rho(1-m^2)\int_S r^2\,d\omega, \tag{1}$$

where $\mathbf{r}$ *is the radius vector drawn from the center to a point on the surface* (S)

of the (outer) bounding ellipsoid

$$\sum_{i=1}^{3} \frac{x_i^2}{a_i^2} = 1. \tag{2}\dagger$$

Proof. By Theorem 1, the attraction at any internal point of a homoeoidal shell is zero. The potential is accordingly a constant throughout the interior and equal to its value at the center. Considering an elementary cone of solid angle $d\omega$ with its vertex at the center, the contribution to $\mathfrak{B}$ by the frustum of the cone intercepted by the homoeoid is

$$G \int_{r_1}^{r_2} \rho r \, d\omega dr = \tfrac{1}{2} G\rho(r_2^2 - r_1^2) \, d\omega, \tag{3}$$

where r_2 and r_1 are the radii of the outer and inner surfaces of the homoeoid in the direction of the cone. Remembering that $r_1 = mr_2$, we obtain the result stated by integrating the contribution (3) over all solid angles (elementary cones).

COROLLARY. *The potential interior to a thin homoeoidal shell of mass M is given by*

$$\mathfrak{B} = \frac{MG}{4\pi a_1 a_2 a_3} \int_S r^2 \, d\omega. \tag{4}$$

Proof. Since the mass M, of the homoeoidal shell considered in Theorem 2, is given by

$$M = \tfrac{4}{3}\pi a_1 a_2 a_3 \, \rho(1-m^3), \tag{5}$$

we can write

$$\mathfrak{B} = \frac{\tfrac{1}{2}MG}{\tfrac{4}{3}\pi a_1 a_2 a_3} \frac{1-m^2}{1-m^3} \int_S r^2 \, d\omega; \tag{6}$$

and the limit of this expression, for a fixed M, as $m \to 1$ is given by equation (4).

We now prove a series of lemmas.

LEMMA 1. *Under the circumstances of Theorem 2*

$$\int_S r^2 \, d\omega = 2\pi a_1 a_2 a_3 \int_0^{\infty} \frac{du}{\Delta}, \tag{7}$$

where

$$\Delta^2 = (a_1^2 + u)(a_2^2 + u)(a_3^2 + u). \tag{8}$$

Proof. In spherical polar coordinates, the equation of S is

$$\frac{1}{r^2} = \frac{\cos^2\vartheta}{a_3^2} + \sin^2\vartheta\left(\frac{\cos^2\varphi}{a_1^2} + \frac{\sin^2\varphi}{a_2^2}\right), \tag{9}$$

---

† The summation convention is suspended in this chapter; and summation when required will be indicated explicitly as in equation (2).

where ϑ is the polar angle measured with respect to the x_3-axis. Accordingly,

$$\int_S r^2 \, d\omega = 8 \int_0^{\pi/2} \int_0^{\pi/2} \frac{\sin\vartheta \, d\vartheta \, d\varphi}{\cos^2\vartheta/a_3^2 + \sin^2\vartheta(\cos^2\varphi/a_1^2 + \sin^2\varphi/a_2^2)}. \tag{10}$$

With the substitution $\qquad\qquad t = \tan\varphi, \tag{11}$

the required integral becomes

$$\int_S r^2 \, d\omega = 8 \int_0^{\pi/2} d\vartheta \sin\vartheta \int_0^{\infty} \frac{dt}{\sin^2\vartheta/a_1^2 + \cos^2\vartheta/a_3^2 + t^2(\sin^2\vartheta/a_2^2 + \cos^2\vartheta/a_3^2)}. \tag{12}$$

The integration over t is elementary and we are left with

$$\int_S r^2 \, d\omega = 4\pi \int_0^{\pi/2} \frac{\sin\vartheta \, d\vartheta}{(\sin^2\vartheta/a_1^2 + \cos^2\vartheta/a_3^2)^{\frac{1}{2}}(\sin^2\vartheta/a_2^2 + \cos^2\vartheta/a_3^2)^{\frac{1}{2}}}$$

$$= 4\pi a_1 a_2 a_3^2 \int_0^{\pi/2} \frac{\sec^2\vartheta \sin\vartheta \, d\vartheta}{(a_1^2 + a_3^2 \tan^2\vartheta)^{\frac{1}{2}}(a_2^2 + a_3^2 \tan^2\vartheta)^{\frac{1}{2}}}. \tag{13}$$

With the transformation of variables

$$u = a_3^2 \tan^2\vartheta, \qquad du = 2a_3^2 \sin\vartheta \sec^3\vartheta \, d\vartheta, \tag{14}$$

we obtain the result stated.

Defining $\qquad\qquad I = a_1 a_2 a_3 \int_0^{\infty} \frac{du}{\Delta}, \tag{15}$

we can express the result of this Lemma as

$$\int_S r^2 \, d\omega = 2\pi I. \tag{16}$$

LEMMA 2. *If l_i $(i = 1, 2, 3)$ are the direction cosines of the radius vector* **r** *joining the center of the ellipsoid to a point on its surface, then*

$$\int_S r^2 l_i^2 \, d\omega = 2\pi a_i^2 A_i, \tag{17}$$

where $\qquad\qquad A_i = a_1 a_2 a_3 \int_0^{\infty} \frac{du}{\Delta(a_i^2 + u)}. \tag{18}$

Proof. Consider, in particular,

$$\int_S r^2 l_3^2 \, d\omega = \int_S r^2 \cos^2\vartheta \, d\omega. \tag{19}$$

Proceeding as in the reduction of the integral in Lemma 1, we observe that the integration over φ is unaffected by the presence of the additional factor $\cos^2\vartheta$. We shall thus be left with

$$\int_S r^2 l_3^2 \, d\omega = 2\pi a_1 a_2 a_3 \int_0^\infty \frac{\cos^2\vartheta}{\Delta} \, du. \tag{20}$$

But by equation (14) $\cos^2\vartheta = a_3^2/(a_3^2+u)$; hence

$$\int_S r^2 l_3^2 \, d\omega = 2\pi a_3^2 A_3. \tag{21}$$

The general result (17) now follows by symmetry.

LEMMA 3.
$$\sum_{i=1}^3 a_i^2 A_i = I \tag{22}$$

and
$$A_i = \frac{I}{a_i^2} - \frac{1}{a_i} \frac{\partial I}{\partial a_i}. \tag{23}$$

The first of these relations follows from summing the result (17) over i and using equation (16); and the second can be directly verified from the definitions.

LEMMA 4.
$$\sum_{i=1}^3 A_i = 2. \tag{24}$$

Proof. With Δ defined as in equation (8),

$$\frac{1}{\Delta} \frac{d\Delta}{du} = \frac{1}{2} \sum_{i=1}^3 \frac{1}{a_i^2+u}. \tag{25}$$

Hence

$$\sum_{i=1}^3 A_i = a_1 a_2 a_3 \int_0^\infty \left(\sum_{i=1}^3 \frac{1}{a_i^2+u} \right) \frac{du}{\Delta} = 2a_1 a_2 a_3 \int_0^\infty \frac{1}{\Delta^2} \frac{d\Delta}{du} \, du$$

$$= -\frac{2a_1 a_2 a_3}{\Delta} \Big|_0^\infty = 2. \tag{26}$$

LEMMA 5.
$$\int_S r^4 l_i^2 \, d\omega = \pi a_i^3 \frac{\partial I}{\partial a_i}. \tag{27}$$

Proof. By differentiating the relation (16) with respect to a_i, we have

$$\int_S r \frac{\partial r}{\partial a_i} \, d\omega = \pi \frac{\partial I}{\partial a_i}. \tag{28}$$

On the other hand, by differentiating equation (9) rewritten in the form

$$\frac{1}{r^2} = \sum_{j=1}^3 \frac{l_j^2}{a_j^2}, \tag{29}$$

we obtain
$$\frac{1}{r^3}\frac{\partial r}{\partial a_i}=\frac{l_i^2}{a_i^3}. \tag{30}$$

Inserting this last relation in equation (28) we obtain the required result.

The integrals representing the A_i's can be expressed in terms of the standard incomplete elliptic integrals

$$E(\theta,\phi)=\int_0^{\phi}(1-\sin^2\theta\sin^2\phi)^{\frac{1}{2}}\,d\phi \quad\text{and}\quad F(\theta,\phi)=\int_0^{\phi}(1-\sin^2\theta\sin^2\phi)^{-\frac{1}{2}}\,d\phi, \tag{31}$$

of the two kinds with the definitions

$$\sin\theta=\left(\frac{a_1^2-a_2^2}{a_1^2-a_3^2}\right)^{\frac{1}{2}}\quad\text{and}\quad\cos\phi=\frac{a_3}{a_1}. \tag{32}$$

We have
$$A_1=\frac{2a_2\,a_3}{a_1^2\sin^3\phi\,\sin^2\theta}[F(\theta,\phi)-E(\theta,\phi)], \tag{33}$$

$$A_2=\frac{2a_2\,a_3}{a_1^2\sin^3\phi\,\sin^2\theta\,\cos^2\theta}\left[E(\theta,\phi)-F(\theta,\phi)\cos^2\theta-\frac{a_3}{a_2}\sin^2\theta\sin\phi\right], \tag{34}$$

and
$$A_3=\frac{2a_2\,a_3}{a_1^2\sin^3\phi\,\cos^2\theta}\left[\frac{a_2}{a_3}\sin\phi-E(\theta,\phi)\right]. \tag{35}$$

The foregoing formulas are directly applicable if $a_1>a_2>a_3$. If $a_1=a_2>a_3$ the integrals defining the A_i's are elementary, and we have

$$A_1=A_2=\frac{(1-e^2)^{\frac{1}{2}}}{e^3}\sin^{-1}e-\frac{1-e^2}{e^2}$$

and
$$A_3=\frac{2}{e^2}-\frac{2(1-e^2)^{\frac{1}{2}}}{e^3}\sin^{-1}e, \tag{36}$$

where
$$e=(1-a_3^2/a_1^2)^{\frac{1}{2}} \tag{37}$$

is the eccentricity of the oblate spheroid. Similarly, if $a_1>a_2=a_3$, we have

$$A_1=\frac{1-e^2}{e^3}\log\frac{1+e}{1-e}-2\frac{1-e^2}{e^2}$$

and
$$A_2=A_3=\frac{1}{e^2}-\frac{1-e^2}{2e^3}\log\frac{1+e}{1-e}, \tag{38}$$

where
$$e=(1-a_3^2/a_1^2)^{\frac{1}{2}} \tag{39}$$

is now the eccentricity of the prolate spheroid.

18. The potential of a homogeneous ellipsoid at an interior point

THEOREM 3. *The potential at an internal point x_i of a solid homogeneous ellipsoid (with semi-axes a_i) is given by*

$$\mathfrak{B}=\pi G\rho\left(I-\sum_{l=1}^{3}A_l x_l^2\right). \tag{40}$$

Proof. Construct an elementary cone of solid angle $d\omega$ with its vertex at x_i. The contribution to the potential at x_i by the matter included between the generators of the cone is given by

$$d\mathfrak{B} = \tfrac{1}{2}G\rho(R_1^2 + R_2^2)\,d\omega, \tag{41}$$

where R_1 and R_2 are the heights of the two half-cones diverging from x_i. The required expression for $\mathfrak{B}$ is obtained by integrating (41) over all solid angles and allowing for the fact that, by the manner of integrating, each elementary cone will be counted twice. Accordingly,

$$\mathfrak{B} = \tfrac{1}{4}G\rho \int_S (R_1^2 + R_2^2)\,d\omega. \tag{42}$$

Let l_i denote the direction cosines of the radius vector, $\mathbf{r}$, drawn from the center parallel to the axis of the elementary cone considered. The coordinates of the points at which the axis of the cone intersects the ellipsoid are

$$x_i + l_i R_1 \quad \text{and} \quad x_i + l_i R_2; \tag{43}$$

and since these points lie on the ellipsoid, it is clear that R_1 and R_2 are the roots of the equation

$$\sum_{i=1}^{3} \left(\frac{x_i + l_i R}{a_i}\right)^2 = 1. \tag{44}$$

Expanding equation (44), and remembering that

$$\frac{1}{r^2} = \sum_{i=1}^{3} \frac{l_i^2}{a_i^2}, \tag{45}$$

we can write

$$\frac{R^2}{r^2} + 2R \sum_{i=1}^{3} \frac{x_i l_i}{a_i^2} + \sum_{i=1}^{3} \frac{x_i^2}{a_i^2} - 1 = 0; \tag{46}$$

and from this equation it follows that

$$R_1^2 + R_2^2 = 4r^4 \left(\sum_{i=1}^{3} \frac{x_i l_i}{a_i^2}\right)^2 + 2r^2 \left(1 - \sum_{i=1}^{3} \frac{x_i^2}{a_i^2}\right). \tag{47}$$

Inserting this last result in equation (42), we obtain

$$\mathfrak{B} = \tfrac{1}{2}G\rho \int_S \left[2r^4 \left(\sum_{i=1}^{3} \frac{x_i l_i}{a_i^2}\right)^2 + r^2 \left(1 - \sum_{i=1}^{3} \frac{x_i^2}{a_i^2}\right)\right] d\omega, \tag{48}$$

or

$$\mathfrak{B} = \tfrac{1}{2}G\rho \int_S \left[2r^4 \sum_{i=1}^{3} \frac{x_i^2 l_i^2}{a_i^4} + r^2 \left(1 - \sum_{i=1}^{3} \frac{x_i^2}{a_i^2}\right)\right] d\omega. \tag{49}$$

By Lemmas 1 and 5, we can now write

$$\mathfrak{V} = \pi G \rho \left[\sum_{i=1}^{3} x_i^2 \left(\frac{1}{a_i} \frac{\partial I}{\partial a_i} - \frac{I}{a_i^2} \right) + I \right]. \tag{50}$$

Now making use of Lemma 3 we obtain the result stated.

19. The potential of a homogeneous ellipsoid at an external point. Maclaurin's and Ivory's theorems

We begin with some elementary results in the geometry of confocal ellipsoids.

Definition. Two points $P\,(x_i)$ and $P'\,(x_i')$ situated, respectively, on the two ellipsoids,

$$\sum_{i=1}^{3} \frac{x_i^2}{a_i^2} = 1 \quad \text{and} \quad \sum_{i=1}^{3} \frac{x_i'^{\,2}}{a_i'^2} = 1, \tag{51}$$

are said to *correspond* if $\qquad x_i/a_i = x_i'/a_i'.$ \hfill (52)

Lemma 6 (*Ivory's Lemma*). *If P, P', and Q, Q' are pairs of corresponding points on two confocal ellipsoids,*

$$\sum_{i=1}^{3} \frac{x_i^2}{a_i^2} = 1 \quad and \quad \sum_{i=1}^{3} \frac{x_i^2}{a_i^2 + \lambda} = 1, \tag{53}$$

then $PQ' = P'Q.$

Proof. Let the two pairs of points considered be (x_i, x_i') and (ξ_i, ξ_i'). Then by definition

$$\frac{x_i}{x_i'} = \frac{\xi_i}{\xi_i'} = \frac{a_i}{(a_i^2 + \lambda)^{\frac{1}{2}}}. \tag{54}$$

Hence,

$$(PQ')^2 - (P'Q)^2 = \sum_{i=1}^{3} \left[(x_i - \xi_i')^2 - (x_i' - \xi_i)^2 \right]$$

$$= \sum_{i=1}^{3} \left\{ \left[x_i - \frac{(a_i^2 + \lambda)^{\frac{1}{2}}}{a_i} \xi_i \right]^2 - \left[\frac{(a_i^2 + \lambda)^{\frac{1}{2}}}{a_i} x_i - \xi_i \right]^2 \right\}$$

$$= \sum_{i=1}^{3} \left[x_i^2 \left(1 - \frac{a_i^2 + \lambda}{a_i^2} \right) + \xi_i^2 \left(\frac{a_i^2 + \lambda}{a_i^2} - 1 \right) \right]$$

$$= -\lambda \left(\sum_{i=1}^{3} \frac{x_i^2}{a_i^2} - \sum_{i=1}^{3} \frac{\xi_i^2}{a_i^2} \right) = 0. \tag{55}$$

Theorem 4. *Consider two infinitesimally thin homogeneous confocal homoeoids of equal volumes and densities. Let P and P' be two corresponding points on two confocal ellipsoids E and E' in the inner and outer* *homoeoidal shells, respectively. Then the potential of the outer homoeoidal shell at P is the same as the potential of the inner homoeoidal shell at P'.*

Proof. Consider two equal elementary volumes at two corresponding points Q and Q' on the two homoeoidal shells; and let the two elementary volumes be further bounded by surfaces of corresponding points.† By Lemma 6, $PQ' = P'Q$. Hence the contribution to the potential at P by the elementary volume at Q' on E' is the same as the contribution to the potential at P' by the elementary volume at Q on E. By summing over all such pairs of elementary volumes in the two shells, we obtain the result stated.

COROLLARY. *The equipotential surfaces external to a thin homoeoid are ellipsoids confocal to the homoeoid.*

Proof. This is an immediate consequence of Theorems 1 and 4. Theorem 1 ensures that the potential of E' on the surface of E is a constant. Therefore (by Theorem 4) the surface at E' is an equipotential surface of E; and since E' is *any* confocal ellipsoid external to E, the corollary stated follows.

THEOREM 5. *The potential of a thin homoeoid of mass M, at a point x_i external to it, is given by*

$$\mathfrak{B} = \tfrac{1}{2}GM \int_\lambda^\infty \frac{du}{[(a_1^2+u)(a_2^2+u)(a_3^2+u)]^{\frac{1}{2}}}, \tag{56}$$

where λ is the ellipsoidal coordinate of x_i defined by the ellipsoid which passes through x_i and is confocal to the homoeoid, i.e. λ is the (algebraically) largest root of the cubic equation

$$\sum_{i=1}^{3} \frac{x_i^2}{a_i^2+\lambda} = 1. \tag{57}$$

† An explanation is needed here. Let $d\sigma$ and $d\sigma'$ be the areas of two triangular elements at P and P' such that the corners are also corresponding points; also, let p and p' be the perpendiculars from the common center of the ellipsoids to the tangent planes at P and P'. The volumes of the tetrahedra (whose bases are $d\sigma$ and $d\sigma'$ and whose common vertex is O) are $\tfrac{1}{3}p\,d\sigma$ and $\tfrac{1}{3}p'\,d\sigma'$. At the same time, it is clear that these volumes are in the ratio $a_1 a_2 a_3 : a_1' a_2' a_3'$ (since the volume of a tetrahedron whose vertex is at the origin and whose triangular base has its corners at $x_i^{(\alpha)}$ ($\alpha = 1, 2, 3$) is given by $\tfrac{1}{6}\|x_i^{(\alpha)}\|$). Accordingly, $p\,d\sigma : p'\,d\sigma' = a_1 a_2 a_3 : a_1' a_2' a_3'$; and since any element of area at P (or P') can be dissected into triangles the relation is generally valid. Since the thickness of a homoeoidal shell is kp where k is a constant, it follows that the volumes of the corresponding elements in the two thin homoeoids are in a constant ratio; and it is clear that this ratio can be none other than the ratio of the entire volumes of the two shells. If the shells are of such thicknesses that they have equal volumes (as in the statement of Theorem 4) then the corresponding volume elements will also have equal volumes; and this last equality is essential to the proof of Theorem 4.

[Note that for an external point such as the one considered in Theorem 5, the ellipsoidal coordinate λ is positive.]

Proof. By Theorem 4, the required potential is equal to the constant potential inside a confocal homoeoidal shell of the same mass M passing through x_i; and by Theorem 2, Corollary, and equations (7) and (8) this is given by

$$\mathfrak{V} = \tfrac{1}{2}GM \int_0^\infty \frac{du'}{[(a_1'^2+u')(a_2'^2+u')(a_3'^2+u')]^{\frac{1}{2}}}. \tag{58}$$

But $a_i'^2 = a_i^2 + \lambda$; therefore, changing the variable of integration to $u = u' + \lambda$ we obtain the result stated.

THEOREM 6 (*Ivory's Theorem*). *Consider two corresponding points P and P' on two equidensity confocal ellipsoids, E and E', with semi-axes a_i and a_i'. Then*

$$\frac{\text{the attraction of } E \text{ at } P' \text{ along } x_i}{\text{the attraction of } E' \text{ at } P \text{ along } x_i} = \frac{a_1 a_2 a_3 a_i'}{a_1' a_2' a_3' a_i}. \tag{59}$$

Proof. Let the equations of the two ellipsoids E and E' be

$$\sum_{i=1}^3 \frac{x_i^2}{a_i^2} = 1 \quad \text{and} \quad \sum_{i=1}^3 \frac{x_i^2}{a_i^2+\lambda} = 1, \tag{60}$$

respectively; and let x_i and x_i' be the two (corresponding) points considered. Since E' passes through x_i', λ is the largest root of the equation

$$\sum_{i=1}^3 \frac{x_i'^2}{a_i^2+\lambda} = 1; \tag{61}$$

also, $$\frac{x_i}{x_i'} = \frac{a_i}{(a_i^2+\lambda)^{\frac{1}{2}}} = \frac{a_i}{a_i'}. \tag{62}$$

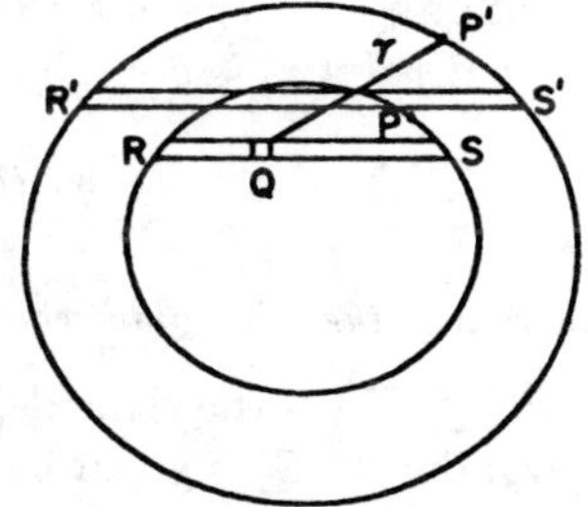

Fig. 4. Illustrating the proof of Ivory's theorem.

Let RS be an elementary rectangular strip in E parallel to the x_1-direction; and let $R'S'$ be the corresponding strip in E' (see Fig. 4). If the cross-sections of the two strips are $dx_2\, dx_3$ and $dx_2'\, dx_3'$, respectively, then, by the manner of their construction,

$$\frac{dx_2\, dx_3}{dx_2'\, dx_3'} = \frac{a_2 a_3}{a_2' a_3'}. \tag{63}$$

If $f'(r) = df/dr$ is the law of force (in its dependence on the distance r) and ρ is the density (assumed to be the same in E and E') then the attraction at P', in the direction x_1, due to the strip RS is given by

$$-\rho\, dx_2\, dx_3 \int_R^S f'(r)\cos\angle P'QS\, dx_1$$

$$= \rho\, dx_2\, dx_3 \int_R^S \frac{df}{dr}\frac{\partial r}{\partial x_1}\, dx_1 = \rho\, dx_2\, dx_3[f(P'S)-f(P'R)]. \tag{64}$$

In the same way, the attraction at P, in the direction x_1, due to the strip $R'S'$ is given by

$$\rho\, dx_2'\, dx_3'[f(PS')-f(PR')]. \tag{65}$$

But by Lemma 6, $P'S = PS'$ and $P'R = PR'$. The two attractions are therefore in the ratio $dx_2\, dx_3 : dx_2'\, dx_3' = a_2\, a_3 : a_2'\, a_3'$ by (63); and this ratio is independent of the particular choice of RS. Consequently by adding the contributions by all the strips (such as RS and $R'S'$) into which the ellipsoids E and E' may be dissected, we obtain the result

$$\frac{\text{the attraction of } E \text{ at } P' \text{ along } x_1}{\text{the attraction of } E' \text{ at } P \text{ along } x_1} = \frac{a_2\, a_3}{a_2'\, a_3'}. \tag{66}$$

The analogous results for the attractions in the other directions follow by symmetry.

THEOREM 7. *The attraction of a solid homogeneous ellipsoid E at an external point x_i' is given by*

$$-2\pi G\rho a_1\, a_2\, a_3\, x_i' \int_\lambda^\infty \frac{du}{\Delta(a_i^2+u)}, \tag{67}$$

where λ is the ellipsoidal coordinate of x_i'.

Proof. With the same definitions as in Theorem 6, the attraction at $P\,(x_i)$ due to E' is given by (cf. Theorem 3)

$$-2\pi G\rho a_1'\, a_2'\, a_3'\, x_i \int_0^\infty \frac{du'}{[(a_1'^2+u')(a_2'^2+u')(a_3'^2+u')]^{\frac{1}{2}}(a_i'^2+u')}. \tag{68}$$

By Theorem 6, the attraction due to E at the external point x_i' is given by

$$-2\pi G\rho a_1'\, a_2'\, a_3'\, x_i'\left[\int_0^\infty \frac{du'}{[(a_1'^2+u')(a_2'^2+u')(a_3'^2+u')]^{\frac{1}{2}}(a_i'^2+u')}\right]\frac{a_1\, a_2\, a_3\, a_i'}{a_1'\, a_2'\, a_3'\, a_i}$$

$$= -2\pi G\rho a_1\, a_2\, a_3\, x_i' \int_0^\infty \frac{du'}{[(a_1'^2+u')(a_2'^2+u')(a_3'^2+u')]^{\frac{1}{2}}(a_i'^2+u')}, \tag{69}$$

where we have made further use of the relation (62). Since $a_i'^2 = a_i^2+\lambda$, we obtain the result (67) by changing the variable of integration to $u = u' + \lambda$.

THEOREM 8 (*Maclaurin's Theorem*). *The attractions of two confocal ellipsoids at a point external to both are proportional to their masses and are in the same direction.*

Proof. Consider the attractions of two different ellipsoids (of densities ρ_1 and ρ_2 and semi-axes a_i and α_i) confocal to one another at some (fixed) point x_i external to both. Then by (69) the attractions are in the ratio $\rho_1 a_1 a_2 a_3 : \rho_2 \alpha_1 \alpha_2 \alpha_3$; the other factors are the same since a'_i, being the semi-axes of the ellipsoid passing through x_i, are the same for both the ellipsoids. And the attractions act in the same direction since all three components are in the same ratio.

THEOREM 9. *The potential of a homogeneous ellipsoid E at an external point x_i is given by*

$$\mathfrak{B} = \pi G \rho a_1 a_2 a_3 \int_\lambda^\infty \frac{du}{\Delta}\left(1 - \sum_{i=1}^{3} \frac{x_i^2}{a_i^2 + u}\right), \tag{70}$$

where
$$\Delta^2 = (a_1^2 + u)(a_2^2 + u)(a_3^2 + u) \tag{71}$$

and λ is the ellipsoidal coordinate of the point considered: that is the positive root of the equation

$$\sum_{i=1}^{3} \frac{x_i^2}{a_i^2 + \lambda} = 1. \tag{72}$$

Proof. By Maclaurin's Theorem, the potential $\mathfrak{B}'$ at x_i due to the ellipsoid E' (confocal to E and passing through x_i, and of the same density) is $(a'_1 a'_2 a'_3 / a_1 a_2 a_3)$ times the potential $\mathfrak{B}$ due to E at the *same* point. But by Theorem 3,

$$\mathfrak{B}' = \pi G \rho a'_1 a'_2 a'_3 \int_0^\infty \frac{du'}{[(a_1'^2 + u')(a_2'^2 + u')(a_3'^2 + u')]^{\frac{1}{2}}}\left(1 - \sum_{i=1}^{3} \frac{x_i^2}{a_i'^2 + u'}\right); \tag{73}$$

therefore

$$\mathfrak{B} = \pi G \rho a_1 a_2 a_3 \int_0^\infty \frac{du'}{[(a_1'^2 + u')(a_2'^2 + u')(a_3'^2 + u')]^{\frac{1}{2}}}\left(1 - \sum_{i=1}^{3} \frac{x_i^2}{a_i'^2 + u'}\right). \tag{74}$$

But $a_i'^2 = a_i^2 + \lambda$; therefore, by changing the variable of integration to $u = u' + \lambda$ we obtain the result stated.

20. The potentials of heterogeneous ellipsoids

We now turn our attention to the potentials of heterogeneous ellipsoids. We shall, in the first instance, restrict our considerations to the special case when the strata of equal density are similar to and concentric with the bounding ellipsoid.

Let a_i denote the semi-axes of the bounding ellipsoid. Under the assumption stated, the strata of equal density are then ellipsoids of semi-axes ma_i where m is a parameter which varies continuously from one to zero (for a solid ellipsoid) or from one to a finite lower limit (for a homoeoidal shell). The equation governing the strata of equal density is

$$\sum_{i=1}^{3} \frac{x_i^2}{a_i^2} = m^2; \tag{75}$$

and we shall let
$$\rho \equiv \rho(m^2). \tag{76}$$

THEOREM 10. *The potential at a point x_i, external to a heterogeneous ellipsoid of the kind described, is given by*

$$\mathfrak{B} = \pi G a_1 a_2 a_3 \int_{\lambda}^{\infty} \frac{du}{\Delta} \int_{m^2(u)}^{1} dm^2 \rho(m^2), \tag{77}$$

where the function $m^2(u)$ is defined by the equation

$$\sum_{i=1}^{3} \frac{x_i^2}{a_i^2+u} = m^2 \tag{78}$$

and λ is the ellipsoidal coordinate of the point x_i considered with respect to the bounding ellipsoid.

Proof. Consider the solid ellipsoid as consisting of a series of infinitesimally thin homoeoidal shells bounded by the ellipsoids of semi-axes ma_i and $(m+dm)a_i$. The mass dM of such a shell is

$$dM(m) = 4\pi a_1 a_2 a_3 \rho(m^2)m^2\, dm. \tag{79}$$

By Theorem 5, the contribution to the potential at an external point x_i by the homoeoidal shell $dM(m)$ is

$$2\pi G a_1 a_2 a_3 \rho(m^2)m^2\, dm \int_{\lambda(m^2)}^{\infty} \frac{dv}{[(m^2a_1^2+v)(m^2a_2^2+v)(m^2a_3^2+v)]^{\frac{1}{2}}}, \tag{80}$$

where $\lambda(m^2)$ is the ellipsoidal coordinate of x_i with respect to the ellipsoid of semi-axes ma_i, i.e. $\lambda(m^2)$ is the positive root of the equation

$$\sum_{i=1}^{3} \frac{x_i^2}{m^2a_i^2+\lambda} = 1. \tag{81}$$

Integrating the contribution (80) over the range of m, we obtain for the total potential the expression

$$\mathfrak{B} = 2\pi G a_1 a_2 a_3 \int_0^1 dm\, m^2 \rho(m^2) \int_{\lambda(m^2)}^{\infty} \frac{dv}{[(m^2 a_1^2 + v)(m^2 a_2^2 + v)(m^2 a_3^2 + v)]^{\frac{1}{2}}}. \tag{82}$$

Writing

$$v = m^2 u \quad \text{and} \quad \lambda(m^2) = m^2 \mu(m^2), \tag{83}$$

where $\mu(m^2)$ is defined implicitly by the equation

$$\sum_{i=1}^{3} \frac{x_i^2}{a_i^2 + \mu} = m^2, \tag{84}$$

we can rewrite equation (82) in the form

$$\mathfrak{B} = \pi G a_1 a_2 a_3 \int_0^1 dm^2 \rho(m^2) \int_{\mu(m^2)}^{\infty} \frac{du}{\Delta}, \tag{85}$$

where Δ has now its usual meaning.

We now invert the order of integrations in equation (85). Noting that

$$\mu \to \infty \quad \text{when} \quad m \to 0$$

and

$$\mu = \lambda \quad (= \text{the ellipsoidal coordinate of } x_i \text{ with respect} \tag{86}$$
$$\text{to the bounding ellipsoid}) \quad \text{when } m = 1,$$

and letting $m^2(u)$ denote the function defined by equation (78), we can write

$$\mathfrak{B} = \pi G a_1 a_2 a_3 \int_{\lambda}^{\infty} \frac{du}{\Delta} \int_{m^2(u)}^{1} dm^2 \rho(m^2). \tag{87}$$

This is the required result.

It is convenient to define the function

$$\psi(m^2) = \int_1^{m^2} dm^2 \rho(m^2) \quad \text{so that} \quad \frac{d\psi}{dm^2} = \rho(m^2). \tag{88}$$

Equation (87) then becomes

$$\mathfrak{B} = \pi G a_1 a_2 a_3 \int_{\lambda}^{\infty} \frac{du}{\Delta} [\psi(1) - \psi(m^2(u))]. \tag{89}$$

Note that for a point on the bounding ellipsoid (for which $\lambda = 0$)

$$\mathfrak{B} = \pi G a_1 a_2 a_3 \int_0^{\infty} \frac{du}{\Delta} [\psi(1) - \psi(m^2(u))]. \tag{90}$$

THEOREM 11. *The potential at any internal point of a homoeoidal shell bounded by the ellipsoids of semi-axes a_i and na_i ($n < 1$) is given by*

$$\mathfrak{V} = \pi G a_1 a_2 a_3 [\psi(1) - \psi(n^2)] \int_0^\infty \frac{du}{\Delta}. \tag{91}$$

Proof. We know that the potential in the interior of an infinitesimal homoeoidal shell is constant; and by Theorem 2, Corollary, the contribution by the homoeoidal shell $dM(m)$ (considered in the proof of Theorem 10) to the potential in its interior is

$$2\pi G a_1 a_2 a_3 \rho(m^2) m^2 \, dm \int_0^\infty \frac{dv}{[(m^2 a_1^2 + v)(m^2 a_2^2 + v)(m^2 a_3^2 + v)]^{\frac{1}{2}}}$$

$$= \pi G a_1 a_2 a_3 \rho(m^2) \, dm^2 \int_0^\infty \frac{du}{\Delta}, \tag{92}$$

where Δ has its usual meaning. The integral appearing as a factor of $\rho(m^2) \, dm^2$ is therefore independent of m; and by integrating this last expression over the present range of m (namely, 1 to n), we obtain equation (91).

THEOREM 12. *The potential at an internal point x_i of a heterogeneous ellipsoid (of the kind we have been considering) is given by*

$$\mathfrak{V} = \pi G a_1 a_2 a_3 \int_0^\infty \frac{du}{\Delta} [\psi(1) - \psi(m^2(u))]. \tag{93}$$

Proof. We may distinguish two contributions to the potential at an internal point: that due to the solid ellipsoid interior to the point x_i considered, and that due to the homoeoidal shell exterior to x_i. Let na_i denote the semi-axes of the equidensity ellipsoid passing through x_i. Then by equations (90) and (91) the two contributions are

$$\pi G a_1 a_2 a_3 \int_0^\infty \frac{du}{\Delta} [\psi(n^2) - \psi(m^2(u))] \tag{94}$$

and $$\pi G a_1 a_2 a_3 [\psi(1) - \psi(n^2)] \int_0^\infty \frac{du}{\Delta}. \tag{95}$$

[Note that for the case under consideration the limits of integration over m^2 in equation (87) are n^2 and $m^2(u)$; and this explains the difference between (90) and (94).] Adding the two contributions (94) and (95), we obtain the required result.

A special case of Theorem 12 which is of considerable importance for the derivation of the potentials $\mathfrak{D}_i$, $\mathfrak{D}_{ij}$, etc. (in § 22) is the following.

Let

$$\rho(m^2) = C(1-m^2)^n = C\left(1 - \sum_{l=1}^{3} \frac{x_l^2}{a_l^2}\right)^n, \tag{96}$$

where C is a constant and n is some positive exponent. In this case (cf. equation (88)),

$$\psi(m^2) = -\frac{C}{n+1}(1-m^2)^{n+1}, \tag{97}$$

and $\psi(m^2(u))$ is given by (cf. equation (78))

$$\psi(m^2(u)) = -\frac{C}{n+1}\left(1 - \sum_{l=1}^{3} \frac{x_l^2}{a_l^2+u}\right)^{n+1} \tag{98}$$

Accordingly, Theorem 12 now gives

$$\mathfrak{B} = \pi G a_1 a_2 a_3 \frac{C}{n+1} \int_0^\infty \frac{du}{\Delta}\left(1 - \sum_{l=1}^{3} \frac{x_l^2}{a_l^2+u}\right)^{n+1} \tag{99}$$

In view of its importance we shall state the result for this special case as a theorem.

THEOREM 13. *For a heterogeneous ellipsoid in which the density distribution has the form*

$$\rho = CE^n, \quad where \quad E = 1 - \sum_{l=1}^{3} \frac{x_l^2}{a_l^2} \tag{100}$$

and C is a constant, the potential at an internal point x_i is given by

$$\mathfrak{B} = \pi G \frac{C}{n+1} a_1 a_2 a_3 \int_0^\infty \frac{du}{\Delta} Q^{n+1}, \tag{101}$$

where

$$Q = 1 - \sum_{l=1}^{3} \frac{x_l^2}{a_l^2+u}. \tag{102}$$

21. The index symbols

We shall show in § 22 below how explicit expressions for the potentials $\mathfrak{D}_i$, $\mathfrak{D}_{ij}$, $\mathfrak{D}_{ijk}$, etc., can be derived very simply with the aid of Theorem 13. But the expressions are most conveniently written down in terms of certain index symbols which we shall now define.

Definition. The index symbols $A_{ijk\ldots}$ and $B_{ijk\ldots}$ are defined by

$$A_{ijk\ldots} = a_1 a_2 a_3 \int_0^\infty \frac{du}{\Delta(a_i^2+u)(a_j^2+u)(a_k^2+u)\ldots} \tag{103}$$

and

$$B_{ijk\ldots} = a_1 a_2 a_3 \int_0^\infty \frac{u\,du}{\Delta(a_i^2+u)(a_j^2+u)(a_k^2+u)\ldots}, \tag{104}$$

where Δ has the same meaning as in equation (5).

As defined the symbols $A_{ijk\ldots}$ and $B_{ijk\ldots}$ are manifestly symmetric in their indices. Also, it is evident that

$$B_{ijkl\ldots} = A_{jkl\ldots} - a_i^2 A_{ijkl\ldots}. \tag{105}$$

Further relations which follow directly from the definitions are

$$a_i^2 A_{ikl\ldots} - a_j^2 A_{jkl\ldots} = +(a_i^2 - a_j^2)B_{ijkl\ldots} \tag{106}$$

and

$$A_{ikl\ldots} - A_{jkl\ldots} = -(a_i^2 - a_j^2)A_{ijkl\ldots}. \tag{107}$$

Relations not as obvious as the foregoing are included in the following Lemma.

LEMMA 7. *The index symbols A_{ijk} of the first four orders satisfy the following relations*:

$$\sum_{l=1}^{3} A_l = 2, \tag{108}$$

$$2A_{ii} + \sum_{l=1}^{3} A_{il} = \frac{2}{a_i^2}, \tag{109}$$

$$2A_{iij} + 2A_{ijj} + \sum_{l=1}^{3} A_{ijl} = \frac{2}{a_i^2 a_j^2}, \tag{110}$$

$$2A_{iijk} + 2A_{ijjk} + 2A_{ijkk} + \sum_{l=1}^{3} A_{ijkl} = \frac{2}{a_i^2 a_j^2 a_k^2}. \tag{111}$$

Proof. Relation (108) has already been established (Lemma 4); and the relations (109)–(111) similarly follow by integrating over the range of u the identities

$$-2\frac{d}{du}\frac{1}{\Delta(a_i^2+u)} = \frac{1}{\Delta(a_i^2+u)}\left(\frac{2}{a_i^2+u} + \sum_{l=1}^{3}\frac{1}{a_l^2+u}\right), \tag{112}$$

$$-2\frac{d}{du}\frac{1}{\Delta(a_i^2+u)(a_j^2+u)} = \frac{1}{\Delta(a_i^2+u)(a_j^2+u)} \times$$

$$\times\left(\frac{2}{a_i^2+u} + \frac{2}{a_j^2+u} + \sum_{l=1}^{3}\frac{1}{a_l^2+u}\right), \tag{113}$$

and

$$-2\frac{d}{du}\frac{1}{\Delta(a_i^2+u)(a_j^2+u)(a_k^2+u)}$$

$$=\frac{1}{\Delta(a_i^2+u)(a_j^2+u)(a_k^2+u)}\left(\frac{2}{a_i^2+u}+\frac{2}{a_j^2+u}+\frac{2}{a_k^2+u}+\sum_{l=1}^{3}\frac{1}{a_l^2+u}\right).$$

$$(114)$$

The relations (108)–(110) can be transformed to obtain further useful identities as follows. Thus, letting $i \neq j \neq k$, we can rewrite equation (109), by making use of the preceding equation (108), in the form

$$3A_{ii}\,a_i^2+A_{ij}\,a_i^2+A_{ik}\,a_i^2 = 2 = A_i+A_j+A_k,\tag{115}$$

or, alternatively,

$$3A_{ii}\,a_i^2+(A_{ij}\,a_i^2+A_i-A_j)+(A_{ik}\,a_i^2+A_i-A_k) = 3A_i.\tag{116}$$

The use of relations which follow from equation (107) now simplifies equation (116) to give

$$3A_{ii}\,a_i^2+A_{ij}\,a_j^2+A_{ik}\,u_k^2 = 3A_i \quad (i \neq j \neq k).\tag{117}$$

Similarly we can deduce from equations (110) and (111) the following relations:

$$5A_{iii}\,a_i^2+A_{iij}\,a_j^2+A_{iik}\,a_k^2 = 5A_{ii},$$
$$3A_{iij}\,a_i^2+3A_{ijj}\,a_j^2+A_{ijk}\,a_k^2 = 5A_{ij},$$
$$7A_{iiii}\,a_i^2+A_{iiij}\,a_j^2+A_{iiik}\,a_k^2 = 7A_{iii},$$
$$5A_{iiij}\,a_i^2+3A_{iijj}\,a_j^2+A_{iijk}\,a_k^2 = 7A_{iij},$$
$$3A_{iijk}\,a_i^2+3A_{ijjk}\,a_j^2+3A_{ijkk}\,a_k^2 = 7A_{ijk}\tag{118}$$

$$(i \neq j \neq k).$$

22. The potentials $\mathfrak{D}_i$, $\mathfrak{D}_{ij}$, and $\mathfrak{D}_{ijk}$

First we state the following Lemma.

LEMMA 8. *If $\mathfrak{B}$ is the gravitational potential due to a continuous distribution of density $\rho(\mathbf{x})$, which is arbitrary except for the requirement that it vanishes on the boundary of the configuration, then the gravitational potential due to the "density distribution" $\partial\rho/\partial x_i$ is given by $\partial\mathfrak{B}/\partial x_i$.*

Proof. We have

$$\frac{\partial\mathfrak{B}}{\partial x_i} = G\int_V \rho(\mathbf{x}')\frac{\partial}{\partial x_i}\frac{1}{|\mathbf{x}-\mathbf{x}'|}\,d\mathbf{x}'$$

$$= -G\int_V \rho(\mathbf{x}')\frac{\partial}{\partial x_i'}\frac{1}{|\mathbf{x}-\mathbf{x}'|}\,d\mathbf{x}'.\tag{119}$$

The Lemma now follows by integrating by parts this last equation and remembering that ρ vanishes on the boundary of V.

It is clear that the essential requirement for the validity of this Lemma is the vanishing of the density on the boundary of the configuration.

COROLLARY. *If $\rho(\mathbf{x})$ and $\operatorname{grad}\rho(\mathbf{x})$ both vanish on the boundary of the configuration, then the gravitational potential induced by the "density" distribution $\partial^2\rho/\partial x_i\,\partial x_j$ is given by $\partial^2\mathfrak{B}/\partial x_i\,\partial x_j$ where $\mathfrak{B}$ is the gravitational potential due to $\rho(\mathbf{x})$.*

It is clear that this Corollary can be extended to still higher partial derivatives of ρ provided ρ and all the necessary lower-order derivatives vanish on the boundary.

THEOREM 14. *The potential*

$$\mathfrak{D}_i(\mathbf{x}) = G\int_V \frac{\rho(\mathbf{x}')x_i'}{|\mathbf{x}-\mathbf{x}'|}\,dx' \tag{120}$$

at an internal point of a homogeneous ellipsoid is given by

$$\frac{\mathfrak{D}_i}{\pi G\rho} = a_i^2\Big(A_i - \sum_{l=1}^{3} A_{il}\,x_l^2\Big)x_i. \tag{121}$$

Proof. According to Theorem 13, the "density" distribution

$$-\tfrac{1}{2}\rho a_i^2\Big(1 - \sum_{l=1}^{3}\frac{x_l^2}{a_l^2}\Big) = -\tfrac{1}{2}\rho a_i^2 E \tag{122}$$

has the potential $\quad \dfrac{\mathfrak{B}}{\pi G\rho} = -\tfrac{1}{4}a_i^2\,a_1\,a_2\,a_3\displaystyle\int_0^{\infty}\frac{du}{\Delta}\,Q^2. \tag{123}$

The distribution (122) clearly vanishes on the boundary of the ellipsoid. Lemma 8 is applicable and we infer that the "density" distribution

$$-\tfrac{1}{2}\rho a_i^2\frac{\partial E}{\partial x_i} = \rho x_i \tag{124}$$

induces the Newtonian potential

$$\frac{\mathfrak{D}_i}{\pi G\rho} = -\tfrac{1}{4}a_i^2\,a_1\,a_2\,a_3\int_0^{\infty}\frac{du}{\Delta}\frac{\partial Q^2}{\partial x_i}$$

$$= a_i^2\,a_1\,a_2\,a_3\int_0^{\infty}\frac{du}{\Delta(a_i^2+u)}\Big(1 - \sum_{l=1}^{3}\frac{x_l^2}{a_l^2+u}\Big)x_i$$

$$= a_i^2\,x_i\Big(A_i - \sum_{l=1}^{3} A_{il}\,x_l^2\Big); \tag{125}$$

and the theorem stated follows.

COROLLARY 1. *The tensor potential $\mathfrak{B}_{ij}$ at an internal point of a homogeneous ellipsoid is given by*

$$\frac{\mathfrak{B}_{ij}}{\pi G\rho} = 2B_{ij}\,x_i\,x_j + a_i^2\,\delta_{ij}\Big(A_i - \sum_{l=1}^{3} A_{il}\,x_l^2\Big). \tag{126}$$

This expression for $\mathfrak{B}_{ij}$ follows from equations (40), (121), and the relation

$$\mathfrak{B}_{ij} = \frac{\partial \mathfrak{D}_i}{\partial x_j} - x_i\frac{\partial \mathfrak{B}}{\partial x_j} \tag{127}$$

derived in Chapter 2 (equation (28)). [Note that in deriving the particular form (126) use has been made of the relation $B_{ij} = A_j - a_i^2 A_{ij}$.]

COROLLARY 2. *The potential energy tensor $\mathfrak{W}_{ij}$ for a homogeneous ellipsoid is given by*

$$\frac{\mathfrak{W}_{ij}}{\pi G\rho} = -2A_i\,I_{ij}, \tag{128}$$

where $\qquad I_{ij} = \tfrac{1}{5}Ma_i^2\,\delta_{ij} \quad (M = mass = \tfrac{4}{3}\pi a_1 a_2 a_3\,\rho) \tag{129}$

is the moment of inertia tensor.

This follows from either of the two representations for $\mathfrak{W}_{ij}$:

$$\mathfrak{W}_{ij} = -\tfrac{1}{2}\int_V \rho\,\mathfrak{B}_{ij}\,d\mathbf{x} \quad \text{or} \quad \int_V \rho x_i\,\frac{\partial \mathfrak{B}}{\partial x_j}\,d\mathbf{x}. \tag{130}$$

THEOREM 15. *The potential*

$$\mathfrak{D}_{ij}(\mathbf{x}) = G\int_V \rho(\mathbf{x}')\,\frac{x_i' x_j'}{|\mathbf{x}-\mathbf{x}'|}\,d\mathbf{x}' \tag{131}$$

at an internal point of a homogeneous ellipsoid is given by

$$\frac{\mathfrak{D}_{ij}}{\pi G\rho} = a_i^2\,a_j^2\Big(A_{ij} - \sum_{l=1}^{3} A_{ijl}\,x_l^2\Big)x_i\,x_j +$$

$$+\tfrac{1}{4}a_i^2\,\delta_{ij}\Big(B_i - 2\sum_{l=1}^{3} B_{il}\,x_l^2 + \sum_{l=1}^{3}\sum_{m=1}^{3} B_{ilm}\,x_l^2\,x_m^2\Big). \tag{132}$$

Proof. We readily verify the identity

$$x_i\,x_j = \tfrac{1}{8}a_i^2\,a_j^2\,\frac{\partial^2 E^2}{\partial x_i\,\partial x_j} + \tfrac{1}{2}a_i^2\,\delta_{ij}\,E. \tag{133}$$

Since $\partial E^2/\partial x_i$ ($i = 1, 2, 3$) vanishes on the boundary of the ellipsoid, Lemma 8, Corollary, is applicable and we can write

$$\frac{\mathfrak{D}_{ij}}{\pi G\rho a_1 a_2 a_3} = \frac{a_i^2\,a_j^2}{24}\int_0^\infty \frac{du}{\Delta}\,\frac{\partial^2 Q^3}{\partial x_i\,\partial x_j} + \tfrac{1}{4}a_i^2\,\delta_{ij}\int_0^\infty \frac{du}{\Delta}\,Q^2. \tag{134}$$

We verify that

$$\frac{\partial^2 Q^3}{\partial x_i\,\partial x_j} = 24\,\frac{x_i x_j}{(a_i^2+u)(a_j^2+u)}\,Q - \frac{6\delta_{ij}}{(a_j^2+u)}\,Q^2 \tag{135}$$

and inserting this relation in equation (134) we obtain

$$\frac{\mathfrak{D}_{ij}}{\pi G\rho} = a_i^2 a_j^2 x_i x_j a_1 a_2 a_3 \int_0^\infty \frac{du}{\Delta(a_i^2+u)(a_j^2+u)}\,Q +$$

$$+ \tfrac{1}{4}a_i^2\,\delta_{ij}\,a_1 a_2 a_3 \int_0^\infty \frac{u\,du}{\Delta(a_i^2+u)}\,Q^2; \tag{136}$$

and expanding this last equation we obtain equation (132).

COROLLARY. *The tensor potential* $\mathfrak{D}_{ij;k}$ *at an internal point is given by*

$\mathfrak{D}_{ij;k} = 2a_k^2 B_{ijk}\, x_i x_j x_k \quad (i \neq j \neq k),$

$\mathfrak{D}_{ij;j} = a_j^2 x_i(-B_{ij}+B_{iij}x_i^2+3B_{ijj}x_j^2+B_{ijk}x_k^2) \quad (i \neq j \neq k),$

$\mathfrak{D}_{ii;j} = a_j^2 x_j\Big[2B_{iij}x_i^2 + a_i^2\Big(A_{ij}-\sum_{l=1}^{3} A_{ijl}x_l^2\Big)\Big] \quad (i \neq j),$

$\mathfrak{D}_{ii;i} = a_i^2 x_i\Big[a_i^2 A_{ii}-2B_{ii}+(4B_{iii}-a_i^2 A_{iii})x_i^2 +$

$$+ \sum_{l \neq i}(2B_{iil}-a_i^2 A_{iil})x_l^2\Big], \tag{137}$$

where a factor $\pi G\rho$ *has been suppressed.*

These expressions follow from inserting in the relation

$$\mathfrak{D}_{ij;k} = \frac{\partial \mathfrak{D}_{ik}}{\partial x_j} - x_i\,\frac{\partial \mathfrak{D}_k}{\partial x_j} \tag{138}$$

the expressions for $\mathfrak{D}_k$ and $\mathfrak{D}_{ik}$ we have derived.

THEOREM 16. *The potential*

$$\mathfrak{D}_{ijk}(\mathbf{x}) = G \int_V \rho(\mathbf{x}')\,\frac{x_i' x_j' x_k'}{|\mathbf{x}-\mathbf{x}'|}\,d\mathbf{x}', \tag{139}$$

at an internal point of a homogeneous ellipsoid, is given by

$$\frac{\mathfrak{D}_{ijk}}{\pi G\rho} = a_i^2 a_j^2 a_k^2\Big(A_{ijk}-\sum_{l=1}^{3} A_{ijkl}x_l^2\Big)x_i x_j x_k +$$

$$+ \tfrac{1}{4}a_i^2 a_j^2 \delta_{jk}\Big(B_{ij}-2\sum_{l=1}^{3} B_{ijl}x_l^2 + \sum_{l=1}^{3}\sum_{m=1}^{3} B_{ijlm}x_l^2 x_m^2\Big)x_i +$$

$$+ \tfrac{1}{4}a_j^2 a_k^2 \delta_{ki}\Big(B_{jk}-2\sum_{l=1}^{3} B_{jkl}x_l^2 + \sum_{l=1}^{3}\sum_{m=1}^{3} B_{jklm}x_l^2 x_m^2\Big)x_j +$$

$$+ \tfrac{1}{4}a_k^2 a_i^2 \delta_{ij}\Big(B_{ki}-2\sum_{l=1}^{3} B_{kil}x_l^2 + \sum_{l=1}^{3}\sum_{m=1}^{3} B_{kilm}x_l^2 x_m^2\Big)x_k. \tag{140}$$

Proof. We readily verify the identity

$$x_i x_j x_k = -\frac{a_i^2 a_j^2 a_k^2}{48} \frac{\partial^3 E^3}{\partial x_i \partial x_j \partial x_k} -$$

$$-\frac{1}{8}\left(a_i^2 a_j^2 \delta_{ki} \frac{\partial E^2}{\partial x_j} + a_j^2 a_k^2 \delta_{ij} \frac{\partial E^2}{\partial x_k} + a_k^2 a_i^2 \delta_{jk} \frac{\partial E^2}{\partial x_i}\right). \qquad (141)$$

Since $\partial^2 E^3/\partial x_i \partial x_j$ and $\partial E^2/\partial x_i$ vanish for all combinations of indices, Lemma 8, Corollary is applicable and we can write

$$\frac{\mathfrak{D}_{ijk}}{\pi G \rho a_1 a_2 a_3} = -\frac{a_i^2 a_j^2 a_k^2}{192} \int_0^\infty \frac{du}{\Delta} \frac{\partial^3 Q^4}{\partial x_i \partial x_j \partial x_k} -$$

$$-\frac{1}{24} \int_0^\infty \frac{du}{\Delta}\left(a_i^2 a_j^2 \delta_{ki} \frac{\partial Q^3}{\partial x_j} + a_j^2 a_k^2 \delta_{ij} \frac{\partial Q^3}{\partial x_k} + a_k^2 a_i^2 \delta_{jk} \frac{\partial Q^3}{\partial x_i}\right). \qquad (142)$$

We readily verify that

$$\frac{\partial^3 Q^4}{\partial x_i \partial x_j \partial x_k} = -192 \frac{x_i x_j x_k}{(a_i^2+u)(a_j^2+u)(a_k^2+u)} Q +$$

$$+48 Q^2\left[\frac{x_i \delta_{jk}}{(a_k^2+u)(a_i^2+u)} + \frac{x_j \delta_{ik}}{(a_i^2+u)(a_j^2+u)} + \frac{x_k \delta_{ij}}{(a_j^2+u)(a_k^2+u)}\right] \qquad (143)$$

and

$$\frac{\partial Q^3}{\partial x_j} = -6 Q^2 \frac{x_i}{a_i^2+u}; \qquad (144)$$

and inserting the foregoing relations in equation (142) we obtain

$$\frac{\mathfrak{D}_{ijk}}{\pi G \rho a_1 a_2 a_3} = a_i^2 a_j^2 a_k^2 x_i x_j x_k \int_0^\infty \frac{Q\, du}{\Delta(a_i^2+u)(a_j^2+u)(a_k^2+u)} +$$

$$+\frac{1}{4} \int_0^\infty Q^2 \frac{u\, du}{\Delta}\left[\frac{a_i^2 a_j^2 x_i \delta_{jk}}{(a_k^2+u)(a_i^2+u)} + \frac{a_j^2 a_k^2 x_j \delta_{ki}}{(a_i^2+u)(a_j^2+u)} + \frac{a_k^2 a_i^2 x_k \delta_{ij}}{(a_j^2+u)(a_k^2+u)}\right]; \qquad (145)$$

and expanding this last equation we obtain equation (140).

23. The first variations of the potential-energy tensors

In Chapter 2, § 15, we have derived general expressions for the first variations of the potential-energy tensors $\mathfrak{W}_{ij}$ and $\mathfrak{W}_{ij;k}$. These variations are needed if we are to make practical use of the linearized form of the virial equations (Chapter 2, equations (152) and (153)) for a treatment of the stability of a configuration. For homogeneous ellipsoids, the required variations can be explicitly evaluated since the tensor potentials $\mathfrak{B}_{ij}$ and $\mathfrak{D}_{ij;k}$, in terms of which they are expressed, are known.

Since $\mathfrak{B}_{ij}$ and $\mathfrak{D}_{ij;k}$ are now polynomials in the coordinates (cf. equations (126) and (137)) it is apparent from the forms of the equations expressing $\delta\mathfrak{W}_{ij}$ and $\delta\mathfrak{W}_{ij;k}$ (Chapter 2, equations (127) and (130)) that these variations can be expressed as linear combinations of the symmetrized quantities (cf. Chapter 2, equations (122)–(125))

$$V_{ijk\ldots} = \int_V \rho\xi_l \frac{\partial}{\partial x_l}(x_i x_j x_k\ldots)\,d\mathbf{x}. \tag{146}$$

(a) The variation $\delta\mathfrak{W}_{ij}$ for homogeneous ellipsoids

We have (Chapter 2, equation (127))

$$\delta\mathfrak{W}_{ij} = -\int_V \rho\xi_l \frac{\partial\mathfrak{B}_{ij}}{\partial x_l}\,d\mathbf{x}. \tag{147}$$

Inserting the expression (126) for $\mathfrak{B}_{ij}$ in the foregoing equation, we obtain, in view of equation (146),

$$\frac{\delta\mathfrak{W}_{ij}}{\pi G\rho} = -2B_{ij}V_{ij} + a_i^2\,\delta_{ij}\sum_{l=1}^{3} A_{ll}V_{ll}. \tag{148}$$

We may note here for later use that the first variations of the particular combinations of $\mathfrak{W}_{ij}$, that occur in Chapter 2, equations (70) and (71), are

$$\frac{\delta\mathfrak{W}_{11} - \delta\mathfrak{W}_{22}}{\pi G\rho} = -(3B_{11} - B_{12})V_{11} + (3B_{22} - B_{12})V_{22} + (B_{23} - B_{13})V_{33} \tag{149}$$

and

$$\frac{\delta\mathfrak{W}_{11} + \delta\mathfrak{W}_{22} - 2\delta\mathfrak{W}_{33}}{\pi G\rho} = -(3B_{11} + B_{12} - 2B_{13})V_{11} -$$
$$-(3B_{22} + B_{12} - 2B_{23})V_{22} + (6B_{33} - B_{13} - B_{23})V_{33}. \tag{150}$$

(b) The variation $\delta\mathfrak{W}_{ij;k}$ for homogeneous ellipsoids

We have (Chapter 2, equation (130))

$$-2\delta\mathfrak{W}_{ij;k} = \int_V \rho\xi_l \frac{\partial}{\partial x_l}(\mathfrak{B}_{ij}x_k + \mathfrak{D}_{ij;k})\,d\mathbf{x}. \tag{151}$$

Inserting the known expressions for $\mathfrak{B}_{ij}$ (equation (126)) and $\mathfrak{D}_{ij;k}$ (equations (137)) in the foregoing equation, we can now express $\delta\mathfrak{W}_{ij;k}$ as linear combinations of V_{ijk}. In view of the complexity of the formulas giving $\mathfrak{D}_{ij;k}$, it is convenient to distinguish the same cases. We find

$$-2\delta\mathfrak{W}_{ij;k} = 2B_{ij;k}V_{ijk}, \tag{152}$$

$$-2\delta\mathfrak{W}_{ij;j} = a_j^2 B_{iij}V_{iii}+(2B_{ij}+3a_j^2 B_{ijj})V_{ijj}+a_j^2 B_{ijk}V_{ikk}-a_j^2 B_{ij}V_i, \tag{153}$$

$$-2\delta\mathfrak{W}_{ii;j} = -a_i^2 A_{ij;j}V_{jjj}-a_i^2 A_{ik;j}V_{jkk}+(2B_{ii;j}-a_i^2 A_{ii;j})V_{iij}+$$
$$+a_i^2(A_i+a_j^2 A_{ij})V_j, \tag{154}$$

$$-2\delta\mathfrak{W}_{ii;i} = [2(B_{ii}+2a_i^2 B_{iii})-a_i^2 A_{ii;i}]V_{iii}+$$
$$+a_i^2(2B_{iij}-A_{ij;i})V_{ijj}+a_i^2(2B_{iik}-A_{ik;i})V_{ikk}+$$
$$+a_i^2(A_i+a_i^2 A_{ii}-2B_{ii})V_i \tag{155}$$

$$(i \neq j \neq k \text{ in the foregoing formulas}),$$

where for brevity we have introduced the further abbreviations

$$A_{ij;k} = A_{ij}+a_k^2 A_{ijk} \quad \text{and} \quad B_{ij;k} = B_{ij}+a_k^2 B_{ijk}. \tag{156}$$

Again for later use, we may note here the first variation of S_{ijj} defined in Chapter 2, equation (84):

$$\delta S_{ijj} = [(2a_i^2+a_j^2)B_{iij}-5a_i^2 B_{iii}-2B_{ii}]V_{iii}+$$
$$+3[B_{jj}+B_{ij}+(a_i^2+2a_j^2)B_{ijj}-a_i^2 B_{iij}]V_{ijj}+$$
$$+[(2a_i^2+a_j^2)B_{ijk}-3a_i^2 B_{iik}]V_{ikk}+[3a_i^2 B_{ii}-(2a_i^2+a_j^2)B_{ij}]V_i \tag{157}$$

$$(i \neq j \neq k).$$

(c) Restrictions on solenoidal displacements

While the expressions for $\delta\mathfrak{W}_{ij}$ and $\delta\mathfrak{W}_{ij;k}$ given in equations (147) and (152)–(155) are of general validity, we shall be concerned, in the later chapters, only with displacements ξ which preserve the same constant density of the undeformed ellipsoids. The displacements should then be divergence free; and this solenoidal requirement on ξ, as we shall now show, is equivalent to an infinite sequence of linear relations among the $V_{ijk...}$'s.

THEOREM 17. *A solenoidal displacement applied to a homogeneous ellipsoid of semi-axes a_i must satisfy the relations represented by*

$$\sum_{q=1}^{3} \frac{V_{ijk...pqq}}{a_q^2} = V_{ijk...p}. \tag{158}$$

Proof. In view of equation (146), we may write

$$\sum_{q=1}^{3} \frac{V_{ijk...pqq}}{a_q^2} = \int_V \rho\xi.\mathrm{grad}\left(x_i x_j x_k...x_p \sum_{q=1}^{1} \frac{x_q^2}{a_q^2}\right) d\mathbf{x}. \tag{159}$$

By an integration by parts, the integral on the right-hand side of equation (159) becomes (since ρ has been assumed to be a constant)

$$\int_S \rho \sum_{q=1}^3 \frac{x_q^2}{a_q^2} x_i x_j x_k \dots x_p \boldsymbol{\xi} . d\mathbf{S} - \int_V \rho \operatorname{div} \boldsymbol{\xi} \sum_{q=1}^3 \frac{x_q^2}{a_q^2} x_i x_j x_k \dots x_p \, d\mathbf{x}$$
$$= \int_S \rho x_i x_j x_k \dots x_p \boldsymbol{\xi} . d\mathbf{S}, \quad (160)$$

since $\sum x_q^2/a_q^2 = 1$ on the surface S of the ellipsoid and $\operatorname{div} \boldsymbol{\xi} = 0$ in the interior. Therefore,

$$\sum_{q=1}^3 \frac{V_{ijk\dots pqq}}{a_q^2} = \int_S \rho x_i x_j x_k \dots x_p \boldsymbol{\xi} . d\mathbf{S}$$
$$= \int_V \rho \boldsymbol{\xi} . \operatorname{grad}(x_i x_j x_k \dots x_p) \, d\mathbf{x} + \int_V \rho \operatorname{div} \boldsymbol{\xi} \, x_i x_j x_k \dots x_p \, d\mathbf{x}$$
$$= V_{ijk\dots p} \quad (161)$$

by a further application of Gauss's theorem.

We may note here, in particular, that for solenoidal displacements, the $V_{ijk\dots}$'s of the second and the third orders must satisfy the relations

$$\sum_{i=1}^3 \frac{V_{ii}}{a_i^2} = 0 \quad (162)$$

and
$$\sum_{j=1}^3 \frac{V_{ijj}}{a_j^2} = V_i. \quad (163)$$

BIBLIOGRAPHICAL NOTES

In treating the classical problem of the potential of a homogeneous ellipsoid, we have preferred to follow the now long-abandoned unsophisticated style and approach of the masters of the nineteenth century who have written on this subject. We have preferred, in fact, to follow Routh where others might have chosen Kellogg. Indeed, in large part the treatment in §§ 17–20 is based on Routh's account in:

E. J. ROUTH, *A Treatise on Analytical Statics*, **2** (Cambridge, England, Cambridge University Press, 1892), 67, 194–231, 239–54.

For essentially similar, but somewhat abbreviated, accounts see:

A. G. WEBSTER, *The Dynamics of Particles and of Rigid, Elastic, and Fluid Bodies* (Leipzig, B. G. Teubner, 1925), pp. 409–24.

A. S. RAMSEY, *An Introduction to the Theory of Newtonian Attraction* (Cambridge, England, Cambridge University Press, 1952), chap. 7.

A standard reference (though it has no bearing on this chapter) is:

O. D. KELLOGG, *Foundations of Potential Theory* (New York, Frederick Ungar, 1929).

§ 22. As stated in § 16, Ferrers was the first to formulate and solve the problem of the potential of a heterogeneous ellipsoid in which the density varies as $x_1^f x_2^g x_3^h$. His results are crucial for the purposes of this book.

> N. M. FERRERS, "On the potentials of ellipsoids, ellipsoidal shells, elliptic laminae, and elliptic rings, of variable densities," *Quart. J. Pure and Appl. Math.*, **14** (1877), 1–22.

The present chapter follows very closely the author's account in Paper XXII in the list on p. 246. Other papers in the list which are related to matters treated in this chapter are Papers VIII, IX, XIII, XVI, and XXXV. A further reference is:

> P. H. ROBERTS, "On the superpotential and supermatrix of a heterogeneous ellipsoid," *Astrophys. J.*, **136** (1962), 1108–14.

4

DIRICHLET'S PROBLEM AND DEDEKIND'S THEOREM

24. Introduction

As we have stated in Chapter 1 (§ 5), it was Dirichlet who first formulated the basic problem of the subject. Dirichlet raised the general question, how a homogeneous fluid mass, under its own gravitation, can maintain at all times an ellipsoidal figure (which may be variable) with internal motions that are, in an inertial frame, linear functions of the position. Formulated in this fashion, the problem clearly distinguishes two frames of reference: an *inertial frame* fixed in space and a *moving frame* whose coordinate axes coincide, at all times, with the principal axes of the ellipsoid. The moving frame will be of variable orientation and the problem of writing the hydrodynamical equations in such moving frames, generally, is of interest in itself. Such moving frames were apparently first considered by Greenhill (1880) though in the particular context of Dirichlet's problem they underlie Riemann's treatment.

In this chapter, we shall begin by considering the general transformation to moving frames and then obtain the form which the virial equations of the second order take in such frames. We then pass on to the basic equations governing Dirichlet's problem and the proof of Dedekind's theorem. This part of our account will be based on an elegant version of Riemann's treatment due to Lebovitz. And finally, we show how the virial equations of the second order, suitably specialized, become identical with the exact equations of Dirichlet's problem. This last identity establishes the particular appropriateness of the virial equations for the treatment of ellipsoidal figures.

25. The hydrodynamical equations in a moving frame

We consider two frames of reference with a common origin: an inertial frame, (X_1, X_2, X_3), and a moving frame, (x_1, x_2, x_3). The orientation of the moving frame, with respect to the inertial frame, will be assumed to be time dependent. Let $\mathbf{T}(t)$ be the linear transformation that relates the coordinates, (X_1, X_2, X_3) and (x_1, x_2, x_3), of a point in the two frames.

Then
$$\mathbf{x} = \mathbf{TX}, \tag{1}$$

or, more explicitly,
$$x_i = T_{ij}X_j. \tag{2}$$

Since $\mathbf{T}$ must represent an orthogonal transformation,
$$\mathbf{TT}^\dagger = 1, \tag{3}$$

where $\mathbf{T}^\dagger$ is the transpose of the matrix $\mathbf{T}$. From equation (3) it follows that the matrix
$$\mathbf{\Omega}^* = \frac{d\mathbf{T}}{dt}\mathbf{T}^\dagger \tag{4}$$

is antisymmetric, since
$$\mathbf{\Omega}^* + \mathbf{\Omega}^{*\dagger} = \frac{d\mathbf{T}}{dt}\mathbf{T}^\dagger + \mathbf{T}\frac{d\mathbf{T}^\dagger}{dt} = \frac{d}{dt}(\mathbf{TT}^\dagger) = 0. \tag{5}$$

The dual, $\mathbf{\Omega}$, of the matrix $\mathbf{\Omega}^*$ is a vector: $\mathbf{\Omega}^*$ and $\mathbf{\Omega}$ are related by
$$\Omega_{ij}^* = \epsilon_{ijk}\Omega_k \quad \text{and} \quad \Omega_i = \tfrac{1}{2}\epsilon_{ijk}\Omega_{jk}^*. \tag{6}$$

Clearly, $\mathbf{\Omega}(t)$ represents a general time-dependent rotation of the (x_1, x_2, x_3)-frame with respect to the inertial frame.

Let $\mathbf{F}(t)$ be any time-dependent vector defined in the inertial frame. Its components, resolved along the instantaneous coordinate axes of the moving frame, are given by
$$\mathbf{F}_{(\mathbf{x})} = \mathbf{TF}, \tag{7}$$

or, equivalently,
$$\mathbf{F} = \mathbf{T}^\dagger\mathbf{F}_{(\mathbf{x})}. \tag{8}$$

Differentiating this last equation with respect to time, we get
$$\frac{d\mathbf{F}}{dt} = \frac{d\mathbf{T}^\dagger}{dt}\mathbf{F}_{(\mathbf{x})} + \mathbf{T}^\dagger\frac{d\mathbf{F}_{(\mathbf{x})}}{dt}; \tag{9}$$

and resolving this equation along the instantaneous coordinate axes of the moving frame, we obtain (on making use of equation (4))
$$\mathbf{T}\frac{d\mathbf{F}}{dt} = -\mathbf{\Omega}^*(\mathbf{TF}) + \frac{d}{dt}(\mathbf{TF}). \tag{10}$$

Since $\mathbf{F}$ is any vector, the operator equation
$$\mathbf{T}\frac{d}{dt} = \left(\frac{d}{dt} - \mathbf{\Omega}^*\right)\mathbf{T} \tag{11}$$

is applicable to any vector defined in the inertial frame.

Applied to the position vector $\mathbf{X}$ and the velocity vector $d\mathbf{X}/dt$, equation (11) gives
$$\mathbf{T}\frac{d\mathbf{X}}{dt} = \left(\frac{d}{dt} - \mathbf{\Omega}^*\right)\mathbf{TX} \tag{12}$$

and
$$\mathbf{T}\frac{d^2\mathbf{X}}{dt^2} = \left(\frac{d}{dt} - \mathbf{\Omega}^*\right)\mathbf{T}\frac{d\mathbf{X}}{dt}. \tag{13}$$

Letting
$$\mathbf{U} = \mathbf{T}\frac{d\mathbf{X}}{dt} \quad \text{and} \quad \mathbf{u} = \frac{d\mathbf{x}}{dt} \tag{14}$$

denote the velocities in the inertial frame (resolved along the instantaneous coordinate axes of the moving frame) and with respect to an observer, at rest in the moving frame, respectively, we can write equations (12) and (13) in the forms

$$\mathbf{U} = \mathbf{u} - \mathbf{\Omega}^*\mathbf{x} \tag{15}$$

and
$$\mathbf{T}\frac{d^2\mathbf{X}}{dt^2} = \frac{d\mathbf{U}}{dt} - \mathbf{\Omega}^*\mathbf{U}. \tag{16}$$

The left-hand side of equation (16) represents the acceleration in the *inertial frame* resolved, however, along the instantaneous directions of the coordinate axes of the moving frame. The hydrodynamical equation governing a fluid, in an inertial frame, therefore, gives

$$\rho\frac{d\mathbf{U}}{dt} - \rho\mathbf{\Omega}^*\mathbf{U} = -\operatorname{grad} p + \rho\operatorname{grad}\mathfrak{B}, \tag{17}$$

where the gradient is evaluated in the coordinates of the moving frame and

$$\frac{d\mathbf{U}}{dt} = \frac{\partial\mathbf{U}}{\partial t} + \mathbf{u}.\operatorname{grad}\mathbf{U}. \tag{18}$$

In view of this last equation, we may rewrite equation (17), in the notation of Cartesian tensors, in the form

$$\rho\frac{\partial U_i}{\partial t} + \rho u_k\frac{\partial U_i}{\partial x_k} = \rho\Omega^*_{im}U_m - \frac{\partial p}{\partial x_i} + \rho\frac{\partial\mathfrak{B}}{\partial x_i}, \tag{19}$$

while the equation of continuity retains its usual form

$$\frac{\partial\rho}{\partial t} + \frac{\partial}{\partial x_k}(\rho u_k) = 0. \tag{20}$$

In equation (19), $\mathbf{U}$ must be expressed in terms of $\mathbf{u}$ by the relation (15), or equivalently by

$$U_i = u_i - \Omega^*_{ik}x_k = u_i + \epsilon_{ijk}\Omega_j x_k. \tag{21}$$

26. The second-order virial equations in a moving frame

To obtain the second-order virial equations in a moving frame, we have simply to multiply equation (17) by x_j and integrate over the volume V occupied by the fluid. The reduction of the different terms proceeds exactly as in § 11*b* (Chapter 2) and we find

$$\frac{d}{dt}\int_V \rho U_i x_j \, d\mathbf{x} = \int_V \rho U_i u_j \, d\mathbf{x} + \int_V \rho \Omega_{im}^* U_m x_j \, d\mathbf{x} +$$
$$+\Pi\delta_{ij}+\mathfrak{W}_{ij}, \qquad (22)$$

where Π and $\mathfrak{W}_{ij}$ have the same meanings as in Chapter 2.

Expressing $\mathbf{U}$ in terms of $\mathbf{u}$ with the aid of equation (21), we obtain after some elementary reductions and rearrangements

$$\frac{d}{dt}\int_V \rho u_i x_j \, d\mathbf{x} - \frac{d}{dt}\int_V \rho \Omega_{im}^* x_m x_j \, d\mathbf{x} = \int_V \rho u_i u_j \, d\mathbf{x} +$$
$$+ \int_V \rho \Omega_{im}^*(u_m x_j - u_j x_m) \, d\mathbf{x} - \int_V \rho \Omega_{im}^* \Omega_{ml}^* x_l x_j \, d\mathbf{x} +$$
$$+\Pi\delta_{ij}+\mathfrak{W}_{ij}. \qquad (23)$$

Alternatively, we may also write

$$\frac{d}{dt}\int_V \rho u_i x_j \, d\mathbf{x} - \frac{d}{dt}(\Omega_{im}^* I_{mj}) = 2\mathfrak{T}_{ij}+\Pi\delta_{ij}+\mathfrak{W}_{ij}-$$
$$-\mathbf{\Omega}_{il}^{*2}I_{lj} + \int_V \rho \Omega_{im}^*(u_m x_j - u_j x_m) \, d\mathbf{x}, \qquad (24)$$

where I_{ij} denotes the moment of inertia tensor.

The particular case when $\mathbf{u}$ is a linear function of $\mathbf{x}$ is of special interest. In this case, writing (cf. Chapter 2, equation (141))

$$u_i = Q_{il} x_l, \qquad (25)$$

we find that equation (24) reduces to the matrix equation

$$\frac{d}{dt}(\mathbf{QI}-\mathbf{\Omega}^*\mathbf{I}) = \mathbf{QIQ}^\dagger+\mathbf{\Omega}^*(\mathbf{QI}-\mathbf{IQ}^\dagger)-\mathbf{\Omega}^{*2}\mathbf{I}+\mathfrak{W}+\Pi\mathbf{1}, \qquad (26)$$

where $\mathfrak{W} = (\mathfrak{W}_{ij})$ and $\mathbf{I} = (I_{ij})$.

27. The Riemann–Lebovitz formulation of Dirichlet's problem

Let $\mathbf{X}(t)$ denote the position of a fluid element in some fixed inertial frame. In Dirichlet's problem, we envisage a homogeneous mass retaining an ellipsoidal figure at all times while $\mathbf{X}(t)$ is a linear function of the coordinates of the element at some chosen initial time $t = 0$.

Let a_1, a_2, and a_3 denote the (variable) semi-axes of the ellipsoid. The constancy of the mass requires that

$$a_1 a_2 a_3 = \text{constant.} \qquad (27)$$

Let the moving frame, considered in § 25, be now so chosen that its coordinate axes coincide with the principal axes of the ellipsoid. There is no loss of generality in our further supposing that at the chosen initial time $t = 0$, the inertial and the moving frames coincide so that

$$X_j(0) = x_j(0). \tag{28}$$

In Dirichlet's problem it is supposed, as we have stated, that $\mathbf{X}(t)$ is a linear function of $\mathbf{x}(0)$. We shall find that it is convenient to express this assumed linear dependence in the form

$$X_i(t) = \sum_{j=1}^{3} P_{ij}(t)\frac{x_j(0)}{a_j(0)}, \tag{29}$$

where $a_j(0)$ is the value of a_j at $t = 0$. The question is now, whether the problem, as it has been formulated, allows a meaningful solution and, in particular, whether an equation governing $\mathbf{P}$ can be derived consistently with the requirements of hydrodynamics.

Letting
$$\mathbf{A} = \begin{vmatrix} a_1 & 0 & 0 \\ 0 & a_2 & 0 \\ 0 & 0 & a_3 \end{vmatrix}, \tag{30}$$

we can express the condition (29) as

$$\mathbf{X}(t) = \mathbf{P}\mathbf{A}_0^{-1}\mathbf{x}(0). \tag{31}$$

In view of this last equation we may write (cf. equation (1))

$$\mathbf{A}^{-1}\mathbf{x} = \mathbf{A}^{-1}\mathbf{T}\mathbf{X} = \mathbf{A}^{-1}\mathbf{T}\mathbf{P}\mathbf{A}_0^{-1}\mathbf{x}(0), \tag{32}$$

or, letting
$$\mathbf{S} = \mathbf{A}^{-1}\mathbf{T}\mathbf{P}, \tag{33}$$

we can write, alternatively,

$$\mathbf{A}^{-1}\mathbf{x} = \mathbf{S}\mathbf{A}_0^{-1}\mathbf{x}(0). \tag{34}$$

In other words, $\mathbf{S}(t)$ *is the linear transformation* (assumed to exist) *that relates $\mathbf{A}^{-1}\mathbf{x}$ at time t to its value at time $t = 0$.*

Consider, in particular, the elements of the fluid that are on the boundary of the ellipsoid. Since there can be no motion normal to a free boundary, these elements, constituting the boundary, must always remain the same. For an element on the boundary of the ellipsoid $\mathbf{A}^{-1}\mathbf{x}$ is a unit vector; accordingly, it must remain a unit vector. From equation (34) it follows that $\mathbf{S}$ must be such that, at all times, it transforms a unit vector into another unit vector. Therefore, $\mathbf{S}$ *must be an orthogonal* matrix, i.e.

$$\mathbf{S}\mathbf{S}^\dagger = 1. \tag{35}$$

Now, even as the orthogonal matrix $\mathbf{T}$ enabled us to define the antisymmetric matrix $\mathbf{\Omega}^*$, so will the orthogonal matrix $\mathbf{S}$ enable us to define the antisymmetric matrix

$$\mathbf{\Lambda}^* = \frac{d\mathbf{S}}{dt}\mathbf{S}^\dagger. \tag{36}$$

And Λ^* is related to its dual Λ by

$$\Lambda_{ij}^* = \epsilon_{ijk}\Lambda_k \quad \text{and} \quad \Lambda_i = \tfrac{1}{2}\epsilon_{ijk}\Lambda_{jk}^*. \tag{37}$$

Thus, Dirichlet's problem reduces to determining whether a matrix $\mathbf{P}$ exists that determines $\mathbf{X}(t)$ by an equation of the form (31) and, at the same time, is expressible in terms of two orthogonal matrices $\mathbf{T}$ and $\mathbf{S}$ in the manner

$$\mathbf{P} = \mathbf{T}^\dagger \mathbf{A} \mathbf{S}, \tag{38}$$

where, further,

$$\frac{d\mathbf{S}}{dt} = \Lambda^*\mathbf{S}, \qquad \frac{d\mathbf{T}}{dt} = \Omega^*\mathbf{T}, \quad \text{and} \quad \mathbf{S}(0) = \mathbf{T}(0) = \mathbf{1}. \tag{39}$$

An immediate consequence of the foregoing relations is that *an interchange in the roles of Λ^* and Ω^* interchanges the roles of $\mathbf{S}$ and $\mathbf{T}$ and transforms $\mathbf{P}$ to its transpose $\mathbf{P}^\dagger$.*

We have already seen that Ω^* is equivalent to a time-dependent rotation. The meaning of Λ^* becomes clearer when we write down the components of the fluid velocity. We have

$$\mathbf{U} = \mathbf{T}\frac{d\mathbf{X}}{dt} = \mathbf{T}\frac{d\mathbf{P}}{dt}\mathbf{A}_0^{-1}\mathbf{x}(0) = \mathbf{T}\frac{d\mathbf{P}}{dt}\mathbf{S}^\dagger\mathbf{A}^{-1}\mathbf{x}, \tag{40}$$

or, using the expression (38) for $\mathbf{P}$, we have

$$\mathbf{U} = \mathbf{T}\left(\mathbf{T}^\dagger\mathbf{A}\frac{d\mathbf{S}}{dt}+\frac{d\mathbf{T}^\dagger}{dt}\mathbf{A}\mathbf{S}+\mathbf{T}^\dagger\frac{d\mathbf{A}}{dt}\mathbf{S}\right)\mathbf{S}^\dagger\mathbf{A}^{-1}\mathbf{x}, \tag{41}$$

or

$$\mathbf{U} = \left(\mathbf{A}\Lambda^*\mathbf{A}^{-1}-\Omega^*+\frac{d\mathbf{A}}{dt}\mathbf{A}^{-1}\right)\mathbf{x}. \tag{42}$$

Comparison of equations (15) and (42) shows that the motion of the fluid consists of a uniform rotation with an angular velocity Ω together with the internal motion

$$\mathbf{u} = \left(\mathbf{A}\Lambda^*\mathbf{A}^{-1}+\frac{d\mathbf{A}}{dt}\mathbf{A}^{-1}\right)\mathbf{x} \tag{43}$$

in the frame in which the orientation of the axes of the ellipsoid remains fixed. The components of $\mathbf{u}$ (in the moving frame) are

$$u_1 = \frac{a_1}{a_2}\Lambda_3 x_2 - \frac{a_1}{a_3}\Lambda_2 x_3 + \frac{1}{a_1}\frac{da_1}{dt}x_1,$$

$$u_2 = \frac{a_2}{a_3}\Lambda_1 x_3 - \frac{a_2}{a_1}\Lambda_3 x_1 + \frac{1}{a_2}\frac{da_2}{dt}x_2,$$

$$u_3 = \frac{a_3}{a_1}\Lambda_2 x_1 - \frac{a_3}{a_2}\Lambda_1 x_2 + \frac{1}{a_3}\frac{da_3}{dt}x_3. \tag{44}$$

This motion is therefore one of *uniform vorticity* $\boldsymbol{\zeta}$ with the components

$$\zeta_k = -\frac{a_i^2+a_j^2}{a_i\,a_j}\Lambda_k \quad (i \neq j \neq k); \tag{45}$$

and superposed on this vortical motion is an *expansion* with the components

$$\frac{1}{a_i}\frac{da_i}{dt}x_i \quad \text{(no summation over repeated indices).} \tag{46}$$

Returning to the equation of motion (17), we first evaluate $d\mathbf{U}/dt$; with $\mathbf{U}$ given by

$$\mathbf{U} = \left[(\mathbf{A}\boldsymbol{\Lambda}^*-\boldsymbol{\Omega}^*\mathbf{A})+\frac{d\mathbf{A}}{dt}\right]\mathbf{A}^{-1}\mathbf{x}, \tag{47}$$

we have

$$\frac{d\mathbf{U}}{dt} = \left[\frac{d}{dt}(\mathbf{A}\boldsymbol{\Lambda}^*-\boldsymbol{\Omega}^*\mathbf{A})+\frac{d^2\mathbf{A}}{dt^2}\right]\mathbf{A}^{-1}\mathbf{x}+$$
$$+\left[(\mathbf{A}\boldsymbol{\Lambda}^*-\boldsymbol{\Omega}^*\mathbf{A})+\frac{d\mathbf{A}}{dt}\right]\frac{d}{dt}(\mathbf{A}^{-1}\mathbf{x}). \tag{48}$$

But by equation (34)

$$\frac{d}{dt}(\mathbf{A}^{-1}\mathbf{x}) = \frac{d\mathbf{S}}{dt}\mathbf{A}_0^{-1}\mathbf{x}(0) = \frac{d\mathbf{S}}{dt}\mathbf{S}^\dagger\mathbf{A}^{-1}\mathbf{x} = \boldsymbol{\Lambda}^*\mathbf{A}^{-1}\mathbf{x}; \tag{49}$$

therefore,

$$\frac{d\mathbf{U}}{dt} = \left[\frac{d^2\mathbf{A}}{dt^2}+\frac{d}{dt}(\mathbf{A}\boldsymbol{\Lambda}^*-\boldsymbol{\Omega}^*\mathbf{A})+\frac{d\mathbf{A}}{dt}\boldsymbol{\Lambda}^*+\mathbf{A}\boldsymbol{\Lambda}^{*2}-\boldsymbol{\Omega}^*\mathbf{A}\boldsymbol{\Lambda}^*\right]\mathbf{A}^{-1}\mathbf{x}. \tag{50}$$

Combining this last result with $-\boldsymbol{\Omega}^*\mathbf{U}$, we obtain

$$\frac{d\mathbf{U}}{dt}-\boldsymbol{\Omega}^*\mathbf{U} = \left[\frac{d^2\mathbf{A}}{dt^2}+\frac{d}{dt}(\mathbf{A}\boldsymbol{\Lambda}^*-\boldsymbol{\Omega}^*\mathbf{A})+\frac{d\mathbf{A}}{dt}\boldsymbol{\Lambda}^*-\boldsymbol{\Omega}^*\frac{d\mathbf{A}}{dt}+\right.$$
$$\left.+\mathbf{A}\boldsymbol{\Lambda}^{*2}+\boldsymbol{\Omega}^{*2}\mathbf{A}-2\boldsymbol{\Omega}^*\mathbf{A}\boldsymbol{\Lambda}^*\right]\mathbf{A}^{-1}\mathbf{x}. \tag{51}$$

On the other hand, we have (cf. Chapter 3, equation (40))

$$\text{grad } \mathfrak{B} = -2\pi G\rho\begin{vmatrix} A_1 & 0 & 0 \\ 0 & A_2 & 0 \\ 0 & 0 & A_3 \end{vmatrix}\mathbf{x} = -2\pi G\rho\mathfrak{A}\mathbf{x} \text{ (say).} \tag{52}$$

The equation of motion (17) therefore gives

$$\rho\left[\frac{d^2\mathbf{A}}{dt^2}+\frac{d}{dt}(\mathbf{A}\boldsymbol{\Lambda}^*-\boldsymbol{\Omega}^*\mathbf{A})+\frac{d\mathbf{A}}{dt}\boldsymbol{\Lambda}^*-\boldsymbol{\Omega}^*\frac{d\mathbf{A}}{dt}+\mathbf{A}\boldsymbol{\Lambda}^{*2}+\boldsymbol{\Omega}^{*2}\mathbf{A}-\right.$$
$$\left.-2\boldsymbol{\Omega}^*\mathbf{A}\boldsymbol{\Lambda}^*+2\pi G\rho\mathfrak{A}\mathbf{A}\right]\mathbf{A}^{-1}\mathbf{x} = -\text{grad}\,p. \tag{53}$$

Since the left-hand side is a homogeneous linear function of the coordinates, the integration of this equation will lead to an expression for p of the form

$$p = p_c(t) + \sum_{i,j=1}^{3} \alpha_{ij}(t)x_i\,x_j, \tag{54}$$

where p_c denotes the central pressure and the α_{ij}'s are certain functions of time. On the other hand, the boundary condition on p requires that it vanishes identically on the surface of the ellipsoid; and a necessary and sufficient condition to ensure it is that p be reducible to the form

$$p = p_c(t)\left(1 - \sum_{i=1}^{3} \frac{x_i^2}{a_i^2}\right). \tag{55}$$

Accordingly,
$$\operatorname{grad} p = -2p_c\,\mathbf{A}^{-2}\mathbf{x}; \tag{56}$$

and inserting this last result in equation (53), we find that we are left with

$$\frac{d^2\mathbf{A}}{dt^2} + \frac{d}{dt}(\mathbf{A}\mathbf{\Lambda}^* - \mathbf{\Omega}^*\mathbf{A}) + \frac{d\mathbf{A}}{dt}\mathbf{\Lambda}^* - \mathbf{\Omega}^*\frac{d\mathbf{A}}{dt} + \mathbf{A}\mathbf{\Lambda}^{*2} + \mathbf{\Omega}^{*2}\mathbf{A} - 2\mathbf{\Omega}^*\mathbf{A}\mathbf{\Lambda}^*$$
$$= -2\pi G\rho\mathfrak{A}\mathbf{A} + \frac{2p_c}{\rho}\mathbf{A}^{-1}. \tag{57}$$

Equation (57) together with the requirement (27) provide a total of ten equations for the ten unknowns a_1, a_2, a_3, p_c, and the components of $\mathbf{\Lambda}$ and $\mathbf{\Omega}$. The fact that we have been able to derive in this manner a consistent set of equations, implies, conversely, that Dirichlet's problem allows solutions that are determined by equations (27) and (57).

Equation (57) is equivalent to a set of equations which Riemann first derived for this problem; the present manner of derivation is due to Lebovitz.

28. Dedekind's theorem

From equation (57), we can deduce Dedekind's theorem; it states that *if a motion determined by*

$$\mathbf{X}(t) = \mathbf{P}(t)\mathbf{A}_0^{-1}\mathbf{x}(0) \tag{58}$$

is admissible under the conditions of Dirichlet's problem, then the motion determined by the transpose $\mathbf{P}^{\dagger}$ of $\mathbf{P}$ is also admissible.

The theorem follows from a consideration of the transposed equation (57). Since $\mathbf{\Lambda}^*$ and $\mathbf{\Omega}^*$ are antisymmetric matrices their transposes are their negatives. Therefore, by transposition equation (57) becomes

$$\frac{d^2\mathbf{A}}{dt^2} + \frac{d}{dt}(-\mathbf{\Lambda}^*\mathbf{A} + \mathbf{A}\mathbf{\Omega}^*) - \mathbf{\Lambda}^*\frac{d\mathbf{A}}{dt} + \frac{d\mathbf{A}}{dt}\mathbf{\Omega}^* + \mathbf{\Lambda}^{*2}\mathbf{A} + \mathbf{A}\mathbf{\Omega}^{*2} - 2\mathbf{\Lambda}^*\mathbf{A}\mathbf{\Omega}^*$$
$$= -2\pi G\rho\mathfrak{A}\mathbf{A} + \frac{2p_c}{\rho}\mathbf{A}^{-1}. \tag{59}$$

We observe that this equation is the same as equation (57) with $\mathbf{\Lambda}^*$ and $\mathbf{\Omega}^*$ interchanged. Therefore, if we have a solution for some $\mathbf{A}$, $\mathbf{\Lambda}^*$, and $\mathbf{\Omega}^*$, then, for the *same* $\mathbf{A}$ there exists another solution in which the roles of $\mathbf{\Lambda}^*$ and $\mathbf{\Omega}^*$ are interchanged. But according to the remarks following equations (38) and (39), interchanging the roles of $\mathbf{\Lambda}^*$ and $\mathbf{\Omega}^*$ is equivalent to replacing $\mathbf{P}$ by its transpose $\mathbf{P}^\dagger$. Dedekind's theorem directly follows from these observations.

We shall call two configurations derived from $\mathbf{P}$ and $\mathbf{P}^\dagger$ the *adjoints* of one another. The velocity field $\mathbf{U}^\dagger$ in the adjoint configuration can be obtained from equation (47) by interchanging $\mathbf{\Lambda}^*$ and $\mathbf{\Omega}^*$. Thus, we obtain

$$\mathbf{U}^\dagger = \left[(\mathbf{A}\mathbf{\Omega}^* - \mathbf{\Lambda}^*\mathbf{A}) + \frac{d\mathbf{A}}{dt}\right]\mathbf{A}^{-1}\mathbf{x}, \tag{60}$$

or

$$\mathbf{U}^\dagger = \left[(\mathbf{A}\mathbf{\Lambda}^* - \mathbf{\Omega}^*\mathbf{A}) + \frac{d\mathbf{A}}{dt}\right]^\dagger \mathbf{A}^{-1}\mathbf{x}. \tag{61}$$

Therefore, *the velocity fields in adjoint configurations are obtained by applying to $\mathbf{A}^{-1}\mathbf{x}$ matrices that are the transposes of one another.*

Suspending the summation convention and letting $i \neq j \neq k$, we find that the typical diagonal and non-diagonal elements of equation (57) are

$$\frac{d^2 a_i}{dt^2} - a_i[(\Lambda_j^2 + \Lambda_k^2) + (\Omega_j^2 + \Omega_k^2)] + 2(a_j \Lambda_k \Omega_k + a_k \Lambda_j \Omega_j)$$
$$= -2\pi G\rho A_i a_i + \frac{2p_c}{\rho a_i}, \tag{62}$$

$$2\frac{d}{dt}(a_i \Lambda_k - a_j \Omega_k) - a_i \frac{d\Lambda_k}{dt} + a_j \frac{d\Omega_k}{dt} + a_i \Lambda_i \Lambda_j + a_j \Omega_j \Omega_i - 2a_k \Lambda_i \Omega_j = 0, \tag{63}$$

and

$$2\frac{d}{dt}(a_i \Omega_k - a_j \Lambda_k) - a_i \frac{d\Omega_k}{dt} + a_j \frac{d\Lambda_k}{dt} + a_i \Omega_i \Omega_j + a_j \Lambda_j \Lambda_i - 2a_k \Omega_i \Lambda_j = 0 \tag{64}$$

$$[i \neq j \neq k \text{ in equations (62), (63), and (64)}].$$

The foregoing equations are clearly unaffected if $(\Lambda_i, \Lambda_j, \Lambda_k)$ and $(\Omega_i, \Omega_j, \Omega_k)$ are replaced by $(\Lambda_i, -\Lambda_j, -\Lambda_k)$ and $(\Omega_i, -\Omega_j, -\Omega_k)$, respectively. Therefore, if we have a solution for some $\mathbf{A}$, $\mathbf{\Lambda}$, and $\mathbf{\Omega}$, we can obtain other solutions by simply changing the signs of any same pair of components of $\mathbf{\Lambda}$ and $\mathbf{\Omega}$. However, unlike the solution that is obtained by interchanging $\mathbf{\Lambda}^*$ and $\mathbf{\Omega}^*$, these other solutions, obtained by changing the signs of some pair of components of $\mathbf{\Lambda}$ and $\mathbf{\Omega}$, are not different from one another in any essential respect.

An important special class of solutions of Dirichlet's problem are those that are *self-adjoint*, i.e. solutions for which $\Lambda^* = \Omega^*$. By identifying Λ^* and Ω^* in equation (57), we obtain

$$\frac{d^2\mathbf{A}}{dt^2} + \frac{d}{dt}(\mathbf{A}\Omega^* - \Omega^*\mathbf{A}) + \frac{d\mathbf{A}}{dt}\Omega^* - \Omega^*\frac{d\mathbf{A}}{dt} + \mathbf{A}\Omega^{*2} + \Omega^{*2}\mathbf{A} - 2\Omega^*\mathbf{A}\Omega^*$$

$$= -2\pi G\rho\mathfrak{A}\mathbf{A} + \frac{2p_c}{\rho}\mathbf{A}^{-1}. \quad (65)$$

Since this equation is the same as its transpose, it follows that equation (65) together with the requirement (27) provide a total of seven equations for the seven unknowns a_1, a_2, a_3, p_c, and the components of Ω; and non-trivial solutions of equations (27) and (65) can be readily constructed (see Chapter 7, § 53 (c)). If one supposes that initial conditions determine the solutions uniquely, then one might conclude that a solution that is initially self-adjoint will always remain self-adjoint. But the possibility cannot be excluded that some self-adjoint solutions of Dirichlet's problem are singular in some sense.

29. The integrals of equation (57)

Equation (57) allows three integrals representing the conservation of energy, angular momentum, and circulation.

The energy integral is obtained by multiplying equations (62), (63), and (64) by

$$\frac{da_i}{dt}, \qquad a_i\Lambda_k - a_j\Omega_k, \quad \text{and} \quad a_i\Omega_k - a_j\Lambda_k \quad (i \neq j \neq k), \qquad (66)$$

respectively, and summing over the different sets $(i, j, k; i \neq j \neq k)$; we find after some lengthy but elementary reductions that we are left with

$$\frac{1}{2}\frac{d}{dt}\left[\sum\left(\frac{da_i}{dt}\right)^2 + \sum(\Lambda_i^2 + \Omega_i^2)(a_j^2 + a_k^2) - 4\sum a_i a_j \Lambda_k \Omega_k \right]$$

$$= -2\pi G\rho \sum A_i a_i \frac{da_i}{dt} + \frac{2p_c}{\rho}\sum\frac{1}{a_i}\frac{da_i}{dt}. \quad (67)$$

In equation (67) the summations, in each case, are over terms that are obtained by a cyclical permutation of the indices. The second term on the right-hand side vanishes† by virtue of the constancy of $a_1 a_2 a_3$; and the first term, on making use of equation (23) of Chapter 3, becomes

$$2\pi G\rho\left(\sum_{i=1}^{3}\frac{\partial I}{\partial a_i}\frac{da_i}{dt} - I\sum_{i=1}^{3}\frac{1}{a_i}\frac{da_i}{dt} \right) = 2\pi G\rho\frac{dI}{dt}. \qquad (68)$$

† More generally, this term is $-2p_c\, d(\rho^{-1})/dt$.

We are thus led to the integral

$$\frac{1}{2}\sum_{i=1}^{3}\left(\frac{da_i}{dt}\right)^2+\tfrac{1}{2}\sum_{i\neq j\neq k}(\Lambda_i^2+\Omega_i^2)(a_j^2+a_k^2)-2\sum_{i\neq j\neq k}a_i\,a_j\,\Lambda_k\Omega_k-2\pi G\rho I$$

$$= \text{constant.} \quad (69)$$

Next, multiplying equations (63) and (64) by

$$a_i[2a_i\,a_j\,\Omega_k-(a_i^2+a_j^2)\Lambda_k] \quad (70)$$

and

$$-a_j[2a_i\,a_j\,\Omega_k-(a_i^2+a_j^2)\Lambda_k], \quad (71)$$

respectively, adding, and summing over the different sets $(i,j,k;\ i\neq j\neq k)$, we deduce that

$$\sum_{i\neq j\neq k}[2a_i\,a_j\,\Omega_k-(a_i^2+a_j^2)\Lambda_k]^2 = \text{constant.} \quad (72)$$

Since the basic equations are invariant to the interchange of $\boldsymbol{\Lambda}^*$ and $\boldsymbol{\Omega}^*$, we can infer the existence of the further integral

$$\sum_{i\neq j\neq k}[2a_i\,a_j\,\Lambda_k-(a_i^2+a_j^2)\Omega_k]^2 = \text{constant.} \quad (73)$$

Now the components of the angular momentum are given by

$$L_i = \epsilon_{ijk}\int_V \rho x_j\,U_k\,dx; \quad (74)$$

and evaluating this expression with the aid of equation (42) we find

$$L_i = \tfrac{1}{5}M[(a_j^2+a_k^2)\Omega_i-2a_j\,a_k\Lambda_i]\quad (i\neq j\neq k). \quad (75)$$

Accordingly, equation (73) represents the conservation of $\mathbf{L}^2$.

The meaning of the integral (72) becomes apparent when we substitute for Λ_k its expression (45) in terms of the vorticity ζ_k; for, with this substitution, equation (72) becomes

$$\sum_{i\neq j\neq k}a_i^2\,a_j^2(2\Omega_k+\zeta_k)^2 = \text{constant.} \quad (76)$$

Now the components of the *circulation* $\mathbf{C}$ (in the inertial frame) are given by

$$C_k = \pi a_i\,a_j(2\Omega_k+\zeta_k)\quad (i\neq j\neq k). \quad (77)$$

Accordingly, equation (76) represents the conservation of $\mathbf{C}^2$.

30. The equivalence to the virial equation

Under the conditions of Dirichlet's problem, the internal motion $\mathbf{u}$ in the ellipsoid (in the frame in which its principal axes remain fixed) is given by equation (43). It will be observed that $\mathbf{u}$ is of the general form (25) considered in § 26 with

$$\mathbf{Q} = \mathbf{A}\boldsymbol{\Lambda}^*\mathbf{A}^{-1}+\frac{d\mathbf{A}}{dt}\mathbf{A}^{-1}. \quad (78)$$

We shall now show that for this special form for $\mathbf{Q}$, the virial equation (26) is identical with Lebovitz's equation (57).

Since for an ellipsoid

$$\mathbf{I} = I_0 \mathbf{A}^2 \quad \text{where} \quad I_0 = \tfrac{1}{5} M = \frac{4\pi}{15} \rho a_1 a_2 a_3, \tag{79}$$

we readily verify that for $\mathbf{Q}$ given by equation (78)

$$\frac{d}{dt}(\mathbf{QI} - \mathbf{\Omega}^* \mathbf{I}) = I_0 \left\{ \frac{d^2 \mathbf{A}}{dt^2}\mathbf{A} + \left(\frac{d\mathbf{A}}{dt}\right)^2 + \left[\frac{d}{dt}(\mathbf{A\Lambda}^* - \mathbf{\Omega}^*\mathbf{A})\right]\mathbf{A} + \right.$$
$$\left. + (\mathbf{A\Lambda}^* - \mathbf{\Omega}^*\mathbf{A})\frac{d\mathbf{A}}{dt} \right\}, \tag{80}$$

$$\mathbf{QIQ}^\dagger = I_0 \left[-\mathbf{A\Lambda}^{*2}\mathbf{A} + \left(\frac{d\mathbf{A}}{dt}\right)^2 - \frac{d\mathbf{A}}{dt}\mathbf{\Lambda}^*\mathbf{A} + \mathbf{A\Lambda}^*\frac{d\mathbf{A}}{dt} \right], \tag{81}$$

and
$$\mathbf{\Omega}^*(\mathbf{QI} - \mathbf{IQ}^\dagger) = 2\mathbf{\Omega}^*\mathbf{A\Lambda}^*\mathbf{A} I_0. \tag{82}$$

Also (cf. Chapter 3, equations (128) and (129))

$$\mathfrak{W} = -2\pi G \rho I_0 \,\mathfrak{A}\mathbf{A}^2. \tag{83}$$

With the foregoing substitutions, equation (26) becomes

$$\left[\frac{d^2\mathbf{A}}{dt^2} + \frac{d}{dt}(\mathbf{A\Lambda}^* - \mathbf{\Omega}^*\mathbf{A}) + \frac{d\mathbf{A}}{dt}\mathbf{\Lambda}^* - \mathbf{\Omega}^*\frac{d\mathbf{A}}{dt} + \mathbf{A\Lambda}^{*2} + \mathbf{\Omega}^{*2}\mathbf{A} - 2\mathbf{\Omega}^*\mathbf{A\Lambda}^* \right]\mathbf{A}$$
$$= -2\pi G\rho\,\mathfrak{A}\mathbf{A}^2 + \frac{\Pi}{I_0}\mathbf{1}. \tag{84}$$

And this equation is clearly equivalent to equation (57). Indeed, it can also be verified that for p given by equation (55)

$$\Pi = \int_V p\,d\mathbf{x} = \frac{2p_c}{\rho} I_0. \tag{85}$$

We have thus shown that *for the solution of Dirichlet's problem, the second-order virial equations provide the necessary and the sufficient conditions.* Therefore, for the treatment of the equilibrium and the stability of the allowed ellipsoidal figures, the virial equations provide all the information that is requisite. And we shall find that they also provide the most convenient means for the treatment of ellipsoidal figures.

BIBLIOGRAPHICAL NOTES

The basic papers of the subject are, of course, those of Dirichlet, Dedekind, and Riemann. Detailed references to them have already been made in Chapter 1 (§ 5).

§ 24. The reference to Greenhill in this section is to his paper:

A. G. GREENHILL, "On the general motion of a liquid ellipsoid under gravitation of its own parts; continuation of a paper on the rotation of a liquid ellipsoid," *Proc. Camb. Phil. Soc.*, **4** (1880), 4–14.

Moving frames of reference are also considered in:

A. B. Basset, *A Treatise on Hydrodynamics*, **1** (Cambridge, England, Deighton, Bell, and Company, 1888; reprint ed. New York, Dover Publications, 1961), 20–25.

A. S. Ramsey, *A Treatise on Hydromechanics. Part II: Hydrodynamics* (London, G. Bell and Sons, 1949), p. 24.

Sir Horace Lamb, *Hydrodynamics* (Cambridge, England, Cambridge University Press, 1932), p. 12.

§§ 27 and 28.   The formulation of Dirichlet's problem in § 27 and the proof of Dedekind's theorem in § 28 are based on Lebovitz's version of Riemann's original treatment.   An outline of his version will be found in Paper XXVII included in the list on p. 247.

For a recent revival of interest in Dirichlet's problem from a group-theoretic point of view, see

Freeman J. Dyson, "Dynamics of a spinning gas cloud," *J. Math. Mech.*, **18** (1968), 91–101.

5

THE MACLAURIN SPHEROIDS

31. Introduction

In this chapter we begin the study of the principal types of ellipsoidal figures that arise, their equilibrium, and their stability. The study will be based on the virial equations derived in Chapter 2. These equations, as we have seen in Chapter 4 (§ 30), provide the information that suffices exactly for a complete analysis of these problems.

This chapter will be devoted to the simplest of these ellipsoidal figures: the Maclaurin spheroids that arise when homogeneous bodies rotate with a uniform angular velocity. We shall consider these figures in some detail since they provide a model for much of the discussion in the remaining chapters.

32. The equilibrium figures

To derive the equilibrium figures, we shall use the second-order virial equations in the forms given in Chapter 2 (§ 12 (a)). With the direction of rotation chosen along the x_3-axis, the relevant equations are (Chapter 2, equation (65))

$$\mathfrak{W}_{ij} + \Omega^2(I_{ij} - \delta_{i3} I_{3j}) = -\delta_{ij} \Pi. \tag{1}$$

The Maclaurin spheroids represent the solution of equation (1) when the configuration is of uniform density and the rotational axis is further assumed to be an axis of symmetry. Under these circumstances, the non-diagonal components of equation (1) are trivially satisfied while the diagonal components give

$$\mathfrak{W}_{11} + \Omega^2 I_{11} = \mathfrak{W}_{33} = -\Pi, \quad \mathfrak{W}_{11} = \mathfrak{W}_{22}, \quad \text{and} \quad I_{11} = I_{22}. \tag{2}$$

Inserting for $\mathfrak{W}_{11}$ and $\mathfrak{W}_{33}$ their values given in Chapter 3, equation (128), we obtain

$$-2A_1 I_{11} + \Omega^2 I_{11} = -2A_3 I_{33}. \tag{3}$$

In equation (3), Ω is measured in the unit $(\pi G \rho)^{\frac{1}{2}}$; this unit for frequency will be used in the remainder of this book (unless explicitly stated otherwise). Equation (3) gives (cf. Chapter 3, equation (106))

$$\Omega^2 = 2\left(A_1 - \frac{a_3^2}{a_1^2} A_3\right) = 2e^2 B_{13}, \tag{4}$$

where
$$e^2 = 1 - a_3^2/a_1^2 \tag{5}$$

defines the eccentricity of the meridional sections. Substituting for A_1 and A_3 their values given in Chapter 3, equations (36), we recover Maclaurin's formula (Chapter 1, equation (13))

$$\Omega^2 = \frac{2(1-e^2)^{\frac{1}{2}}}{e^3}(3-2e^2)\sin^{-1}e - \frac{6}{e^2}(1-e^2). \tag{6}$$

Indeed, it is to be noted that equation (4) is identical with the equation (Chapter 1, equation (12)) derived by Maclaurin on the basis of Newton's arguments involving the canals.

TABLE I

The variations of Ω^2 and L along the Maclaurin sequence

e	$\Omega^2/\pi G\rho$	$L/(GM^3\bar{a})^{\frac{1}{2}}$	e	$\Omega^2/\pi G\rho$	$L/(GM^3\bar{a})^{\frac{1}{2}}$	e	$\Omega^2/\pi G\rho$	$L/(GM^3\bar{a})^{\frac{1}{2}}$
0	0	0	0·75	0·31947	0·25792	0·91	0·44507	0·41563
0·10	0·00534	0·02539	0·80	0·36316	0·29345	0·92	0·44816	0·43302
0·15	0·01204	0·03829	0·81	0·37190	0·30153	0·93	0·44933	0·45254
0·20	0·02146	0·05144	0·81267	0·37423	0·30375	0·94	0·44785	0·47480
0·25	0·03363	0·06491	0·82	0·38059	0·31001	0·95	0·44264	0·50074
0·30	0·04862	0·07882	0·83	0·38917	0·31893	0·95289	0·44022	0·50912
0·35	0·06647	0·09329	0·84	0·39761	0·32835	0·96	0·43193	0·53194
0·40	0·08727	0·10846	0·85	0·40583	0·33833	0·97	0·41257	0·57123
0·45	0·11108	0·12450	0·86	0·41378	0·34895	0·98	0·37802	0·62486
0·50	0·13799	0·14163	0·87	0·42136	0·36029	0·99	0·31030	0·71209
0·55	0·16807	0·16013	0·88	0·42845	0·37247	0·995	0·24371	0·79443
0·60	0·20135	0·18037	0·89	0·43490	0·38563	0·999	0·12540	0·97380
0·65	0·23783	0·20286	0·90	0·44053	0·39994	0·9999	0·04286	1·22633
0·70	0·27734	0·22834						

In Table I we list the values of Ω^2 for various values of the eccentricity e. The dependence of Ω^2 on e is further illustrated in Fig. 5.

Table I also includes the angular momentum L measured in the unit $(GM^3\bar{a})^{\frac{1}{2}}$ where $\bar{a} = (a_1^2 a_3)^{\frac{1}{3}}$ is the radius of a sphere of the same mass M as the spheroid; it is given by

$$\frac{L}{(GM^3\bar{a})^{\frac{1}{2}}} = \frac{\sqrt{3}}{5}\left(\frac{a_1}{\bar{a}}\right)^2 \Omega \quad [\bar{a} = (a_1^2 a_3)^{\frac{1}{3}}]. \tag{7}$$

The variation of this quantity along the sequence is illustrated in Fig. 6.

We observe that while the angular momentum (for a given mass) increases monotonically along the sequence, the angular velocity does not: the fact that Ω^2 attains a maximum along the sequence emerged already from Thomas Simpson's first calculations in 1743 (see p. 4).

The maximum of Ω^2 occurs at an eccentricity where

$$\frac{d\Omega^2}{de} = \frac{2}{e^3}(9-2e^2) - \frac{2(9-8e^2)}{e^4(1-e^2)^{\frac{1}{2}}}\sin^{-1}e = 0. \tag{8}$$

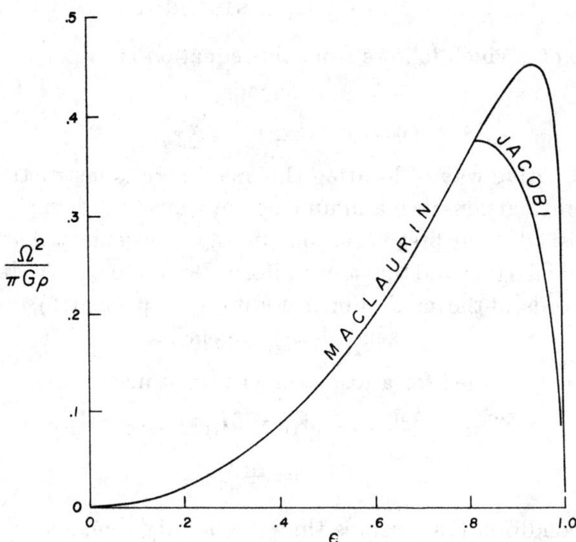

Fig. 5. The square of the angular velocity (in the unit $\pi G\rho$) along the Maclaurin and the Jacobian sequences. The abscissa, in both cases, is the eccentricity of the (1, 3)-section.

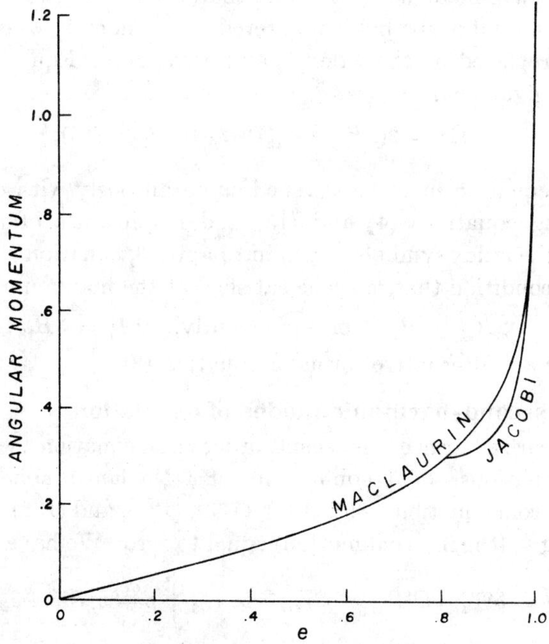

Fig. 6. The angular momentum (in the unit $(GM^3\bar{a})^{\frac{1}{2}}$) along the Maclaurin and the Jacobian sequences. The abscissa in both cases is the eccentricity of the (1, 3-)section.

The value of e which follows from this equation is

$$e = 0\cdot92995, \tag{9}$$

where
$$\Omega^2 = 0\cdot449331 = \Omega^2_{\mathrm{max}}. \tag{10}$$

An alternative way of locating this maximum is instructive. At the point where Ω^2 attains its maximum, not only must the defining equations (2) be satisfied, their first variations, for a displacement that preserves the spheroidal figure and the same uniform density, must also be satisfied. In other words, at the maximum, in addition to equation (2), the equation

$$\delta\mathfrak{W}_{11}+\Omega^2 V_{11} = \delta\mathfrak{W}_{33} \tag{11}$$

must also be satisfied for a displacement for which

$$\delta\mathfrak{W}_{11} = \delta\mathfrak{W}_{22}, \qquad V_{11} = \delta I_{11} = \delta I_{22} = V_{22}, \tag{12}$$

and
$$V_{11}+V_{22} = -\frac{a_1^2}{a_3^2}V_{33}. \tag{13}$$

The last condition (13) ensures that the density remains constant and unchanged (cf. Chapter 3, equation (162)). By making use of Chapter 3, equation (150), we now obtain

$$\Omega^2 V_{11} = \delta\mathfrak{W}_{33}-\delta\mathfrak{W}_{11} = 2(2B_{11}-B_{13})V_{11}-(3B_{33}-B_{13})V_{33}, \tag{14}$$

where we have made use of the fact that, since $a_1 = a_2$ for a Maclaurin spheroid, an index symbol is unaltered if the index 2, wherever it may occur, is replaced by the index 1, and conversely. Equations (13) and (14) now give

$$\Omega^2 = 2(2B_{11}-B_{13})+2\frac{a_3^2}{a_1^2}(3B_{33}-B_{13}); \tag{15}$$

and this equation must be satisfied simultaneously with equation (4). Combining equations (4) and (15) and making use of the relations between the index symbols (given in Chapter 3, equation (117)), we find that the condition that must be satisfied at the maximum of Ω^2 is

$$2B_{11} = B_{13}, \quad\text{or equivalently,}\quad 2A_1 = 3B_{13}; \tag{16}$$

and these are alternative forms of equation (8).

33. The second-harmonic modes of oscillation

The linearized form of the second-order virial equations, which governs small oscillations about equilibrium of a Maclaurin spheroid, can be obtained from equations (89), (131), (133), (137), and (152) of Chapter 2 by setting u_i (the internal motion) equal to zero. We have

$$\frac{d^2 V_{i;j}}{dt^2} = \delta\mathfrak{W}_{ij}+\Omega^2(V_{ij}-\delta_{i3}V_{3j})+2\Omega\epsilon_{il3}\int_V \rho\frac{\partial\xi_l}{\partial t}x_j\,d\mathbf{x}+\delta_{ij}\delta\Pi. \tag{17}$$

We shall now suppose that the Lagrangian displacement $\xi(\mathbf{x}, t)$ is of the form

$$\xi(\mathbf{x}, t) = e^{\lambda t}\xi(\mathbf{x}), \tag{18}$$

where λ is a characteristic-value parameter to be determined. For ξ of this chosen form, equation (17) becomes

$$\lambda^2 V_{i;j} - 2\lambda\Omega\epsilon_{il3} V_{l;j} = \delta\mathfrak{W}_{ij} + \Omega^2(V_{ij} - \delta_{i3} V_{3j}) + \delta_{ij}\delta\Pi. \tag{19}$$

The nine equations, which equation (19) represents, fall into two non-combining groups, of four and five equations, distinguished by their parity (i.e. oddness or evenness) with respect to the index 3. It is convenient to have these equations written out explicitly. The equations odd in the index 3 are

$$\lambda^2 V_{3;1} = \delta\mathfrak{W}_{31} = -2B_{13} V_{13}, \tag{20}$$

$$\lambda^2 V_{3;2} = \delta\mathfrak{W}_{32} = -2B_{13} V_{23}, \tag{21}$$

$$\lambda^2 V_{1;3} - 2\lambda\Omega V_{2;3} = \delta\mathfrak{W}_{13} + \Omega^2 V_{13} = (-2B_{13} + \Omega^2)V_{13}, \tag{22}$$

and $\qquad \lambda^2 V_{2;3} + 2\lambda\Omega V_{1;3} = \delta\mathfrak{W}_{23} + \Omega^2 V_{23} = (-2B_{13} + \Omega^2)V_{23}, \tag{23}$

where we have substituted for $\delta\mathfrak{W}_{ij}$ in accordance with equation (148) of Chapter 3 and allowed for the present equality of B_{13} and B_{23}; also, we have suppressed a common factor $\pi G\rho$ consistently with the unit, $(\pi G\rho)^{\frac{1}{2}}$, in which we are presently measuring frequency. And the equations even in the index 3 are

$$\lambda^2 V_{3;3} = \delta\mathfrak{W}_{33} + \delta\Pi, \tag{24}$$

$$\lambda^2 V_{1;1} - 2\lambda\Omega V_{2;1} = \delta\mathfrak{W}_{11} + \Omega^2 V_{11} + \delta\Pi, \tag{25}$$

$$\lambda^2 V_{2;2} + 2\lambda\Omega V_{1;2} = \delta\mathfrak{W}_{22} + \Omega^2 V_{22} + \delta\Pi, \tag{26}$$

$$\lambda^2 V_{1;2} - 2\lambda\Omega V_{2;2} = \delta\mathfrak{W}_{12} + \Omega^2 V_{12} = (-2B_{11} + \Omega^2)V_{12}, \tag{27}$$

$$\lambda^2 V_{2;1} + 2\lambda\Omega V_{1;1} = \delta\mathfrak{W}_{12} + \Omega^2 V_{12} = (-2B_{11} + \Omega^2)V_{12}; \tag{28}$$

and these equations must be supplemented by the condition

$$\frac{V_{11}}{a_1^2} + \frac{V_{22}}{a_1^2} + \frac{V_{33}}{a_3^2} = 0 \tag{29}$$

required by the solenoidal character of ξ (see Chapter 3, § 23 (c)).

(a) The transverse-shear modes

We shall consider first the equations odd in the index 3. By adding equations (20) and (22) and similarly equations (21) and (23), we obtain

$$(\lambda^2 + 4B_{13} - \Omega^2)V_{13} - 2\lambda\Omega V_{23} + 2\lambda\Omega V_{3;2} = 0 \tag{30}$$

and $\qquad (\lambda^2 + 4B_{13} - \Omega^2)V_{23} + 2\lambda\Omega V_{13} - 2\lambda\Omega V_{3;1} = 0. \tag{31}$

Eliminating $V_{3;1}$ and $V_{3;2}$ from the foregoing equations with the aid of equations (20) and (21), we have

$$\lambda(\lambda^2+4B_{13}-\Omega^2)V_{13}-2\Omega(\lambda^2+2B_{13})V_{23} = 0 \tag{32}$$

and

$$\lambda(\lambda^2+4B_{13}-\Omega^2)V_{23}+2\Omega(\lambda^2+2B_{13})V_{13} = 0. \tag{33}$$

If V_{13} and V_{23} are not to vanish identically, we must have

$$\lambda^2(\lambda^2+4B_{13}-\Omega^2)^2+4\Omega^2(\lambda^2+2B_{13})^2 = 0. \tag{34}$$

Writing

$$\lambda = i\sigma \tag{35}$$

(so that a real σ implies stability), we can factorize equation (34) to give

$$\sigma(\sigma^2-4B_{13}+\Omega^2)-2\Omega(\sigma^2-2B_{13}) = 0 \tag{36}$$

and a similar equation with $-\Omega$ in place of Ω. Equation (36) can be further factorized to give

$$(\sigma-\Omega)(\sigma^2-\sigma\Omega-4B_{13}) = 0; \tag{37}$$

and the roots of this equation are

$$\sigma = \Omega \quad \text{and} \quad \sigma = \tfrac{1}{2}[\Omega\pm(16B_{13}+\Omega^2)^{\frac{1}{2}}]. \tag{38}$$

By replacing Ω by $-\Omega$ in the foregoing, we obtain further roots; these additional roots simply correspond to the fact that equation (34) is even in λ so that if λ is a root then so is $-\lambda$.

The three frequencies given by equations (38) are all real; they, therefore, represent stable modes of oscillation. It should also be noted that the solution for these odd modes does not depend on the divergence condition (29).

When λ^2 is a characteristic root, equations (32) and (33) give

$$\frac{V_{13}}{V_{23}} = \frac{2\Omega(\lambda^2+2B_{13})}{\lambda(\lambda^2+4B_{13}-\Omega^2)}, \tag{39}$$

or, in view of equation (34) satisfied by λ^2,

$$V_{13} = \pm iV_{23}. \tag{40}$$

But the ratio of equations (20) and (21) requires that

$$V_{3;1}/V_{3;2} = V_{13}/V_{23}; \tag{41}$$

accordingly, we must have

$$V_{3;1}/V_{3;2} = V_{1;3}/V_{2;3} = \pm i. \tag{42}$$

It now follows from equations (20) and (21) that

$$\frac{V_{1;3}}{V_{3;1}} = \frac{V_{2;3}}{V_{3;2}} = \frac{\sigma^2}{2B_{13}}-1. \tag{43}$$

(b) *The toroidal modes*

Turning to the equations even in the index 3, we can combine equations (25)–(28) to give the pair of equations

$$\lambda^2 V_{12} + \lambda\Omega(V_{11} - V_{22}) = 2(-2B_{11} + \Omega^2)V_{12} \tag{44}$$

and

$$\tfrac{1}{2}\lambda^2(V_{11} - V_{22}) - 2\lambda\Omega V_{12} = \delta\mathfrak{W}_{11} - \delta\mathfrak{W}_{22} + \Omega^2(V_{11} - V_{22})$$
$$= (-2B_{11} + \Omega^2)(V_{11} - V_{22}), \tag{45}$$

where in writing the alternative form for the right-hand side of equation (45) we have made use of equation (149) in Chapter 3. Rearranging equations (44) and (45), we have

$$[\lambda^2 + 2(2B_{11} - \Omega^2)]V_{12} + \lambda\Omega(V_{11} - V_{22}) = 0 \tag{46}$$

and

$$[\lambda^2 + 2(2B_{11} - \Omega^2)](V_{11} - V_{22}) - 4\lambda\Omega V_{12} = 0; \tag{47}$$

and this pair of equations leads to the characteristic equation

$$[\lambda^2 + 2(2B_{11} - \Omega^2)]^2 + 4\lambda^2\Omega^2 = 0. \tag{48}$$

Again, writing $\lambda - i\sigma$, we can factorize equation (48) to give

$$\sigma^2 - 2\sigma\Omega - 2(2B_{11} - \Omega^2) = 0, \tag{49}$$

and a similar equation with $\bar{\Omega}$ in place of Ω. The roots of equation (49) are

$$\sigma = \Omega \pm (4B_{11} - \Omega^2)^{\frac{1}{2}}. \tag{50}$$

When λ^2 is a characteristic root, it follows from equations (46)–(48) that

$$\frac{V_{11} - V_{22}}{V_{12}} = \frac{4\lambda\Omega}{\lambda^2 + 2(2B_{11} - \Omega^2)} = \pm 2i. \tag{51}$$

For these modes

$$V_{33} = 0, \tag{52}$$

and the divergence condition (29) gives

$$V_{11} = -V_{22}. \tag{53}$$

On the other hand, from equations (27) and (28) we obtain (on subtraction)

$$\lambda^2(V_{1;2} - V_{2;1}) = \lambda\Omega(V_{11} + V_{22}). \tag{54}$$

Accordingly, if $\lambda \neq 0$ and $V_{11} = -V_{22}$, we must have

$$V_{1;2} = V_{2;1}. \tag{55}$$

Hence, for these *toroidal modes*,

$$V_{1;1} = -V_{2;2} = \pm i V_{1;2}, \qquad V_{1;2} = V_{2;1}, \tag{56}$$

and the remaining $V_{i;j}$'s are all zero.

We observe that according to equation (50) when

$$\Omega^2 = 2B_{11}, \quad \sigma = 0 \text{ is a non-trivial root.} \tag{57}$$

But σ does *not* become complex at this point; it becomes complex when

$$\Omega^2 = 4B_{11} \quad \text{and} \quad \sigma = \Omega \text{ is a double root.} \tag{58}$$

Thus, *while the Maclaurin spheroid is neutral to a deformation belonging to this mode at* $\Omega^2 = 2B_{11}$, *it becomes unstable, by overstable oscillations with a frequency* Ω, *at* $\Omega^2 = 4B_{11}$. It is found that these two points occur at

$$e = 0{\cdot}812670 \quad \text{where} \quad \Omega^2 = 2B_{11} = 0{\cdot}374230$$

and
$$e = 0{\cdot}952887 \quad \text{where} \quad \Omega^2 = 4B_{11} = 0{\cdot}440220. \tag{59}$$

We shall show in Chapter 6 that the point $\Omega^2 = 2B_{11}$ is a *point of bifurcation*: at this point the Jacobian sequence branches off. It is also the point where, as we shall show in § 37 below, instability can be induced if some dissipative mechanism is operative. In contrast, at the point $\Omega^2 = 4B_{11}$ dynamical instability sets in; this point was first located by Riemann by a method to which we shall refer in Chapter 7, § 53 (*a*).

(c) The pulsation mode

So far, we have accounted for five normal modes. A sixth mode remains to be determined.

Combining equations (24)–(26) in such a manner that $\delta\Pi$ is eliminated, we have (cf. Chapter 3, equation (150))

$$\tfrac{1}{2}\lambda^2(V_{11}+V_{22})+2\lambda\Omega(V_{1;2}-V_{2;1})-\lambda^2 V_{33}$$
$$= \delta\mathfrak{W}_{11}+\delta\mathfrak{W}_{22}-2\delta\mathfrak{W}_{33}+\Omega^2(V_{11}+V_{22})$$
$$= (-4B_{11}+2B_{13}+\Omega^2)(V_{11}+V_{22})+(6B_{33}-2B_{13})V_{33}. \tag{60}$$

If we exclude the root $\lambda = 0$, we can eliminate $(V_{1;2}-V_{2;1})$ with the aid of equation (54) since we must now suppose that $(V_{11}+V_{22}) \neq 0$. We thus obtain

$$\lambda^2[\tfrac{1}{2}(V_{11}+V_{22})-V_{33}] = (-4B_{11}+2B_{13}-\Omega^2)(V_{11}+V_{22})+$$
$$+2(3B_{33}-B_{13})V_{33}. \tag{61}$$

We can now eliminate $(V_{11}+V_{22})$ from this equation by making use of the divergence condition (29). In this manner we obtain for the oscillation frequency the equation

$$\sigma^2\left(\frac{1}{2}+\frac{a_3^2}{a_1^2}\right) = (4B_{11}-2B_{13}+\Omega^2)+2\frac{a_3^2}{a_1^2}(3B_{33}-B_{13}). \tag{62}$$

An alternative form of this equation, obtained by substituting for Ω^2 its expression (4), is

$$\sigma^2\left(\frac{1}{2}+\frac{a_3^2}{a_1^2}\right) = 4B_{11}+\frac{a_3^2}{a_1^2}(6B_{33}-4B_{13}). \tag{63}$$

And for this pulsation mode the only non-vanishing $V_{i;j}$'s are

$$V_{1;1} = V_{2;2} = -\frac{a_1^2}{2a_3^2}V_{3;3} \quad \text{and} \quad V_{1;2} = -V_{2;1} = \pm 2i\frac{\Omega}{\sigma}V_{1;1}. \qquad (64)$$

In Table II, the characteristic frequencies belonging to the different modes are listed; and their variations along the sequence are further illustrated in Figs. 7a and 7b.

TABLE II

The characteristic frequencies of oscillation of the Maclaurin spheroid

e	Transverse-shear modes $\sigma_o/(\pi G\rho)^{\frac{1}{2}}$		Toroidal modes $\sigma_e/(\pi G\rho)^{\frac{1}{2}}$		Pulsation modes $\sigma_p/(\pi G\rho)^{\frac{1}{2}}$
0	1·03280	1·03280	1·03280	1·03280	1·03280
0·2	1·11156	0·96509	1·16273	0·86978	1·04569
0·3	1·15547	0·93498	1·21513	0·77416	1·06209
0·4	1·20254	0·90713	1·25786	0·66704	1·08552
0·5	1·25272	0·88124	1·28865	0·54570	1·11631
0·6	1·30555	0·85682	1·30319	0·40574	1·15438
0·7	1·35937	0·83274	1·29270	0·23943	1·19788
0·8	1·40841	0·80578	1·23624	0·03099	1·23792
0·81	1·41246	0·80263	1·22632	0·00664	1·24075
0·81207†	1·41350	0·80176	1·22340	0	1·24143
0·82	1·41624	0·79932	1·21527	0·01856	1·24314
0·83	1·41967	0·79584	1·20296	0·04471	1·24501
0·85	1·42527	0·78821	1·17382	0·10028	1·24669
0·90	1·42633	0·76261	1·06057	0·26688	1·22978
0·92	1·41687	0·74742	0·98559	0·35331	1·20698
0·94	1·39558	0·72636	0·86841	0·47002	1·16581
0·95289‡	1·37084	0·70735	$0·66349 \pm 0i$		1·12302
0·96	1·35102	0·69381	$0·65722 \pm 0·14762i$		1·09541
0·98	1·24641	0·63158	$0·61483 \pm 0·27887i$		0·93919
0·99	1·12160	0·56455	$0·55705 \pm 0·30598i$		0·78555
0·999	0·70871	0·35459	$0·35412 \pm 0·23529i$		0·40898

† Point of bifurcation.
‡ Point of onset of dynamical instability.

(d) The proper solutions for the displacements

In some ways it is remarkable that the nine linearized virial equations of the second order have enabled the exact solution of the characteristic-value problem, associated with these second-harmonic oscillations, without any simplifying or approximative assumptions. It is found that this is also the case for the third- and fourth-harmonic oscillations: the eighteen third-order and the thirty fourth-order virial equations similarly enable the exact solution of the associated characteristic-value problems. (The demonstration with respect to the third-order equations will be found in Chapter 6, § 42 (b); for the demonstration with respect

to the fourth-order equations, see Paper XXXV in the list on p. 246.)
It is apparent that the same must hold for all the higher-order oscillations
as well.

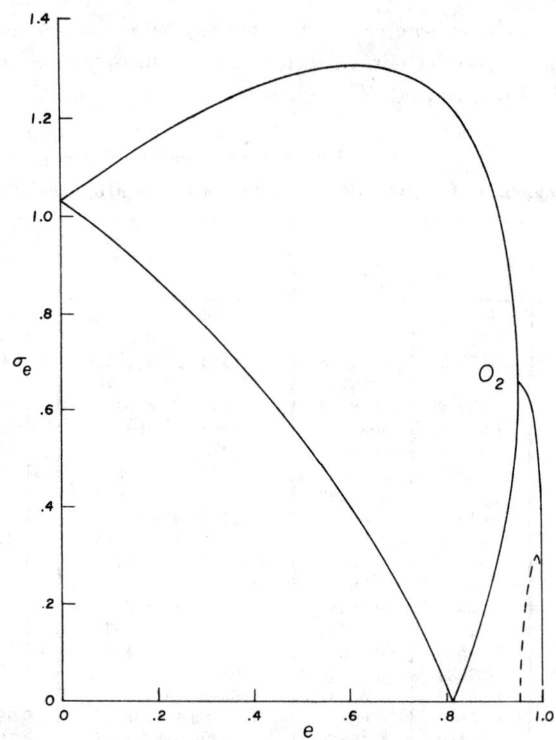

Fig. 7a. The characteristic frequencies (in the unit $(\pi G\rho)^{\frac{1}{2}}$) of the two
even modes of second-harmonic oscillation of the Maclaurin spheroid.
The point of bifurcation, where the Jacobian sequence branches off,
occurs at $e = 0\cdot8127$. At O_2 ($e = 0\cdot9529$) the Maclaurin spheroid becomes dynamically unstable. The real and the imaginary parts of
the frequency, beyond O_2, are shown by the full-line and the dashed
curves, respectively.

The fact, that the virial equations of the different orders *exactly* suffice
to solve the characteristic-value problems of the different harmonics,
must mean that the specification of the $V_{i;jk...}$'s similarly suffices to determine the proper solution ξ_i, and conversely. This reciprocal relationship
between the ξ_i's and the $V_{i;jk...}$'s will clearly be present if the ξ_i's are
homogeneous polynomials in the coordinates x_i: linear for the second-
harmonic oscillations, quadratic for the third-harmonic oscillations, and
so on. It would, therefore, appear that the general forms of the displace-

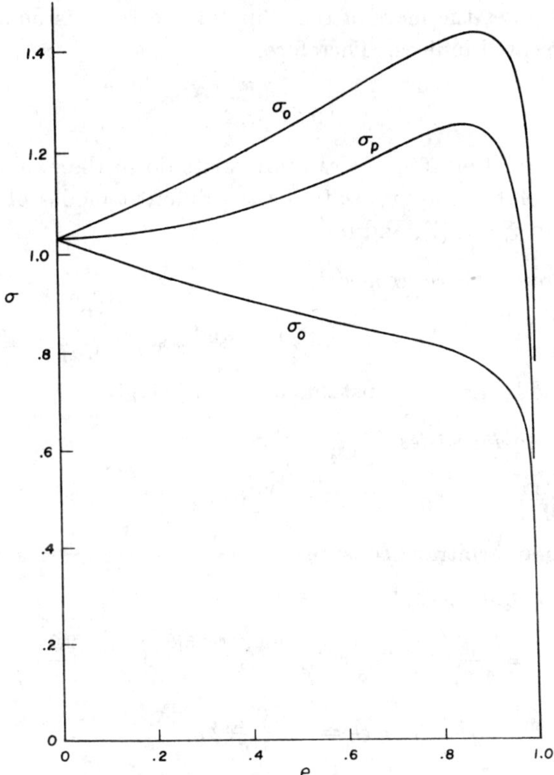

FIG. 7b. The characteristic frequencies (in the unit $(\pi G\rho)^{\frac{1}{2}}$) of the pulsation mode and the two odd modes of second-harmonic oscillation of the Maclaurin spheroid. The respective curves are labeled σ_p and σ_o.

ments ξ_i, appropriate for the second- and the third-harmonic oscillations, are

$$\xi_i = \sum_{m=1}^{3} L_{i;m}\, x_m \quad \text{(for second-harmonic oscillations)} \tag{65}$$

and

$$\xi_i = \sum_{m,n=1}^{3} L_{i;mn}\, x_m\, x_n \quad \text{(for third-harmonic oscillations)}, \tag{66}$$

where $L_{i;m}$ and $L_{i;mn}$ are constants. The nine and the eighteen constants in these forms for ξ_i will in fact be uniquely determined by the specification of the nine $V_{i;j}$'s and the eighteen $V_{i;jk}$'s; and conversely. Thus, considering in particular the second-harmonic modes,

$$V_{i;j} = \sum_{m=1}^{3} L_{i;m} \int_V \rho x_m x_j\, d\mathbf{x} = \tfrac{1}{5} M L_{i;j}\, a_j^2, \tag{67}$$

where M denotes the mass of the ellipsoid and there is no summation over the repeated indices. Therefore,

$$L_{i;j} = \frac{5V_{i;j}}{Ma_j^2}.$$ (68)

Using the relation (68), we can now write down the explicit form of the proper solutions appropriate for the different modes of oscillation considered in §§ (a), (b), and (c).

(i) *The transverse-shear modes*:

$$\xi_1 = \frac{5V_{1;3}}{Ma_3^2}x_3, \qquad \xi_2 = \pm i\frac{5V_{1;3}}{Ma_3^2}x_3, \quad \text{and} \quad \xi_3 = \frac{5V_{3;1}}{Ma_1^2}(x_1 \pm ix_2),$$ (69)

where $V_{1;3}$ and $V_{3;1}$ are two constants whose ratio is given by equation (43).

(ii) *The toroidal modes*:

$$\xi_1 = \frac{5V_{1;1}}{Ma_1^2}(x_1 \pm ix_2), \qquad \xi_2 = \frac{5V_{1;1}}{Ma_1^2}(-x_2 \pm ix_1), \quad \text{and} \quad \xi_3 = 0,$$ (70)

where $V_{1;1}$ is an arbitrary constant.

(iii) *The pulsation mode*:

$$\xi_1 = \frac{5V_{1;1}}{Ma_1^2}\left(x_1 \pm 2i\frac{\Omega}{\sigma}x_2\right), \qquad \xi_2 = \frac{5V_{1;1}}{Ma_1^2}\left(x_2 \mp 2i\frac{\Omega}{\sigma}x_1\right),$$

and $$\xi_3 = -\frac{10V_{1;1}}{Ma_1^2}x_3,$$ (71)

where $V_{1;1}$ is an arbitrary constant.

34. A necessary condition for the occurrence of a point of bifurcation

In § 33 we saw that the Maclaurin spheroid allows a deformation, by a displacement of the form (70), with respect to which it is neutral when $\Omega^2 = 2B_{11}$. It will be shown explicitly in Chapter 6, § 39 that the Jacobian sequence branches off from the Maclaurin sequence at this point. On this account, the point $\Omega^2 = 2B_{11}$ is called a *point of bifurcation*. We shall now formulate this notion of a point of bifurcation more generally.

Consider a sequence of equilibrium configurations arranged "linearly" with respect to some parameter (such as the eccentricity e along the Maclaurin sequence). Suppose that as we follow such a sequence we arrive at a point where a parting of the ways occurs. At such a point, as Darwin has picturesquely described, "a physical evolutionist must, as it were, change carriages in the evolutionary journey." More precisely, at the parting of the ways, we can distinguish two distinct equilibrium configurations where there was only one before. We say then that we have arrived at a point of bifurcation. A point of bifurcation defined in this manner presupposes that the existence of equilibrium configurations along both prongs of the fork has been established.

The occurrence of a point of bifurcation implies that a non-trivial Lagrangian displacement ξ exists such that the deformation of the configuration by this displacement leaves its equilibrium unaffected. The displacement in question is, in fact, the one which will deform the configuration from the "shape" it has in one of the branches to the shape it has in the other. From the existence of a displacement ξ, that leaves the equilibrium unaffected, we may draw two inferences: *first*, that at the point of bifurcation the equilibrium configuration must be characterized by a proper mode of oscillation with respect to which it is neutral; and *second*, that if J is *any* integral property which vanishes by virtue of equilibrium, then its first variation, induced by the displacement ξ, must also vanish at the point of bifurcation. We shall now formulate the second of these two inferences more precisely.

The integral properties J with which we shall be concerned are those which occur in the equilibrium virial equations. These are all of the general form

$$J = \int_V \rho(\mathbf{x})Q_1(\mathbf{x})\,d\mathbf{x} + \int_V \int_V \rho(\mathbf{x})\rho(\mathbf{x}')Q_2(\mathbf{x}, \mathbf{x}')\,d\mathbf{x}d\mathbf{x}', \qquad (72)$$

where $Q_1(\mathbf{x})$ and $Q_2(\mathbf{x}, \mathbf{x}')$ are functions which are defined for all points $\mathbf{x}$ and pairs of points $(\mathbf{x}, \mathbf{x}')$, respectively, in the volume V occupied by the fluid. The first variation of J induced by a displacement ξ is (cf. Chapter 2, equation (120))

$$\delta J = \int_V \rho(\mathbf{x})\xi_l(\mathbf{x})\frac{\partial Q_1(\mathbf{x})}{\partial x_l}\,d\mathbf{x} +$$

$$+ \int_V \int_V \rho(\mathbf{x})\rho(\mathbf{x}')\left[\xi_l(\mathbf{x})\frac{\partial}{\partial x_l} + \xi_l(\mathbf{x}')\frac{\partial}{\partial x_l'}\right]Q_2(\mathbf{x}, \mathbf{x}')\,d\mathbf{x}d\mathbf{x}'. \quad (73)$$

The assertion is that *if*

$$J = 0 \quad \textit{by virtue of equilibrium} \tag{74}$$

then $\quad \delta J = 0 \quad$ *at a point of bifurcation for some non-trivial* ξ. $\qquad$ (75)

We now define a *neutral point* along a sequence of equilibrium configurations as *a point where a characteristic frequency, belonging to some proper normal mode of oscillation, vanishes.* Clearly, a point of bifurcation must also be a neutral point; but a neutral point need not necessarily be a point of bifurcation.

The principal conclusion to be drawn from the foregoing discussion is the following. If J is any integral property that vanishes by virtue of the conditions for equilibrium, then a necessary condition for the occurrence of a neutral point (and, *a fortiori*, for a point of bifurcation) is that a non-trivial Lagrangian displacement $\boldsymbol{\xi}$ exists for which $\delta J = 0$. For this criterion to be useful it is necessary that δJ does not vanish identically for the forms of $\boldsymbol{\xi}$ that may be relevant: the property $\delta J = 0$ cannot then discriminate the neutral point from any other point. However, to exhibit and isolate a point of bifurcation along a sequence, it does not matter if J itself vanishes identically (or trivially) provided its appropriate first variation does not vanish equally trivially. Thus, the virial relation

$$J = \mathfrak{W}_{12} + \Omega^2 I_{12} = 0 \tag{76}$$

provides an example: it vanishes trivially along the Maclaurin sequence; but its first variation does not vanish identically along the sequence; and, as we shall see presently in § 35, the requirement that δJ vanish isolates, in fact, the point of bifurcation $\Omega^2 = 2B_{11}$.

35. The isolation of the point $\Omega^2 = 2B_{11}$ from a consideration of the virial relations

We shall illustrate the ideas described in § 34 by showing how the point of bifurcation along the Maclaurin sequence, where a non-axisymmetric sequence might branch off, can be isolated from a consideration of the integral properties provided by the second-order virial relations (Chapter 2, equations (68)–(71)). The first variations of these integral properties give

$$\delta \mathfrak{W}_{13} = \delta \mathfrak{W}_{23} = 0, \qquad \delta I_{13} = V_{13} = 0, \qquad \delta I_{23} = V_{23} = 0, \tag{77}$$

$$\delta \mathfrak{W}_{12} + \Omega^2 V_{12} = 0, \tag{78}$$

$$\delta \mathfrak{W}_{11} - \delta \mathfrak{W}_{22} + \Omega^2 (V_{11} - V_{22}) = 0, \tag{79}$$

and $$\delta \mathfrak{W}_{11} + \delta \mathfrak{W}_{22} - 2 \delta \mathfrak{W}_{33} + \Omega^2 (V_{11} + V_{22}) = 0, \tag{80}$$

as conditions that must be satisfied at a neutral point.

By making use of equations (148)–(150) of Chapter 3 and remembering the present equality of B_{11}, B_{12}, and B_{22} as well as of B_{13} and B_{23}, we find that equations (77)–(80) reduce to

$$V_{13} = V_{23} = 0, \tag{81}$$

$$(\Omega^2 - 2B_{11}) V_{12} = 0, \tag{82}$$

$$(\Omega^2 - 2B_{11})(V_{11} - V_{22}) = 0, \tag{83}$$

and
$$[\Omega^2 - 2(2B_{11} - B_{13})](V_{11} + V_{22}) + 2(3B_{33} - B_{13})V_{33} = 0, \tag{84}$$

where it may be noted that equation (84), when $V_{11} = V_{22}$, is the same as equation (14) used to isolate the maximum of Ω^2.

The further condition that the Lagrangian displacement leaves the initial constant density unchanged requires that we supplement equations (81)–(84) by equation (29).

At a point of bifurcation, where a new sequence of configurations without rotational symmetry is presumed to branch off, we may require that

$$V_{11} \neq V_{22}. \tag{85}$$

With this stipulation, a non-trivial solution of equations (82)–(84) is possible only if

$$\Omega^2 = 2B_{11}; \tag{86}$$

and at a point where this equality is satisfied, we can have

$$V_{13} = V_{23} = V_{33} = 0, \qquad V_{12} \neq 0, \quad \text{and/or} \quad V_{11} = -V_{22}. \tag{87}$$

A displacement that is compatible with the requirements enumerated in (87), and deforms the spheroid into a tri-axial ellipsoid, clearly exists. The possibility of a sequence of tri-axial ellipsoids branching off from the Maclaurin sequence at the point $\Omega^2 = 2B_{11}$ is, therefore, established.

36. The stable part of the Maclaurin sequence as a curve of bifurcation in four different ways

First, we shall clarify further the concepts underlying the location, in § 35, of the point of bifurcation along the Maclaurin sequence.

The Maclaurin spheroid rotating with a uniform angular velocity Ω_{Mc} (say), when viewed from a frame of reference rotating with that same angular velocity, appears as in hydrostatic equilibrium with no internal motions. An infinitesimal Lagrangian displacement that deforms it into a tri-axial ellipsoid is

$$\xi_1 = \alpha x_2, \qquad \xi_2 = \beta x_1, \quad \text{and} \quad \xi_3 = 0, \tag{88}$$

where α and β are two (infinitesimal) constants. Such a deformation will in general destroy the hydrostatic equilibrium, *as observed in the chosen frame*: for example, the first variation of the integral property (76), namely,

$$(\Omega^2_{\mathrm{Mc}} - 2B_{11})V_{12} \quad \text{with} \quad V_{12} = \tfrac{1}{5}M(\alpha + \beta)a_1^2 \neq 0, \tag{89}$$

will not vanish for an arbitrarily selected member of the Maclaurin sequence. Consequently, the deformation of the spheroid by the displacement (88) must result in a state of continuing motions (*again, as observed in the chosen frame*) except when $\Omega^2_{\mathrm{Mc}} = 2B_{11}$.

The foregoing remarks emphasize why a point of bifurcation must also be a neutral point. A Lagrangian displacement, which deforms the spheroid into a tri-axial ellipsoid, must also be a proper solution of an associated characteristic-value problem; otherwise, we cannot be certain that all the conditions requisite for equilibrium are satisfied. However, we have shown in § 33 (b) that when $\Omega_{\mathrm{Mc}}^2 = 2B_{11}$, the mode belonging to the characteristic value

$$\sigma = \Omega_{\mathrm{Mc}} - (4B_{11} - \Omega_{\mathrm{Mc}}^2)^{\frac{1}{2}} \tag{90}$$

does indeed become neutral; and the proper solution (70) to which σ belongs does deform the spheroid into an ellipsoid.

When the reasons for the occurrence of a point of bifurcation at $\Omega_{\mathrm{Mc}}^2 = 2B_{11}$ are stated in the foregoing manner, the question naturally occurs whether a selected normal mode of oscillation, of some chosen Maclaurin spheroid, cannot be "neutralized" by viewing it from a frame of reference rotating with a suitably adjusted angular velocity $\Omega \neq \Omega_{\mathrm{Mc}}$. We shall now examine how one may accomplish such a neutralization.

We suppose then that a Maclaurin spheroid is observed from a frame of reference rotating with an angular velocity Ω different from Ω_{Mc} but about the same axis. We shall call the frame, from which the spheroid is viewed, the *observer's frame*. In the observer's frame, the Maclaurin spheroid will appear as having stationary internal motions with the uniform vorticity

$$\zeta = 2(\Omega_{\mathrm{Mc}} - \Omega) \tag{91}$$

about the x_3-axis. Since the transverse sections are circular, the motion associated with ζ is purely rotational and is exactly the difference between Ω and Ω_{Mc}. The components of the internal motion, in the observer's frame, are

$$u_1 = -(\Omega_{\mathrm{Mc}} - \Omega)x_2, \qquad u_2 = +(\Omega_{\mathrm{Mc}} - \Omega)x_1, \quad \text{and} \quad u_3 = 0. \tag{92}$$

The virial equations governing small oscillations under these more general circumstances can be written down† by making use of equations (141)–(145) and (152) of Chapter 2; and they can be solved *for the characteristic frequencies as they will be recognized in the observer's frame.* However, the required characteristic frequencies can be obtained, without detailed calculations, by a method which, though somewhat indirect, is instructive.

Consider the oscillations of a Maclaurin spheroid, as in § 33, in the frame rotating with the same angular velocity Ω_{Mc}. Since we are dealing

---

† In Chapter 7, § 49 and § 52 the relevant equations are in fact written down and solved under the very general requirements of the Riemann ellipsoids.

here with a slightly perturbed homogeneous spheroid, its internal state
can be completely described by the motion of its bounding surface and,
more particularly, by its normal component to the surface, namely,

$$X = \xi_j \frac{\partial}{\partial x_j}\left(\frac{x_1^2+x_2^2}{a_1^2}+\frac{x_3^2}{a_3^2}\right) = 2\frac{\xi_1 x_1+\xi_2 x_2}{a_1^2}+2\frac{\xi_3 x_3}{a_3^2}. \tag{93}$$

By substituting for ξ its expressions, given in equations (69)–(71),
appropriate for the different modes, we obtain

$$X_{\text{trans}} = A_{\text{trans}}\, x_3(x_1^2+x_2^2)^{\frac{1}{2}}e^{i\sigma_o t+i\varphi}, \tag{94}$$

$$X_{\text{tor}} = A_{\text{tor}}(x_1^2+x_2^2)e^{i\sigma_e t+2i\varphi}, \tag{95}$$

and
$$X_{\text{puls}} = A_{\text{puls}}\left(\frac{x_1^2+x_2^2}{a_1^2}-2\frac{x_3^2}{a_3^2}\right)e^{i\sigma_p t}, \tag{96}$$

where A_{trans}, A_{tor}, and A_{puls} are constants and σ_o, σ_e, and σ_p are the
characteristic frequencies of oscillation given by equations (38), (50), and
(63), respectively. Also, φ in equations (94)–(96) denotes the azimuthal
angle measured in the equatorial plane and in the same sense as Ω_{Mc}.

We now inquire how the motion of the surface described by equations
(94)–(96), in the frame rotating with the angular velocity Ω_{Mc}, will appear
in the observer's frame rotating with an angular velocity Ω different
from Ω_{Mc}.

Let φ_0 denote the azimuthal angle in the observer's frame. It is
related to φ by
$$\varphi_0 = \varphi-(\Omega-\Omega_{\text{Mc}})t. \tag{97}$$

Accordingly, the motion of the surface in the observer's frame will be
described by the equations

$$X_{\text{trans}} = A_{\text{trans}}\, x_3(x_1^2+x_2^2)^{\frac{1}{2}}\exp\{i[\sigma_o+(\Omega-\Omega_{\text{Mc}})]t+i\varphi_0\}, \tag{98}$$

$$X_{\text{tor}} = A_{\text{tor}}(x_1^2+x_2^2)\exp\{i[\sigma_e+2(\Omega-\Omega_{\text{Mc}})]t+2i\varphi_0\}, \tag{99}$$

and
$$X_{\text{puls}} = A_{\text{puls}}\left(\frac{x_1^2+x_2^2}{a_1^2}-2\frac{x_3^2}{a_3^2}\right)\exp(i\sigma_p t). \tag{100}$$

From these equations it follows that while the motions accompanying
the pulsation mode are described independently of the choice of the
observer's frame, this is not the case with the transverse-shear and the
toroidal modes. In the observer's frame, these latter modes will be
attributed to the *characteristic frequencies*

$$\sigma_o(\Omega) = \sigma_o+(\Omega-\Omega_{\text{Mc}}) \tag{101}$$

and
$$\sigma_e(\Omega) = \sigma_e+2(\Omega-\Omega_{\text{Mc}}), \tag{102}$$

where, to avoid ambiguity, we have distinguished the σ's by the angular velocities of the frames to which they refer. Inserting for σ_o and σ_e their values given in equations (38) and (50), we obtain

$$2\sigma_o(\Omega) = 2\Omega - \Omega_{\mathrm{Mc}} \pm (16 B_{13} + \Omega_{\mathrm{Mc}}^2)^{\frac{1}{2}} \tag{103}$$

and

$$\sigma_e(\Omega) = 2\Omega - \Omega_{\mathrm{Mc}} \pm (4 B_{11} - \Omega_{\mathrm{Mc}}^2)^{\frac{1}{2}}. \tag{104}$$

From equations (103) and (104) it follows that the transverse-shear and the toroidal modes of a chosen Maclaurin spheroid can be neutralized by viewing it in frames that are rotating with the angular velocities

$$\Omega = \tfrac{1}{2}[\Omega_{\mathrm{Mc}} \mp (16 B_{13} + \Omega_{\mathrm{Mc}}^2)^{\frac{1}{2}}] \tag{105}$$

and

$$\Omega = \tfrac{1}{2}[\Omega_{\mathrm{Mc}} \mp (4 B_{11} - \Omega_{\mathrm{Mc}}^2)^{\frac{1}{2}}]. \tag{106}$$

In agreement with our earlier result, we infer from equation (104) that, with the choice of the minus sign, $\sigma_e = 0$ for $\Omega = \Omega_{\mathrm{Mc}}$ and $\Omega_{\mathrm{Mc}}^2 = 2 B_{11}$; but we also notice that, with the choice of the plus sign, $\sigma_e = 0$ for $\Omega = 0$ and the same value $\Omega_{\mathrm{Mc}}^2 = 2 B_{11}$. These *two* possibilities, for neutralizing the one or the other of the two toroidal modes at $\Omega_{\mathrm{Mc}}^2 = 2 B_{11}$, correspond to the fact that at this point two distinct sequences branch off: the Jacobian and the Dedekind (cf. Chapter 1, § 5; also Chapter 6 where the two sequences are studied in detail).

Returning to equations (105) and (106), we observe that while equation (105) provides real values for Ω (for the choice of the observer's frame) along the entire Maclaurin sequence, equation (106) provides real values for Ω only for $\Omega_{\mathrm{Mc}}^2 \leqslant 4 B_{11}$, i.e. only for the stable part of the sequence. We may, therefore, conclude that *every point of the dynamically stable part of the Maclaurin sequence is a point of bifurcation in four different ways; and the points of the remaining unstable part of the sequence are points of bifurcation in two different ways.* As we shall see in greater detail in Chapter 7 (§§ 48 (*b*) and 51 (*c*)) the four types of ellipsoidal figures that branch off from the Maclaurin sequence are the *Riemann ellipsoids*.

Finally, we may note that if the Maclaurin spheroids are viewed from a frame rotating with the angular velocity $\tfrac{1}{2}\Omega_{\mathrm{Mc}}$, the characteristic frequencies (attributed to the transverse-shear and the toroidal modes) become

$$\sigma_o(\tfrac{1}{2}\Omega_{\mathrm{Mc}}) = \pm\tfrac{1}{2}(16 B_{13} + \Omega_{\mathrm{Mc}}^2)^{\frac{1}{2}} \tag{107}$$

and

$$\sigma_e(\tfrac{1}{2}\Omega_{\mathrm{Mc}}) = \pm(4 B_{11} - \Omega_{\mathrm{Mc}}^2)^{\frac{1}{2}}. \tag{108}$$

For this choice of the observer's frame, the vorticity $\zeta = \Omega_{\mathrm{Mc}}$ and (cf. Chapter 4, equation (45))

$$\Lambda(\tfrac{1}{2}\Omega_{\mathrm{Mc}}) = -\frac{a_1 a_2}{a_1^2 + a_2^2}\,\zeta = -\tfrac{1}{2}\Omega_{\mathrm{Mc}}. \tag{109}$$

Also, it can be verified that in this frame the directions of the principal axes of the ellipsoid are fixed. *The motions accompanying these oscillatory modes are therefore self-adjoint* in the sense of Dedekind's theorem (Chapter 4, § 28 and Chapter 7, § 53).

37. The effect of viscous dissipation on the stability of the Maclaurin spheroid

In § 33, we have seen that while the Maclaurin spheroid allows a neutral mode of oscillation at $\Omega^2 = 2B_{11}$, it does not become unstable at this point; it becomes unstable only at a subsequent point where $\Omega^2 = 4B_{11}$. These results were deduced from an analysis based on the assumption that the fluid is inviscid. We shall now show that if this assumption is relaxed and the fluid is considered viscid, the Maclaurin spheroid becomes unstable already at $\Omega^2 = 2B_{11}$. Instability which thus manifests itself only if some dissipative mechanism is operative is called *secular instability*. It differs from "ordinary" or dynamical instability, which sets in independently of any dissipative mechanism, in that the e-folding time of the instability is directly proportional to the efficiency of the dissipative process that is operating.

In considering the secular instability of the Maclaurin spheroid, we shall suppose that we are dealing with stresses derived from ordinary viscosity defined in terms of a *coefficient of kinematic viscosity ν* (of dimensions $cm^2\ sec^{-1}$). The treatment will be based on the linearized version of the second-order virial equations suitably generalized to include the presence of viscous stresses; and the problem will be solved in a *low Reynolds-number approximation*, i.e. in an approximation in which the effects arising from viscous dissipation are considered small and taken into account in the first order.

The analysis in this section derives from an investigation due to Rosenkilde.

(a) *The second-order virial equations allowing for viscous dissipation*

The Navier–Stokes equation governing the motion of a viscous fluid can be written in the form

$$\rho\,\frac{du_i}{dt} = \rho\frac{\partial \mathfrak{B}}{\partial x_i} - \frac{\partial p}{\partial x_i} + \frac{\partial P_{ik}}{\partial x_k}, \tag{110}$$

where

$$P_{ik} = \rho\nu\left(\frac{\partial u_i}{\partial x_k} + \frac{\partial u_k}{\partial x_i} - \frac{2}{3}\frac{\partial u_l}{\partial x_l}\delta_{ik}\right) \tag{111}$$

defines the stress due to viscosity. Associated with equations (110) and (111) is the boundary condition that *on a free surface the normal component of the total stress*, namely,

$$(-p\delta_{ik}+P_{ik})n_k \tag{112}$$

(where **n** denotes the unit outward normal), *must vanish.*

From equation (110) we can derive the appropriate form of the second-order virial equation in the usual manner by multiplying it by x_j and integrating over the volume V occupied by the fluid. The reductions proceed exactly as in § 11 (*a*) except that we must now include the additional term in the viscous stress. We have

$$\int_V x_j\left(-\frac{\partial p}{\partial x_i}+\frac{\partial P_{ik}}{\partial x_k}\right) d\mathbf{x} = \int_S x_j(-p\delta_{ik}+P_{ik})n_k\, dS+$$
$$+\delta_{ij}\int_V p\, d\mathbf{x} - \int_V P_{ij}\, d\mathbf{x}. \tag{113}$$

The integrated part vanishes by virtue of the boundary condition that must be satisfied on a free surface. We thus obtain (cf. Chapter 2, equation (47))

$$\frac{d}{dt}\int_V \rho u_i x_j\, d\mathbf{x} = 2\mathfrak{T}_{ij}+\mathfrak{W}_{ij}+\delta_{ij}\,\Pi-\mathfrak{P}_{ij}, \tag{114}$$

where we have written
$$\mathfrak{P}_{ij} = \int_V P_{ij}\, d\mathbf{x}. \tag{115}$$

We may call $\mathfrak{P}_{ij}$ the *shear-energy tensor.*

The form which equation (113) takes in a rotating frame can be readily written down. Since the velocity defined in the rotating frame differs from the velocity defined in the inertial frame only by $\mathbf{x}\times\mathbf{\Omega}$, the expressions for $\mathfrak{P}_{ik}$ in the two frames are, formally, the same. We, therefore, have (cf. Chapter 2, equation (63))

$$\frac{d}{dt}\int_V \rho u_i x_j\, d\mathbf{x} = 2\mathfrak{T}_{ij}+\mathfrak{W}_{ij}+\Omega^2(I_{ij}-\delta_{i3}\,I_{3j})+$$
$$+2\epsilon_{il3}\Omega\int_V \rho u_l x_j\, d\mathbf{x}+\delta_{ij}\,\Pi-\mathfrak{P}_{ij}, \tag{116}$$

where we have supposed that the direction of $\mathbf{\Omega}$ is along the x_3-axis.

We turn next to the linearized form of equation (116) which will describe small departures from an initial flow. All the terms except the one in $\mathfrak{P}_{ij}$ have already been considered in Chapter 2, § 15. And the required first variation of $\mathfrak{P}_{ij}$ can be written down with the aid of equations (97) and (119) of Chapter 2. We have

$$\delta\mathfrak{P}_{ij} = \int_V \rho\left[\nu\left(\frac{\partial\Delta u_i}{\partial x_j}+\frac{\partial\Delta u_j}{\partial x_i}-\frac{2}{3}\frac{\partial\Delta u_l}{\partial x_l}\delta_{ij}\right.\right.-$$

$$\left.\left.-\frac{\partial\xi_k}{\partial x_j}\frac{\partial u_i}{\partial x_k}-\frac{\partial\xi_k}{\partial x_i}\frac{\partial u_j}{\partial x_k}+\frac{2}{3}\frac{\partial\xi_k}{\partial x_l}\frac{\partial u_l}{\partial x_k}\delta_{ij}\right)+\Delta\nu\left(\frac{\partial u_i}{\partial x_j}+\frac{\partial u_j}{\partial x_i}-\frac{2}{3}\frac{\partial u_l}{\partial x_l}\delta_{ij}\right)\right]\,d\mathbf{x},$$

$$(117)$$

where (cf. Chapter 2, equation (89))

$$\Delta\mathbf{u} = \frac{d\boldsymbol{\xi}}{dt} = \frac{\partial\boldsymbol{\xi}}{\partial t}+u_j\frac{\partial\boldsymbol{\xi}}{\partial x_j}. \tag{118}$$

In case the initial state is stationary and is without any internal motions, the foregoing expression for $\delta\mathfrak{P}_{ij}$ simplifies considerably: we have

$$\delta\mathfrak{P}_{ij} = \int_V \rho\nu\frac{\partial}{\partial t}\left(\frac{\partial\xi_i}{\partial x_j}+\frac{\partial\xi_j}{\partial x_i}-\frac{2}{3}\frac{\partial\xi_l}{\partial x_l}\delta_{ij}\right)\,d\mathbf{x}. \tag{119}$$

The corresponding linearized version of the virial equation (116) is (cf. equation (17))

$$\frac{d^2V_{i;j}}{dt^2} = \delta\mathfrak{W}_{ij}+\Omega^2(V_{ij}-\delta_{i3}V_{3j})+2\Omega\epsilon_{il3}\int_V \rho\frac{\partial\xi_l}{\partial t}x_j\,d\mathbf{x} +$$

$$+\delta_{ij}\,\delta\Pi-\int_V \rho\nu\frac{\partial}{\partial t}\left(\frac{\partial\xi_i}{\partial x_j}+\frac{\partial\xi_j}{\partial x_i}-\frac{2}{3}\frac{\partial\xi_l}{\partial x_l}\delta_{ij}\right)\,d\mathbf{x}. \tag{120}$$

Finally, if we suppose that the Lagrangian displacement is of the form (18) and is also divergence free, equation (120) becomes

$$\lambda^2V_{i;j}-2\lambda\Omega\epsilon_{il3}V_{l;j} = \delta\mathfrak{W}_{ij}+\Omega^2(V_{ij}-\delta_{i3}V_{3j})+\delta_{ij}\,\delta\Pi-\delta\mathfrak{P}_{ij}, \tag{121}$$

where

$$\delta\mathfrak{P}_{ij} = \lambda\int_V \rho\nu\left(\frac{\partial\xi_i}{\partial x_j}+\frac{\partial\xi_j}{\partial x_i}\right)\,d\mathbf{x}. \tag{122}$$

(b) The low Reynolds-number approximation

While equations (121) and (122) are exact, the accomplishment of their solution, in the manner of the equations of the inviscid problem in § 33, requires that $\delta\mathfrak{P}_{ij}$ be also expressed in terms of the $V_{i;j}$'s. In general this is clearly not possible. However, in a low Reynolds-number approximation, in which the effects arising from viscous dissipation are considered as small perturbations on the inviscid flow, we may evaluate $\delta\mathfrak{P}_{ij}$ in terms of the proper solution valid in the inviscid limit. As we have seen in § 33 (d), the proper solution, in this limit, is a linear function of the coordinates and can be expressed in terms of the $V_{i;j}$'s; thus (cf. equations (65) and (68))

$$\xi_i = \sum_{k=1}^{3} \frac{5V_{i;k}}{Ma_k^2} x_k. \tag{123}$$

For ξ_i given by this equation

$$\delta\mathfrak{P}_{ij} = 5\lambda\nu\left(\frac{V_{i;j}}{a_j^2} + \frac{V_{j;i}}{a_i^2}\right). \tag{124}$$

With this expression for $\delta\mathfrak{P}_{ij}$, equation (121), as in the inviscid limit, involves only the $V_{i;j}$'s; and its solution can be accomplished similarly.

(c) *The effect of viscous dissipation on the toroidal modes*

We shall consider only the even modes. In place of equations (25)–(28), we now have

$$\lambda^2 V_{1;1} - 2\lambda\Omega V_{2;1} = \delta\mathfrak{W}_{11} + \Omega^2 V_{11} + \delta\Pi - \frac{5\lambda\nu}{a_1^2} V_{11}, \tag{125}$$

$$\lambda^2 V_{2;2} + 2\lambda\Omega V_{1;2} = \delta\mathfrak{W}_{22} + \Omega^2 V_{22} + \delta\Pi - \frac{5\lambda\nu}{a_1^2} V_{22}, \tag{126}$$

$$\lambda^2 V_{1;2} - 2\lambda\Omega V_{2;2} = -(2B_{11} - \Omega^2)V_{12} - \frac{5\lambda\nu}{a_1^2} V_{12}, \tag{127}$$

and

$$\lambda^2 V_{2;1} + 2\lambda\Omega V_{1;1} = -(2B_{11} - \Omega^2)V_{12} - \frac{5\lambda\nu}{a_1^2} V_{12}. \tag{128}$$

By combining these equations in the same manner as in § 31 (*b*), we now find, in place of equations (46) and (47),

$$\left[\lambda^2 + 2(2B_{11} - \Omega^2) + \frac{10\lambda\nu}{a_1^2}\right]V_{12} + \lambda\Omega(V_{11} - V_{22}) = 0 \tag{129}$$

and

$$\left[\lambda^2 + 2(2B_{11} - \Omega^2) + \frac{10\lambda\nu}{a_1^2}\right](V_{11} - V_{22}) - 4\lambda\Omega V_{12} = 0; \tag{130}$$

and this pair of equations leads to the characteristic equation (cf. equation (49))

$$\sigma^2 - 2\sigma\Omega - 2(2B_{11} - \Omega^2) - \frac{10\sigma\nu}{a_1^2} i = 0, \tag{131}$$

where we have written $\lambda = i\sigma$.

Since we are working in a low Reynolds-number approximation, we may write

$$\sigma = \sigma_0 + \nu\Delta\sigma + O(\nu^2), \tag{132}$$

where σ_0 is a characteristic frequency in the inviscid limit. With this substitution, we obtain from equation (131)

$$\Delta\sigma = i\frac{5\sigma_0}{a_1^2(\sigma_0 - \Omega)}. \tag{133}$$

For the mode that becomes neutral† at $\Omega^2 = 2B_{11}$

$$\sigma_0 = \Omega - (4B_{11} - \Omega^2)^{\frac{1}{2}}, \tag{134}$$

where the square root (for $\Omega^2 < 4B_{11}$) is to be taken with the positive sign. Equation (133) now gives

$$\Delta\sigma = i\,\frac{5}{a_1^2}\,\frac{(4B_{11}-\Omega^2)^{\frac{1}{2}}-\Omega}{(4B_{11}-\Omega^2)^{\frac{1}{2}}}. \tag{135}$$

We observe, then, that for this mode

$$i\nu\Delta\sigma = -\,\frac{5\nu}{a_1^2}\,\frac{(4B_{11}-\Omega^2)^{\frac{1}{2}}-\Omega}{(4B_{11}-\Omega^2)^{\frac{1}{2}}}. \tag{136}$$

From equation (136) it follows that *while this mode is damped prior to the neutral point at $\Omega^2 = 2B_{11}$, it is amplified in the interval*

$$4B_{11} > \Omega^2 > 2B_{11}.$$

Consequently, even the slightest viscosity will induce instability in the interval between the neutral point and the point of onset of the dynamical instability. However, the e-folding time of the instability is proportional to a_1^2/ν and is given by

$$\tau = \frac{a_1^2}{\nu}\,\frac{(4B_{11}-\Omega^2)^{\frac{1}{2}}}{5[\Omega-(4B_{11}-\Omega^2)^{\frac{1}{2}}]}. \tag{137}$$

Table III provides a brief listing of the constant of proportionality in this relation.

TABLE III

The e-folding time τ of the secular instability
(τ is listed in the unit a_1^2/ν)

e	τ	e	τ
0·812670	∞	0·90	0·2974
0·82	6·4464	0·92	0·1790
0·84	1·5537	0·94	0·0848
0·86	0·7899	0·95	0·0330
0·88	0·4743	0·952887	0

It is clear from equation (136) that the instability induced by viscosity is *not* operative beyond the point of dynamical instability at $\Omega^2 = 4B_{11}$. And the singularity the solution exhibits at $\Omega^2 = 4B_{11}$ is not real: it arises solely from the inadmissibility of the original substitution (132) in

---

† A qualification "in the absence of viscosity" is not necessary here since $\Delta\sigma$ vanishes simultaneously with σ_0.

the neighborhood of this point; for equation (131) without this substitution gives

$$\sigma = \Omega \pm \left(\frac{5\Omega\nu}{a_1^2}\right)^{\frac{1}{2}}(1+i)+O(\nu) \quad \text{at} \quad \Omega^2 = 4B_{11}. \tag{138}$$

BIBLIOGRAPHICAL NOTES

The present chapter attempts to bring together, in the context of the Maclaurin spheroids, methods and ideas scattered through the papers listed on p. 246. Thus, the manner of locating the maximum of Ω^2 along the Maclaurin sequence is derived from Papers XVI and XXII; the treatment of the oscillation of the Maclaurin spheroid is based on Lebovitz's Paper II as well as on the account given in Paper XXII; the ideas described in §§ 34–36 are developed in Papers IX, XVI, XXV (§ 9), and XXVIII (§ 5); and finally the analysis in § 37 is derived from Rosenkilde's Paper XXXVII.

§ 37. For an alternative treatment of the effect of viscous dissipation on the stability of the Maclaurin spheroid see:

P. H. ROBERTS and K. STEWARTSON, "On the stability of a Maclaurin spheroid of small viscosity," *Astrophys. J.*, **137** (1963), 777–90.

6

THE JACOBI AND THE DEDEKIND ELLIPSOIDS

38. Introduction

THIS chapter is devoted, principally, to the study of the two adjoint sequences of Jacobi and Dedekind. Instability along these two sequences sets in by a mode of oscillation belonging to the third harmonics; and along the Jacobian sequence, Poincaré's renowned sequence of pear-shaped configurations branches off at the point of onset of the instability. We shall find that the various problems associated with equilibrium and stability along these two sequences, often considered as "difficult" in the past, find their solutions naturally by the methods developed in the earlier chapters.

In this chapter we also include results bearing on the third harmonic oscillations of the Maclaurin spheroid since they are most readily obtained by specializing the equations appropriate for the Jacobi ellipsoid.

39. The Jacobi ellipsoids: the equilibrium figures

We return to equation (1) of Chapter 5 and ask if it allows ellipsoidal figures, with three unequal axes, as solutions under the same underlying assumption of homogeneity.

Explicitly written out, the second-order virial equation gives

$$\mathfrak{W}_{11}+\Omega^2 I_{11} = \mathfrak{W}_{22}+\Omega^2 I_{22} = \mathfrak{W}_{33}; \tag{1}$$

or, inserting the known expressions for the potential energy and the moment of inertia tensors, we have

$$\Omega^2 a_1^2 - 2A_1 a_1^2 = \Omega^2 a_2^2 - 2A_2 a_2^2 = -2A_3 a_3^2. \tag{2}$$

Adding $2a_1^2 a_2^2 A_{12}$ to each of the three sides of the triangle of equalities (2), we obtain

$$a_1^2(\Omega^2 - 2B_{12}) = a_2^2(\Omega^2 - 2B_{12}) = 2(A_{12} a_1^2 a_2^2 - A_3 a_3^2). \tag{3}$$

These equalities will allow solutions with $a_1 \neq a_2$, if and only if

$$a_1^2 a_2^2 A_{12} = a_3^2 A_3 \tag{4}$$

and

$$\Omega^2 = 2B_{12}. \tag{5}$$

Equation (4) is a geometric restriction on the ellipsoid: it determines a unique relation between the ratios of the axes, a_2/a_1 and a_3/a_1, in order that equilibrium may at all be possible; it is therefore the distinguishing feature of the Jacobian figures. And the value of Ω^2 that is to be associated with each Jacobian figure is determined by equation (5).

It will be noticed that equations (4) and (5) are exactly in the forms that they were first derived by Jacobi (cf. Chapter 1, equations (18) and (21)).

The "first" member of the Jacobian sequence is a spheroid determined by the equations

$$a_1^4 A_{11} = a_3^2 A_3 \tag{6}$$

and

$$\Omega^2 = 2B_{11}. \tag{7}$$

This spheroid is also a member of the Maclaurin sequence: for, along the Maclaurin sequence (cf. Chapter 5, equation (4))

$$\Omega^2 = 2\left(1 - \frac{a_3^2}{a_1^2}\right)B_{13}, \tag{8}$$

and the simultaneous satisfaction of equations (7) and (8) requires

$$(a_1^2 - a_3^2)B_{13} = (a_1^2 - a_3^2)(A_1 - a_3^2 A_{13}) = a_1^2 B_{11} = a_1^2(A_1 - a_1^2 A_{11}), \tag{9}$$

or,

$$a_1^4 A_{11} = a_3^2(a_1^2 - a_3^2)A_{13} + a_3^2 A_1; \tag{10}$$

and this equation is the same as equation (6). Therefore, the Jacobian sequence, determined by equations (4) and (5), bifurcates from the Maclaurin sequence, determined by equation (8), from the point where $\Omega^2 = 2B_{11}$. This result is in agreement with what has been established in Chapter 5, § 35, namely, that at the point where $\Omega^2 = 2B_{11}$, the Maclaurin spheroid can be quasi-statically deformed into a neighboring tri-axial ellipsoid.

There is an alternative way of deriving formula (5) that clarifies its origin. Suppose that an ellipsoid with unequal axes ($a_1 \neq a_2 \neq a_3$), rotating with a uniform angular velocity Ω, about the x_3-axis, is a permissible figure of equilibrium. Then in the frame of reference rotating with the angular velocity Ω, it cannot clearly make any difference how we choose the direction from which we measure the azimuthal angle φ. In other words, the equilibrium of the ellipsoid cannot be affected by the displacement

$$\xi_1 = \alpha x_2, \quad \xi_2 = \beta x_1, \quad \text{and} \quad \xi_3 = 0, \tag{11}$$

where α and β are two infinitesimal constants, since its only effect is to rotate the orientation of the ellipsoid in the (x_1, x_2)-plane by the angle

$$\delta\varphi = \frac{\alpha a_2^2 + \beta a_1^2}{a_2^2 - a_1^2}. \tag{12}$$

For the displacement (11),

$$V_{12} = \delta I_{12} = \delta \int_V \rho x_1 x_2 \, d\mathbf{x} = \int_V \rho(\xi_1 x_2 + \xi_2 x_1) \, d\mathbf{x}$$

$$= \alpha I_{22} + \beta I_{11} \neq 0; \qquad (13)$$

and the first variation of the virial relation

$$\mathfrak{W}_{12} + \Omega^2 I_{12} = 0, \qquad (14)$$

namely (cf. Chapter 3, equation (148))

$$\delta\mathfrak{W}_{12} + \Omega^2 V_{12} = (\Omega^2 - 2B_{12})V_{12} = 0, \qquad (15)$$

cannot be satisfied unless Ω^2 is given by equation (5). *Equation (5) is, therefore, the expression of the invariance of the equilibrium of the ellipsoid to a choice of the orientation of the coordinate axes in the equatorial plane.*

TABLE IV

The properties of the Jacobi ellipsoids

a_2/a_1	a_3/a_1	$\Omega^2/(\pi G\rho)$	$L/(GM^3\bar{a})^{\frac{1}{2}}$	a_2/a_1	a_3/a_1	$\Omega^2/(\pi G\rho)$	$L/(GM^3\bar{a})^{\frac{1}{2}}$
1·00	0·582724	0·374230	0·303751	0·48	0·372384	0·302642	0·369473
0·96	0·570801	0·373987	0·303959	0·44	0·349632	0·287267	0·385940
0·92	0·558330	0·373190	0·304602	0·40	0·325609	0·269678	0·406073
0·88	0·545263	0·371785	0·305749	0·36	0·300232	0·249693	0·430872
0·84	0·531574	0·369697	0·307467	0·32	0·273419	0·227153	0·461750
0·80	0·517216	0·366837	0·309837	0·28	0·245083	0·201946	0·500777
0·76	0·502147	0·363114	0·312956	0·24	0·215143	0·174052	0·551140
0·72	0·486322	0·358424	0·316938	0·20	0·183524	0·143610	0·618069
0·68	0·469689	0·352649	0·321923	0·16	0·150166	0·111044	0·710927
0·64	0·452194	0·345665	0·328081	0·12	0·115038	0·077281	0·848770
0·60	0·433781	0·337330	0·335618	0·08	0·078166	0·044168	1·079302
0·56	0·414386	0·327493	0·344796	0·04	0·039688	0·015415	1·58276
0·52	0·393944	0·315989	0·355941	0	0	0	∞

In Table IV we list the properties of the Jacobi ellipsoids; and in Figs. 5 and 6 (see p. 79), the variations of Ω^2 and

$$\frac{L}{(GM^3\bar{a})^{\frac{1}{2}}} = \frac{\sqrt{3}}{10} \frac{a_1^2 + a_2^2}{\bar{a}^2} \Omega \qquad [\bar{a} = (a_1 a_2 a_3)^{\frac{1}{3}}] \qquad (16)$$

(cf. Chapter 5, equation (7)) along the Jacobian sequence are contrasted with those along the Maclaurin sequence.

40. The bifurcation of Poincaré's sequence of pear-shaped configurations from the Jacobian sequence

As has been explained in Chapter 5, § 34, at a point of bifurcation, some proper mode of oscillation must become neutral; also, the first variation of any integral property (that obtains by virtue of equilibrium) resulting from a quasi-static deformation by the proper displacement, belonging to the neutral mode, must vanish.

Since a displacement $\boldsymbol{\xi}$ that deforms an ellipsoid into a pear-shaped figure must be quadratic in the coordinates, none of the integral properties provided by the *second-order* virial equations can discriminate the associated point of bifurcation: their first variations for such displacements vanish identically. However, the relations (78), (79), and (83) of Chapter 2, provided by the *third-order* virial equations, *will* discriminate neutral points belonging to such quadratic displacements. Indeed, *a necessary and sufficient condition for the occurrence of a neutral point, belonging to a displacement quadratic in the coordinates, is that the equations*

$$\delta\mathfrak{W}_{ij;k}+\delta\mathfrak{W}_{ik;j}+\Omega^2(V_{ijk}-\delta_{i3}V_{3jk}) = 0, \tag{17}$$

$$2\delta\mathfrak{W}_{ij;j}+\Omega^2(V_{ijj}-\delta_{i3}V_{3jj}) = 0, \tag{18}$$

and $\qquad \delta S_{ijj}+\Omega^2[V_{iii}-3V_{ijj}-\delta_{i3}(V_{3ii}-V_{3jj})+2\delta_{j3}V_{3ij}] = 0 \tag{19}$

$[i \neq j \neq k$; and no summation over repeated indices in equations

(17), (18), and (19)]

are satisfied for some non-trivial set of the V_{ijk}*'s.*

Expressions for $\delta\mathfrak{W}_{ij;k}$ and δS_{ijj} have been given in equations (152)–(157) of Chapter 3; and these expressions involve the same V_{ijk}'s that occur explicitly in equations (17)–(19). An examination of equations (17)–(19) and the expressions for $\delta\mathfrak{W}_{ij;k}$ and δS_{ijj} shows, in fact, that the equations fall into four non-combining groups distinguished by their oddness or evenness with respect to the indices 1, 2, and 3. Thus, there are three equations (derived from equation (17)) which are odd in all three indices and three sets of four equations each (derived from equations (18) and (19)) which are odd in one of the three indices and even in the remaining two. For deformations which preserve the symmetry about the $(1,3)$- and the $(1,2)$-planes and destroy only the fore and aft symmetry about the $(2,3)$-plane, we need consider only the set of four relations which are odd in the index 1 and even in the indices 2 and 3; the V_{ijk}'s even in the index 1 will vanish identically for such deformations and the two sets, even in the index 1, will provide no useful information in the context. The equations we have to consider, then, are

$$\delta J_1 = -2\delta\mathfrak{W}_{12;2}-\Omega^2 V_{122} = 0, \tag{20}$$

$$\delta J_2 = -2\delta\mathfrak{W}_{13;3}-\Omega^2 V_{133} = 0, \tag{21}$$

$$\delta J_3 = \delta S_{122}+\Omega^2(V_{111}-3V_{122}) = 0, \tag{22}$$

and $\qquad \delta J_4 = \delta S_{133}+\Omega^2(V_{111}-V_{133}) = 0. \tag{23}$

At a point of bifurcation, these four equations must be satisfied non-trivially.

We first observe that the expressions for $\delta\mathfrak{W}_{12;2}$, $\delta\mathfrak{W}_{13;3}$, δS_{122}, and δS_{133} involve, in addition to V_{111}, V_{122}, and V_{133}, the first moment,

$$V_1 = \int_V \rho \xi_1 \, d\mathbf{x}, \tag{24}$$

as well. We shall assume, quite generally, that

$$V_i = 0 \quad (i = 1, 2, 3). \tag{25}$$

The meaning of this assumption is that we are restricting ourselves to deformations that keep the center of mass invariant. Since the only motion that the center of mass of a self-gravitating mass is capable of is a uniform motion (cf. Chapter 2, equations (43) and the remarks following) no generality is lost by making this assumption.

Inserting, then, the appropriate expressions for $\delta\mathfrak{W}_{ij;j}$ and δS_{ijj} that occur in equations (20)–(23), we shall obtain a set of four linear homogeneous equations for V_{111}, V_{122}, and V_{133}. These equations may be written in the forms

$$\delta J_i = \langle i \mid 111 \rangle V_{111} + \langle i \mid 122 \rangle V_{122} + \langle i \mid 133 \rangle V_{133} \quad (i = 1,\dots, 4), \tag{26}$$

where $\langle i \mid 111 \rangle$, etc., are certain matrix elements that are known.

If we should require, as indeed we must, that the Lagrangian displacement in question is also solenoidal, then we must supplement equations (26) by the further condition (cf. Chapter 3, equation (163))

$$\frac{V_{111}}{a_1^2} + \frac{V_{122}}{a_2^2} + \frac{V_{133}}{a_3^2} = 0. \tag{27}$$

We may now conclude that *the existence of a non-trivial neutral mode of oscillation, belonging to the third harmonics and such as to deform a Jacobi ellipsoid into a pear-shaped figure with no fore and aft symmetry about its longest axis, requires that for some member of the Jacobian sequence the 4×3-matrix representing equation (26) (or, the 5×3-matrix, if equation (27) is also included) is at most of rank 2.*

Considering first the 4×3-matrix representing equation (26) and letting Δ_i denote the determinant of order 3 obtained by omitting the ith row of the matrix, we find that the Δ_i's do indeed vanish simultaneously at a certain uniquely determinate point along the Jacobian sequence. This fact is manifest from Table V in which the values of Δ_i are listed for Jacobi ellipsoids in the neighborhood of the common neutral point. That the characteristic vector, $(V_{111}, V_{122}, V_{133})$, belonging to the neutral point so determined, satisfies also the solenoidal condition (27) is apparent from the fact that the determinant Δ_{12}, of equation (27) and equation (26) for $i = 1$ and 2, vanishes simultaneously with the other determinants (see Table V in which the values of Δ_{12} are also listed).

The fact that the 5×3-matrix representing equations (26) and (27) is of rank 2 at a uniquely determinate point along the Jacobian sequence,

TABLE V

The location of the point of bifurcation by the virial relations

a_2/a_1	0·431	0·432	0·433	0·434
Δ_1	$-0\cdot0001179$	$-0\cdot0000223$	$+0\cdot00007377$	$+0\cdot0001705$
Δ_2	$-0\cdot0001791$	$-0\cdot0000339$	$+0\cdot00011227$	$+0\cdot0002597$
Δ_3	$+0\cdot00001002$	$+0\cdot00000191$	$-0\cdot000006323$	$-0\cdot00001468$
Δ_4	$+0\cdot00001002$	$+0\cdot00000190$	$-0\cdot000006324$	$-0\cdot00001468$
Δ_{12}	$-0\cdot01754$	$-0\cdot00330$	$+0\cdot01083$	$+0\cdot02487$

should be capable of a direct analytical demonstration. Such a demonstration has not been attempted; but an indirect demonstration will be given in § (*a*) below.

Finally, we may note that by interpolation among the values listed in Table V, it is found that the neutral point along the Jacobian sequence for deformation into a pear-shaped figure occurs where

$$a_2/a_1 = 0\cdot432232, \quad a_3/a_1 = 0\cdot345069, \quad \text{and} \quad \Omega^2 = 0\cdot284030. \qquad (28)$$

(These values are slightly different from those of Darwin which are generally quoted.)

(*a*) A direct determination of the point of bifurcation

The question to which we shall address ourselves is whether a Jacobi ellipsoid can be deformed, quasi-statically, into a pear-shaped figure without violating *any* of the conditions for equilibrium. Quite generally, the underlying problem can be formulated as follows.

For a uniformly rotating homogeneous configuration, the equation of hydrostatic equilibrium integrates to give

$$\frac{p}{\rho} = \mathfrak{B}(\mathbf{x}) + \tfrac{1}{2}\Omega^2(x_1^2 + x_2^2) + \text{constant}. \qquad (29)$$

And if
$$S(\mathbf{x}) = 0 \qquad (30)$$

is the equation of the bounding surface, then on this surface the pressure must vanish identically for some suitably chosen value for the constant of integration in equation (29).

For an undeformed Jacobi ellipsoid

$$S = S_J = \sum_{i=1}^{3} \frac{x_i^2}{a_i^2} - 1 = 0; \qquad (31)$$

and when the conditions (4) and (5) are satisfied

$$\left(\frac{p}{\rho}\right)_J = a_3^2 A_3\left(1 - \sum_{i=1}^{3} \frac{x_i^2}{a_i^2}\right), \tag{32}$$

where a factor $\pi G\rho$ has been suppressed consistently with our assumption of measuring Ω^2 in the unit $\pi G\rho$.

We consider now a deformation of the Jacobi ellipsoid by a Lagrangian displacement $\boldsymbol{\xi}$ and seek the conditions that the deformation will not violate any of the conditions for equilibrium.

By the displacement $\boldsymbol{\xi}$, the equation of the bounding surface becomes

$$S = S_J - \xi_i\frac{\partial S_J}{\partial x_i} = \sum_{i=1}^{3} \frac{x_i^2}{a_i^2} - 1 - 2\sum_{i=1}^{3} \frac{\xi_i x_i}{a_i^2} = 0, \tag{33}$$

while $\mathfrak{B}(\mathbf{x})$ and Ω^2 are changed by the amounts

$$\delta\mathfrak{B} = G\delta\int_V \frac{\rho(\mathbf{x}')}{|\mathbf{x}-\mathbf{x}'|}\, d\mathbf{x}' = G\int_V \rho(\mathbf{x}')\xi_i(\mathbf{x}')\frac{\partial}{\partial x_i'}\frac{1}{|\mathbf{x}-\mathbf{x}'|}\, d\mathbf{x}'$$

$$= -G\frac{\partial}{\partial x_i}\int_V \frac{\rho(\mathbf{x}')\xi_i(\mathbf{x}')}{|\mathbf{x}-\mathbf{x}'|}\, d\mathbf{x}' \tag{34}$$

and

$$\delta\Omega^2 = \delta\left(\frac{\mathfrak{W}_{33}-\mathfrak{W}_{11}}{I_{11}}\right) = \frac{\delta\mathfrak{W}_{33}}{I_{11}} - \frac{\delta\mathfrak{W}_{11}}{I_{11}} - \Omega_J^2\frac{V_{11}}{I_{11}}. \tag{35}$$

With $\delta\mathfrak{B}$ and $\delta\Omega^2$ determined in terms of $\boldsymbol{\xi}$ by the foregoing equations, the pressure in the deformed configuration is given by

$$\frac{p}{\rho} = \mathfrak{B}_J(\mathbf{x}) + \tfrac{1}{2}\Omega_J^2(x_1^2+x_2^2) + \delta\mathfrak{B}(\mathbf{x}) + \tfrac{1}{2}\delta\Omega^2(x_1^2+x_2^2) + \text{constant}; \tag{36}$$

or, in view of equation (32),

$$\frac{p}{\rho} = a_3^2 A_3\left(1 - \sum_{i=1}^{3} \frac{x_i^2}{a_i^2}\right) + \delta\mathfrak{B}(\mathbf{x}) + \tfrac{1}{2}\delta\Omega^2(x_1^2+x_2^2) + \text{constant}. \tag{37}$$

In this last equation the "constant" is of order $|\boldsymbol{\xi}|$.

The boundary condition requires that the pressure given by equation (37) vanishes on the *deformed surface* (33). Accordingly, making use of equation (33), we may write

$$\left(\frac{p}{\rho}\right)_S = -2a_3^2 A_3\sum_{i=1}^{3} \frac{\xi_i x_i}{a_i^2} + \delta\mathfrak{B}(\mathbf{x}) + \tfrac{1}{2}\delta\Omega^2(x_1^2+x_2^2) + \text{constant}$$

$$= 0. \tag{38}$$

Since the expression for $(p/\rho)_S$ given in equation (38) is of order $|\boldsymbol{\xi}|$, it will suffice if it vanishes on the original *undeformed ellipsoid*. It is this last requirement that leads to the desired conditions on $\boldsymbol{\xi}$.

We shall apply the method that we have described to the problem on hand. First, we observe that the most general quadratic displacement, that will deform an incompressible ellipsoid into a pear-shaped figure with no fore and aft symmetry, should be expressible as a linear combination of the following five:

$$\boldsymbol{\xi}^{(0)} = (1, 0, 0), \quad \boldsymbol{\xi}^{(1)} = (x_1^2, -2x_1 x_2, 0), \quad \boldsymbol{\xi}^{(2)} = (x_2^2, 0, 0),$$
$$\boldsymbol{\xi}^{(3)} = (x_3^2, 0, 0), \quad \text{and} \quad \boldsymbol{\xi}^{(4)} = (0, x_1 x_2, -x_1 x_3). \tag{39}$$

Even though these five solenoidal displacements are linearly independent as vectors, they are *not linearly independent modulo the ellipsoid* in the sense that no non-trivial combination of them will fail to deform the boundary $S_J = 0$. Thus, the displacements

$$S_2 \boldsymbol{\xi}^{(2)} + S_3 \boldsymbol{\xi}^{(3)} \quad \text{and} \quad S_4 \boldsymbol{\xi}^{(4)}, \tag{40}$$

where S_2, S_3, and S_4 are arbitrary constants, deform the ellipsoidal boundary into the surfaces

$$S_J - 2x_1 \left(S_2 \frac{x_2^2}{a_1^2} + S_3 \frac{x_3^2}{a_1^2} \right) = 0$$

and

$$S_J - 2S_4 x_1 \left(\frac{x_2^2}{a_2^2} - \frac{x_3^2}{a_3^2} \right) = 0, \tag{41}$$

respectively. Therefore, the displacement

$$\boldsymbol{\xi}^{(4)} - \frac{a_1^2}{a_2^2} \boldsymbol{\xi}^{(2)} + \frac{a_1^2}{a_3^2} \boldsymbol{\xi}^{(3)} \tag{42}$$

will leave the surface $S_J = 0$ unaffected; and we may write

$$\boldsymbol{\xi}^{(4)} \equiv \frac{a_1^2}{a_2^2} \boldsymbol{\xi}^{(2)} - \frac{a_1^2}{a_3^2} \boldsymbol{\xi}^{(3)} \quad (\mathrm{mod}\, S_J = 0). \tag{43}$$

It will, therefore, suffice to consider for $\boldsymbol{\xi}$ a linear combination of the four displacements $\boldsymbol{\xi}^{(0)}, \dots, \boldsymbol{\xi}^{(3)}$. We shall write

$$\boldsymbol{\xi} = - \sum_{i=0}^{3} S_i \boldsymbol{\xi}^{(i)} \tag{44}$$

where we have introduced a minus sign for later convenience.

By the displacement (44), the boundary of the ellipsoid will be deformed into the surface

$$S = \sum_{i=1}^{3} \frac{x_i^2}{a_i^2} - 1 + 2x_1 \left[\frac{S_0}{a_1^2} + \frac{S_1}{a_1^2} x_1^2 + \left(\frac{S_2}{a_1^2} - 2\frac{S_1}{a_2^2} \right) x_2^2 + \frac{S_3}{a_1^2} x_3^2 \right] = 0. \tag{45}$$

As we have noted earlier (cf. equation (25)) we may, without loss of generality, impose on $\boldsymbol{\xi}$ the condition

$$V_1 = \int_V \rho \xi_1 \, d\mathbf{x} = 0. \tag{46}$$

This condition gives

$$5S_0 + S_1 a_1^2 + S_2 a_2^2 + S_3 a_3^2 = 0. \tag{47}$$

By making use of the general formulas (34) and (35), we now find that

$$\delta\Omega^2 = 0 \tag{48}$$

and

$$\delta\mathfrak{B} = \sum_{i=0}^{3} S_i \, \delta\mathfrak{B}^{(i)}, \tag{49}$$

where, by equations (40) and (131) of Chapter 3,

$$\delta\mathfrak{B}^{(0)} = \frac{\partial}{\partial x_1}\left(I - \sum_{l=1}^{3} A_l x_l^2\right) = -2A_1 x_1, \tag{50}$$

and

$$\delta\mathfrak{B}^{(1)} = \frac{\partial \mathfrak{D}_{11}}{\partial x_1} - 2\frac{\partial \mathfrak{D}_{12}}{\partial x_2}, \quad \delta\mathfrak{B}^{(2)} = \frac{\partial \mathfrak{D}_{22}}{\partial x_1}, \quad \text{and} \quad \delta\mathfrak{B}^{(3)} = \frac{\partial \mathfrak{D}_{33}}{\partial x_1}. \tag{51}$$

Inserting the expressions for $\mathfrak{D}_{ij}$ given in equation (132) of Chapter 3, we find

$$\delta\mathfrak{B}^{(1)} = (-4A_{111} a_1^4 + B_{111} a_1^2)x_1^3 + (-2A_{112} a_1^4 + B_{112} a_1^2 + 6A_{122} a_1^2 a_2^2)x_1 x_2^2 +$$
$$+ (-2A_{113} a_1^4 + B_{113} a_1^2)x_1 x_3^2 + (2A_{11} a_1^4 - B_{11} a_1^2 - 2a_1^2 a_2^2 A_{12})x_1, \tag{52}$$
$$\delta\mathfrak{B}^{(2)} = B_{112} a_2^2 x_1^3 + (B_{122} a_2^2 - 2A_{122} a_2^4)x_1 x_2^2 + B_{123} a_2^2 x_1 x_3^2 - B_{12} a_2^2 x_1, \tag{53}$$
$$\delta\mathfrak{B}^{(3)} = B_{113} a_3^2 x_1^3 + B_{123} a_3^2 x_1 x_2^2 + (B_{133} a_3^2 - 2A_{133} a_3^4)x_1 x_3^2 - B_{13} a_3^2 x_1. \tag{54}$$

With the foregoing explicit expressions for the various quantities, equation (38) gives

$$\left(\frac{p}{\rho}\right)_S = x_1\left\{\left[\left(B_{111} a_1^2 - 4A_{111} a_1^4 + 2\frac{a_3^2}{a_1^2}A_3\right)S_1 + B_{112} a_2^2 S_2 + B_{113} a_3^2 S_3\right]x_1^2 + \right.$$

$$+ \left[\left(B_{112} a_1^2 - 2A_{112} a_1^4 + 6A_{122} a_1^2 a_2^2 - 4\frac{a_3^2}{a_2^2}A_3\right)S_1 + \right.$$

$$+ \left(B_{122} a_2^2 - 2A_{122} a_2^4 + 2\frac{a_3^2}{a_1^2}A_3\right)S_2 + B_{123} a_3^2 S_3\bigg]x_2^2 +$$

$$+ \left[(B_{113} a_1^2 - 2A_{113} a_1^4)S_1 + B_{123} a_2^2 S_2 + \right.$$

$$+ \left(B_{133} a_3^2 - 2A_{133} a_3^4 + 2\frac{a_3^2}{a_1^2}A_3\right)S_3\bigg]x_3^2 +$$

$$+ \left(-2A_1 + 2\frac{a_3^2}{a_1^2}A_3\right)S_0 + (-B_{11} a_1^2 + 2A_{11} a_1^4 - 2A_{12} a_1^2 a_2^2)S_1 -$$

$$\left. - B_{12} a_2^2 S_2 - B_{13} a_3^2 S_3\right\} + \text{constant}$$

$$= x_1(Q_1 x_1^2 + Q_2 x_2^2 + Q_3 x_3^2 + Q_0) + \text{constant (say)}, \tag{55}$$

where $Q_0, \ldots, Q_3$ are known linear combinations of $S_0, \ldots, S_3$.

We have already remarked that it will suffice if $(p/\rho)_S$, now given by equation (55), vanishes on the original undeformed ellipsoid. To satisfy this boundary condition, it is clearly necessary and sufficient that

$$-Q_0 = Q_1 a_1^2 = Q_2 a_2^2 = Q_3 a_3^2. \tag{56}$$

Equation (56) together with equation (47) provide a system of four linear homogeneous equations for the constants $S_0,..., S_3$. In order that these four constants may not all vanish identically, we must require that the determinant of the system vanishes. This condition on the determinant of equations (47) and (56) determines, *uniquely and without any ambiguity*, the point along the Jacobian sequence where the Jacobian ellipsoid may be quasi-statically deformed into a pear-shaped figure without violating any of the conditions required for equilibrium. And we find that the point occurs where

$$a_2/a_1 = 0\cdot432232, \quad a_3/a_1 = 0\cdot345069, \quad \text{and} \quad \Omega^2 = 0\cdot284030, \tag{57}$$

in exact agreement with the determination (28) via the virial relations.

Also, we find that the pear-shaped figure at the point of bifurcation is obtained by deforming the Jacobi ellipsoid at that point by the displacement

$$\xi = S_0(1 - 14\cdot180x_2^2 - 19\cdot744x_3^2)\mathbf{1}_{x_1}, \tag{58}$$

where S_0 is an arbitrary constant and $\mathbf{1}_{x_1}$ is a unit vector in the direction of x_1. (Note that the displacement $\xi^{(I)}$ is absent from the proper displacement (58). This fact can be established directly from the relevant equations; but we omit the proof.) This direct determination of the displacement suffices to establish why the virial relations (20)–(23), together with the solenoidal condition (27), must agree in determining the same neutral point.

Finally, in Fig. 8 (taken from a paper by Darwin) we illustrate the relative dimensions of the critical Jacobi ellipsoid and its deformation into a pear-shaped figure by the displacement (58). As Darwin has remarked, "Comparison with M. Poincaré's sketch shows that the figure is considerably longer than he supposed."

41. The points along the Maclaurin sequence where sequences of pear-shaped configurations bifurcate

The relations (20)–(23) can be used to isolate further neutral points along the Maclaurin sequence, beyond the first where the Jacobian sequence bifurcates. These further neutral points belong to the third-harmonic modes of oscillation.

The expressions for $\delta\mathfrak{W}_{ij;j}$ and δS_{ijj} given in equations (153) and

(157) of Chapter 3 simplify for a Maclaurin spheroid since a_1 and a_2 are now equal and any index symbol is unaltered if index 2, wherever it may occur, is replaced by 1, and conversely. Thus, we find that

$$\delta S_{122} = -2(B_{11}+a_1^2\,B_{111})(V_{111}-3V_{122}), \tag{59}$$

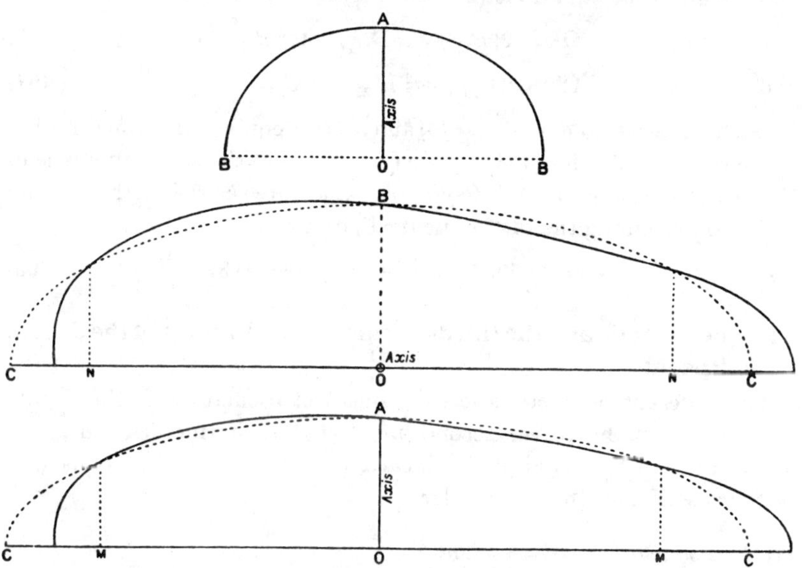

FIG. 8. Darwin's illustration of the critical Jacobian figure from which the pear-shaped sequence branches off; the three elliptical sections shown are (from top to bottom) in the (3, 2)-, (2, 1)-, and (1, 3)-planes. The full-line curves show the pear-shaped deformation of the ellipsoidal figure (the dotted curves) by the displacement (58) to which it is neutral.

and equation (22) gives

$$\delta J_3 = [\Omega^2-2(B_{11}+a_1^2\,B_{111})](V_{111}-3V_{122}). \tag{60}$$

Accordingly, a neutral point occurs where

$$\Omega^2 = 2(B_{11}+a_1^2\,B_{111}) \tag{61}$$

for a displacement for which

$$V_{111} \neq 3V_{122} \quad \text{and} \quad V_{133} = 0. \tag{62}$$

It is found that equation (61) is satisfied where

$$e = 0{\cdot}89926 \quad \text{and} \quad \Omega^2 = 0{\cdot}44015. \tag{63}$$

Considering next the relations (20), (21), and (23), we should set in them

$$V_{111} = 3V_{122}, \tag{64}$$

in order that there may be no inconsistency with relation (22) and equation (60). Also, under these same circumstances, the solenoidal condition (27) gives

$$V_{122} = \tfrac{1}{3}V_{111} = -\frac{a_1^2}{4a_3^2}V_{133};$$ (65)

and we find that equations (20) and (21) reduce to

$$\Omega^2 = 2(B_{11}+3a_1^2 B_{111})-4a_3^2 B_{113}$$ (66)

and

$$\Omega^2 = 2B_{13}+3a_3^2 B_{133}-a_1^2 B_{113}.$$ (67)

It can be shown that equations (66) and (67) are equivalent to one another—the right-hand sides of these two equations are the same by virtue of the relations among the index symbols—and agree as well with equation (23) in determining the further neutral point at

$$e = 0\cdot 96937 \quad \text{where} \quad \Omega^2 = 0\cdot 414132.$$ (68)

42. The second- and the third-harmonic oscillations of the Jacobi ellipsoid

The different characteristic frequencies of oscillation of the Jacobi ellipsoid belonging to the second and the third harmonics can all be determined with the aid of the linearized forms of the virial equations of the second and the third orders.

(a) The second-harmonic oscillations

The treatment of the second-harmonic oscillations is straightforward; it can be carried out exactly as for the Maclaurin spheroids in Chapter 5, § 33, and indeed with only very minor modifications in that analysis.

The relevant equations of the problem are equations (20)–(28) of Chapter 5 with the only modifications that in equations (21) and (23) B_{23} should be written in place of B_{13}, and in equations (27) and (28) B_{12} in place of B_{11}. (The latter substitution in equations (27) and (28) makes the right-hand sides of these equations vanish by virtue of the formula determining Ω^2.)

Considering first the odd modes and proceeding exactly as in § 33 (a), we find that equation (32) is unaltered, while in equation (33), B_{23} replaces B_{13}. Therefore, instead of equation (34) of Chapter 5, we now have

$$\lambda^2(\lambda^2+4B_{13}-\Omega^2)(\lambda^2+4B_{23}-\Omega^2)+4\Omega^2(\lambda^2+2B_{13})(\lambda^2+2B_{23}) = 0,$$ (69)

or, writing $\lambda^2 = -\sigma^2$, we have

$$\sigma^2(\sigma^2+\Omega^2-4B_{13})(\sigma^2+\Omega^2-4B_{23})-4\Omega^2(\sigma^2-2B_{13})(\sigma^2-2B_{23}) = 0.$$ (70)

Besides the root $\sigma^2 = \Omega^2$ which equation (70) manifestly allows, we have

$$\sigma^2 = 2B_{13}+2B_{23}+\tfrac{1}{2}\Omega^2\pm[(2B_{13}+2B_{23}+\tfrac{1}{2}\Omega^2)^2-16B_{13}B_{23}]^{\frac{1}{2}}. \tag{71}$$

Turning next to the even modes, we first observe that since the right-hand sides of equations (27) and (28) of Chapter 5 are now zero (as we have noted earlier),

$$\lambda = 0 \text{ is a double root;} \tag{72}$$

and if we exclude these roots, we must have

$$\lambda V_{1;2} = \Omega V_{22} \quad \text{and} \quad \lambda V_{2;1} = -\Omega V_{11}. \tag{73}$$

Next, eliminating $\delta\Pi$ from equations (24)–(26) of Chapter 5, we obtain

$$\tfrac{1}{2}\lambda^2(V_{11}-V_{22})-2\lambda\Omega V_{12} = \delta\mathfrak{W}_{11}-\delta\mathfrak{W}_{22}+\Omega^2(V_{11}-V_{22}) \tag{74}$$

and

$$\tfrac{1}{2}\lambda^2(V_{11}+V_{22})-\lambda^2 V_{33}+2\lambda\Omega(V_{1;2}-V_{2;1})$$
$$= \delta\mathfrak{W}_{11}+\delta\mathfrak{W}_{22}-2\delta\mathfrak{W}_{33}+\Omega^2(V_{11}+V_{22}). \tag{75}$$

Making use of the relations (73), we find that the foregoing equations can be reduced to the forms

$$(\tfrac{1}{2}\lambda^2 \mid \Omega^2)(V_{11}-V_{22}) = \delta\mathfrak{W}_{11}-\delta\mathfrak{W}_{22} \tag{76}$$

and $\qquad (\tfrac{1}{2}\lambda^2+\Omega^2)(V_{11}+V_{22})-\lambda^2 V_{33} = \delta\mathfrak{W}_{11}+\delta\mathfrak{W}_{22}-2\delta\mathfrak{W}_{33}. \tag{77}$

Explicit expressions for the quantities on the right-hand sides of equations (76) and (77) have been given in equations (149) and (150) of Chapter 3. These expressions are linear combinations of V_{11}, V_{22}, and V_{33}; therefore, equations (76) and (77) together with the solenoidal condition,

$$\frac{V_{11}}{a_1^2}+\frac{V_{22}}{a_2^2}+\frac{V_{33}}{a_3^2} = 0, \tag{78}$$

provide a set of three linear homogeneous equations for V_{11}, V_{22}, and V_{33}. The resulting characteristic equation for $\lambda^2 = -\sigma^2$ can be brought to the form

$$\begin{vmatrix} \Omega^2+3B_{11}-B_{13}-\tfrac{1}{2}\sigma^2 & B_{12}-B_{23} & \Omega^2+3(B_{11}-B_{33})+B_{12}-B_{23} \\[4pt] B_{12}-B_{13} & \Omega^2+3B_{22}-B_{23}-\tfrac{1}{2}\sigma^2 & \Omega^2+3(B_{22}-B_{33})+B_{12}-B_{13} \\[4pt] \dfrac{1}{a_1^2} & \dfrac{1}{a_2^2} & \dfrac{1}{a_1^2}+\dfrac{1}{a_2^2}+\dfrac{1}{a_3^2} \end{vmatrix} = 0. \tag{79}$$

The variations of the characteristic roots, determined by equations (71) and (79), along the Jacobian sequence are illustrated in Fig. 9.

(b) *The third-harmonic oscillations*

The linearized version of the third-order virial equations has been written down in Chapter 2 (equations (132), (134), (140), and (153)). Since in the present context no internal motions are present in the initial

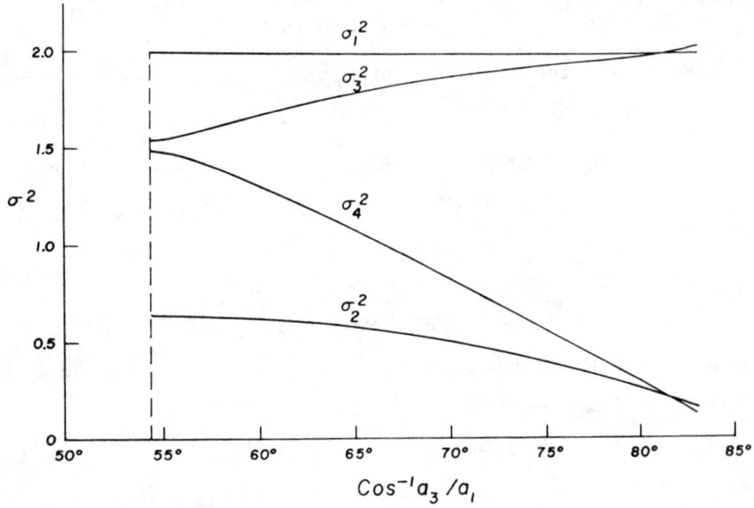

FIG. 9. The squares of the characteristic frequencies (in the unit $\pi G\rho$) of the odd (σ_1^2 and σ_2^2) and the even (σ_3^2 and σ_4^2) modes of second-harmonic oscillation of the Jacobi ellipsoid.

unperturbed state, the equations are considerably simplified and we are left with

$$\frac{d^2 V_{i;jk}}{dt^2} = \delta\mathfrak{W}_{ij;k} + \delta\mathfrak{W}_{ik;j} + \Omega^2(V_{ijk} - \delta_{i3}V_{3jk}) +$$

$$+ 2\epsilon_{il3}\Omega \int_V \rho \frac{\partial \xi_l}{\partial t} x_j x_k \, d\mathbf{x} + \delta_{ij}\delta\Pi_k + \delta_{ik}\delta\Pi_j. \quad (80)$$

We shall now suppose that the Lagrangian displacement $\boldsymbol{\xi}(\mathbf{x}, t)$ is of the form

$$\boldsymbol{\xi}(\mathbf{x}, t) = \boldsymbol{\xi}(\mathbf{x})e^{\lambda t}, \quad (81)$$

where λ is a parameter whose characteristic values are to be determined. For $\boldsymbol{\xi}$ of this chosen form, equation (80) becomes

$$\lambda^2 V_{i;jk} - 2\lambda\Omega\epsilon_{il3}V_{l;jk} = \delta\mathfrak{W}_{ij;k} + \delta\mathfrak{W}_{ik;j} + \Omega^2(V_{ijk} - \delta_{i3}V_{3jk})$$

$$+ \delta_{ij}\delta\Pi_k + \delta_{ik}\delta\Pi_j. \quad (82)$$

Equation (82) represents a total of eighteen equations. These eighteen equations fall into two non-combining groups of ten and eight equations, respectively, distinguished by their parity (i.e. evenness or oddness) with respect to the index 3. It is convenient to have these equations written out explicitly. The equations even in the index 3 are

$$\lambda^2 V_{3;31} - \delta\mathfrak{W}_{13;3} - \delta\mathfrak{W}_{33;1} = \delta\Pi_1,$$
$$\lambda^2 V_{3;32} - \delta\mathfrak{W}_{23;3} - \delta\mathfrak{W}_{33;2} = \delta\Pi_2,$$
$$\lambda^2 V_{1;11} - 2\lambda\Omega V_{2;11} - \Omega^2 V_{111} - 2\delta\mathfrak{W}_{11;1} = 2\delta\Pi_1,$$
$$\lambda^2 V_{2;22} + 2\lambda\Omega V_{1;22} - \Omega^2 V_{222} - 2\delta\mathfrak{W}_{22;2} = 2\delta\Pi_2,$$
$$\lambda^2 V_{1;22} - 2\lambda\Omega V_{2;22} - \Omega^2 V_{122} - 2\delta\mathfrak{W}_{12;2} = 0,$$
$$\lambda^2 V_{2;11} + 2\lambda\Omega V_{1;11} - \Omega^2 V_{211} - 2\delta\mathfrak{W}_{12;1} = 0,$$
$$\lambda^2 V_{1;12} - 2\lambda\Omega V_{2;12} - \Omega^2 V_{112} - \delta\mathfrak{W}_{11;2} - \delta\mathfrak{W}_{12;1} = \delta\Pi_2,$$
$$\lambda^2 V_{2;12} + 2\lambda\Omega V_{1;12} - \Omega^2 V_{122} - \delta\mathfrak{W}_{22;1} - \delta\mathfrak{W}_{12;2} = \delta\Pi_1,$$
$$\lambda^2 V_{1;33} - 2\lambda\Omega V_{2;33} - \Omega^2 V_{133} - 2\delta\mathfrak{W}_{13;3} = 0,$$
$$\lambda^2 V_{2;33} + 2\lambda\Omega V_{1;33} - \Omega^2 V_{233} - 2\delta\mathfrak{W}_{23;3} = 0. \tag{83}$$

And the equations odd in the index 3 are

$$\lambda^2 V_{3;11} - 2\delta\mathfrak{W}_{13;1} = 0,$$
$$\lambda^2 V_{3;22} - 2\delta\mathfrak{W}_{23;2} = 0,$$
$$\lambda^2 V_{3;12} - \delta\mathfrak{W}_{13;2} - \delta\mathfrak{W}_{23;1} = 0,$$
$$\lambda^2 V_{3;33} - 2\delta\mathfrak{W}_{33;3} = 2\delta\Pi_3,$$
$$\lambda^2 V_{1;13} - 2\lambda\Omega V_{2;13} - \Omega^2 V_{113} - \delta\mathfrak{W}_{11;3} - \delta\mathfrak{W}_{13;1} = \delta\Pi_3,$$
$$\lambda^2 V_{2;23} + 2\lambda\Omega V_{1;23} - \Omega^2 V_{223} - \delta\mathfrak{W}_{22;3} - \delta\mathfrak{W}_{23;2} = \delta\Pi_3,$$
$$\lambda^2 V_{1;23} - 2\lambda\Omega V_{2;23} - \Omega^2 V_{123} - \delta\mathfrak{W}_{12;3} - \delta\mathfrak{W}_{10;2} = 0,$$
$$\lambda^2 V_{2;13} + 2\lambda\Omega V_{1;13} - \Omega^2 V_{123} - \delta\mathfrak{W}_{21;3} - \delta\mathfrak{W}_{23;1} = 0. \tag{84}$$

For reasons that we have already explained, we may, without any loss of generality, supplement the foregoing equations by the three additional conditions (25) which express the invariance of the center of mass to the deformation considered.

Considering first equations (83), we find that by suitably combining them so as to eliminate $\delta\Pi_1$ and $\delta\Pi_2$ as well as the unsymmetrized $V_{i;jk}$'s (in favor of the symmetrized V_{ijk}'s), we can reduce them to the following four equations:

$$(\lambda^2 - 3\Omega^2)(V_{122} - \tfrac{1}{3}V_{111}) + 2\lambda\Omega(V_{112} - \tfrac{1}{3}V_{222}) + \delta S_{122} = 0, \tag{85}$$
$$(\lambda^2 - 3\Omega^2)(V_{112} - \tfrac{1}{3}V_{222}) - 2\lambda\Omega(V_{122} - \tfrac{1}{3}V_{111}) + \delta S_{112} = 0, \tag{86}$$
$$\lambda(\lambda^2 + 4\Omega^2)[(\lambda^2 - \Omega^2)V_{133} - \tfrac{1}{3}(\lambda^2 + \Omega^2)V_{111} + \delta S_{133}] +$$
$$+ 2\Omega(\lambda^2 + 4\Omega^2)(\Omega^2 V_{112} + 2\delta\mathfrak{W}_{12;1}) - 2\Omega\lambda^2(\Omega^2 V_{233} + 2\delta\mathfrak{W}_{23;3}) +$$
$$+ 4\Omega^2\lambda(\Omega^2 V_{133} + 2\delta\mathfrak{W}_{13;3}) = 0, \tag{87}$$

$$\lambda(\lambda^2+4\Omega^2)[(\lambda^2-\Omega^2)V_{233}-\tfrac{1}{3}(\lambda^2+\Omega^2)V_{222}+\delta S_{233}]-$$
$$-2\Omega(\lambda^2+4\Omega^2)(\Omega^2 V_{122}+2\delta\mathfrak{W}_{12;2})+2\Omega\lambda^2(\Omega^2 V_{133}+2\delta\mathfrak{W}_{13;3})+$$
$$+4\Omega^2\lambda(\Omega^2 V_{233}+2\delta\mathfrak{W}_{23;3}) = 0. \quad (88)$$

From equations (153) and (157) of Chapter 3, it is apparent that the $\delta\mathfrak{W}_{ij;j}$'s and δS_{ijj}'s that occur in equations (85)–(88) are linear combinations of the same six V_{ijk}'s as are already present. Therefore, equations (85)–(88) together with the two solenoidal conditions,

$$\frac{V_{111}}{a_1^2}+\frac{V_{122}}{a_2^2}+\frac{V_{133}}{a_3^2} = 0 \qquad (89)$$

and
$$\frac{V_{211}}{a_1^2}+\frac{V_{222}}{a_2^2}+\frac{V_{233}}{a_3^2} = 0, \qquad (90)$$

provide a system of six linear homogeneous equations for the six V_{ijk}'s. The requirement that the determinant of the system vanish leads to a characteristic equation for λ^2. The characteristic equation is found to be of degree seven in λ^2; but the manner of the reductions has introduced a spurious root $\lambda^2 = -4\Omega^2$ so that only six of the seven roots are genuine.†

In a similar fashion, equations (84), suitably combined so as to eliminate $\delta\Pi_3$ and the unsymmetrized $V_{i;jk}$'s, lead to the following three equations:

$$\lambda(\lambda^2-2\Omega^2)V_{123}+\lambda^2\Omega(V_{113}-V_{223})+\lambda\delta S_{123}-2\Omega(\delta\mathfrak{W}_{13;1}-\delta\mathfrak{W}_{23;2}) = 0, \quad (91)$$
$$\lambda^4(V_{113}+V_{223}-\tfrac{2}{3}V_{333})+\lambda^2[2\Omega^2(V_{113}+V_{223})+\delta S_{113}+\delta S_{223}]+$$
$$+4\lambda\Omega(\delta\mathfrak{W}_{13;2}-\delta\mathfrak{W}_{23;1})-8\Omega^2(\delta\mathfrak{W}_{13;1}+\delta\mathfrak{W}_{23;2}) = 0, \quad (92)$$
$$\lambda^3(V_{113}-V_{223})-4\lambda^2\Omega V_{123}+\lambda[2\Omega^2(V_{223}-V_{113})+\delta S_{113}-\delta S_{223}]+$$
$$+4\Omega(\delta\mathfrak{W}_{13;2}+\delta\mathfrak{W}_{23;1}) = 0, \quad (93)$$

where in equation (91) (cf. Chapter 2, equation (84))

$$\delta S_{123} = -2\delta\mathfrak{W}_{12;3}-2\delta\mathfrak{W}_{23;1}-2\delta\mathfrak{W}_{31;2}$$
$$= 2[B_{12}+B_{23}+B_{31}+(a_1^2+a_2^2+a_3^2)B_{123}]V_{123}. \qquad (94)$$

Equations (91)–(93) supplemented by the further solenoidal condition

$$\frac{V_{113}}{a_1^2}+\frac{V_{223}}{a_2^2}+\frac{V_{333}}{a_3^2} = 0, \qquad (95)$$

provide a system of four linear homogeneous equations for the four V_{ijk}'s that are odd in the index 3. The requirement that the determinant of the system vanish leads to a second characteristic equation for λ^2.

---

† For an explicit clarification of this remark, as well as for some further details of the derivations (here drastically curtailed), see the original paper (Paper XIII in the list on p. 245) from which the present account is abstracted.

This characteristic equation is of degree five in λ^2, so that we have altogether eleven characteristic roots that belong to the third harmonics.

The characteristic equations which follow from equations (85)–(90) and (91)–(95) have been solved for a number of Jacobi ellipsoids. Their variations along the Jacobian sequence are illustrated in Figs. 10a, 10b,

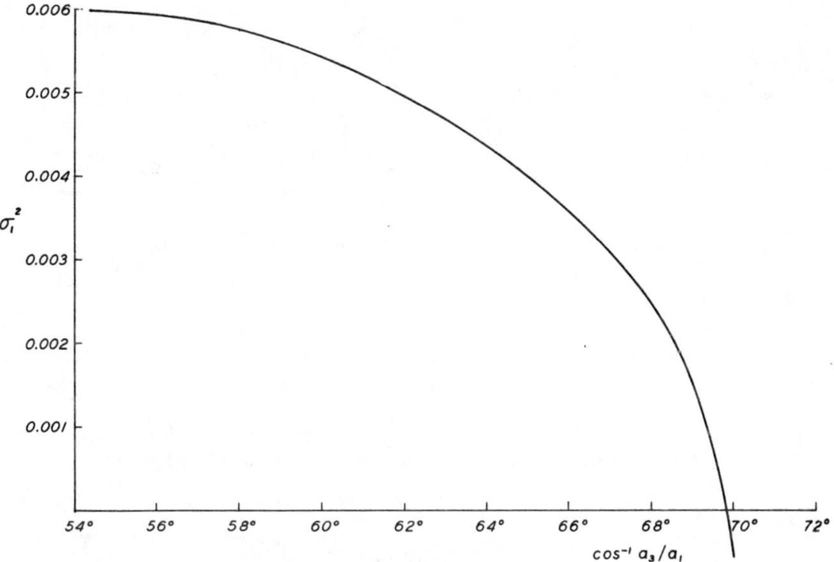

FIG. 10a. The square of the characteristic frequency (in the unit $\pi G\rho$) of oscillation of the Jacobi ellipsoid belonging to the mode that becomes unstable at the point where the pear-shaped sequence bifurcates.

and 10c. We observe that *the Jacobi ellipsoid is unstable only with respect to the one mode that becomes neutral at the point where the pear-shaped configurations bifurcate.* In this respect the behavior along the Jacobian sequence is different from that along the Maclaurin sequence: at the point where the Jacobian sequence bifurcates, the Maclaurin spheroid does not become dynamically unstable; it becomes only secularly unstable. This difference in the behavior along the Jacobian and the Maclaurin sequences is contrary to some earlier expectations of Poincaré and Darwin; it was first conclusively established by Cartan (by methods very different from the ones adopted in this book).

43. The third-harmonic oscillations of the Maclaurin spheroid

In Chapter 5, we have considered the second-harmonic oscillations of the Maclaurin spheroid. The specialization of the analysis of § 42 for

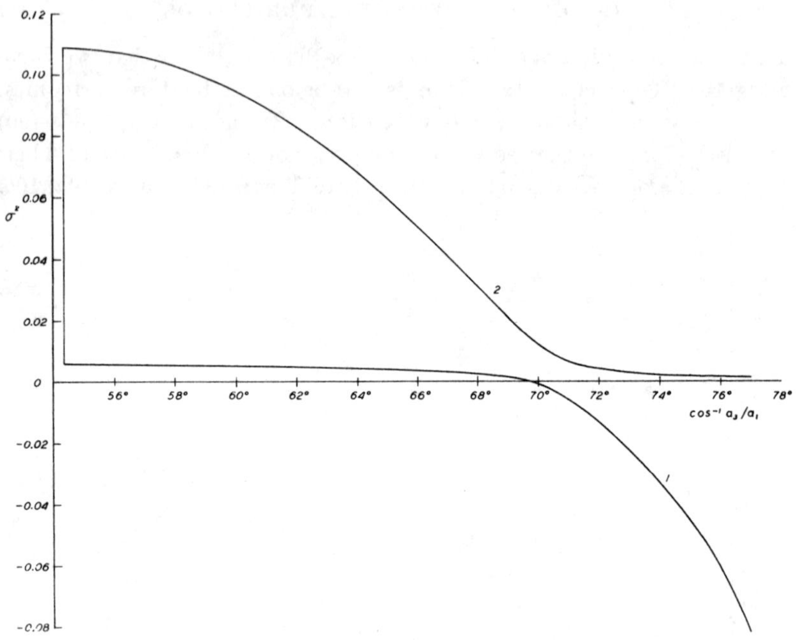

FIG. 10*b*. The squares of the characteristic frequencies (in the unit $\pi G\rho$) of the two lowest even modes of third-harmonic oscillation of the Jacobi ellipsoid.

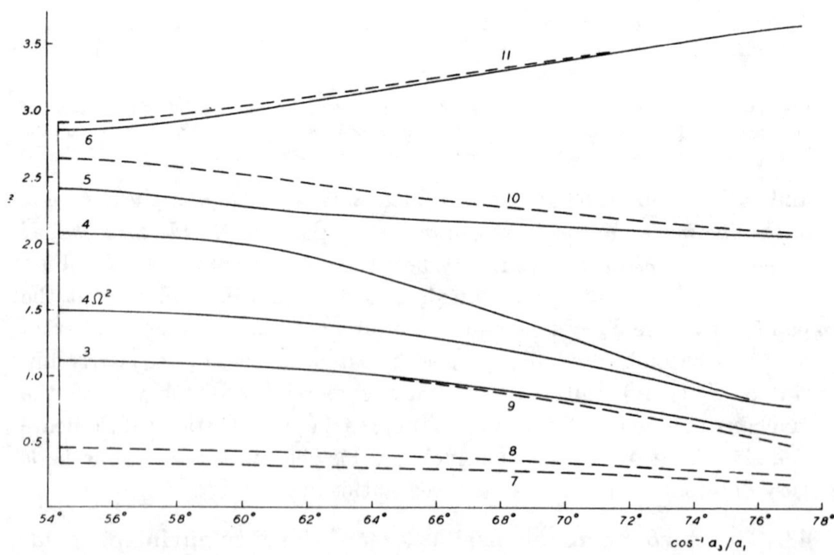

FIG. 10c. The squares of the characteristic frequencies (in the unit $\pi G\rho$) of the remaining nine modes of third-harmonic oscillation of the Jacobi ellipsoid; and $4\Omega^2$. The full-line curves refer to the even and the dashed curves to the odd modes.

the case $a_1 = a_2$ enables us to solve the corresponding problem for the third-harmonic oscillations of the Maclaurin spheroid; and the solution has some interest.

First, we observe that according to equation (157) of Chapter 3 (cf. also equation (59))

$$\delta S_{122} = 6(B_{11}+a_1^2 B_{111})(V_{122}-\tfrac{1}{3}V_{111}) \tag{96}$$

and
$$\delta S_{112} = 6(B_{11}+a_1^2 B_{111})(V_{112}-\tfrac{1}{3}V_{222}), \tag{97}$$

when $a_1 = a_2$. Therefore, equations (85) and (86), specialized for the Maclaurin spheroid, become

$$[\lambda^2-3\Omega^2+6(B_{11}+a_1^2 B_{111})](V_{122}-\tfrac{1}{3}V_{111})+2\lambda\Omega(V_{112}-\tfrac{1}{3}V_{222}) = 0 \tag{98}$$

and
$$[\lambda^2-3\Omega^2+6(B_{11}+a_1^2 B_{111})](V_{112}-\tfrac{1}{3}V_{222})-2\lambda\Omega(V_{122}-\tfrac{1}{3}V_{111}) = 0. \tag{99}$$

These equations can, therefore, be considered independently of the divergence conditions and of the remaining pair of equations.

Equations (98) and (99) lead to the characteristic equation

$$[\lambda^2-3\Omega^2+6(B_{11}+a_1^2 B_{111})]^2+4\lambda^2\Omega^2 = 0, \tag{100}$$

or, writing $\lambda^2 = -\sigma^2$, we can factorize the equation to give

$$\sigma^2-2\Omega\sigma+[3\Omega^2-6(B_{11}+a_1^2 B_{111})] = 0, \tag{101}$$

and a similar equation with $-\sigma$ in place of σ. The roots of equation (101) are
$$\sigma = \Omega\pm[6(B_{11}+a_1^2 B_{111})-2\Omega^2]^{\frac{1}{2}}. \tag{102}$$

Accordingly, this mode of oscillation becomes neutral when

$$\Omega^2 = 2(B_{11}+a_1^2 B_{111}) \quad (\sigma = 0) \tag{103}$$

and unstable when

$$\Omega^2 > 3(B_{11}+a_1^2 B_{111}) \quad (\sigma \text{ complex}). \tag{104}$$

The result (103) is in agreement with what was found in § 41 (see equation 61)).

The point where neutrality occurs via this mode has already been determined (cf. equation (63)); and the point where instability sets in is found to occur where

$$e = 0\cdot96696 \quad \text{and} \quad \Omega^2 = 0\cdot41972. \tag{105}$$

Returning to the two remaining equations (87) and (88), we can now put
$$V_{111} = 3V_{122} \quad \text{and} \quad V_{222} = 3V_{112}, \tag{106}$$

in order that the roots given by equation (100) may be excluded and there is no inconsistency with equations (98) and (99). With the substitutions (106), equations (87) and (88) become

$$\lambda(\lambda^2+4\Omega^2)[(\lambda^2-\Omega^2)V_{133}-(\lambda^2+\Omega^2)V_{122}+\delta S_{133}]+$$
$$+2\Omega(\lambda^2+4\Omega^2)(\Omega^2 V_{112}+2\delta\mathfrak{W}_{12;1})-2\Omega\lambda^2(\Omega^2 V_{233}+2\delta\mathfrak{W}_{23;3})+$$
$$+4\Omega^2\lambda(\Omega^2 V_{133}+2\delta\mathfrak{W}_{13;3}) = 0 \quad (107)$$

and

$$\lambda(\lambda^2+4\Omega^2)[(\lambda^2-\Omega^2)V_{233}-(\lambda^2+\Omega^2)V_{112}+\delta S_{233}]-$$
$$-2\Omega(\lambda^2+4\Omega^2)(\Omega^2 V_{122}+2\delta\mathfrak{W}_{12;2})+2\Omega\lambda^2(\Omega^2 V_{133}+2\delta\mathfrak{W}_{13;3})+$$
$$+4\Omega^2\lambda(\Omega^2 V_{233}+2\delta\mathfrak{W}_{23;3}) = 0. \quad (108)$$

The solenoidal conditions (89) and (90), in view of equations (106) and the fact that $a_1 = a_2$, now give

$$V_{122} = -\frac{a_1^2}{4a_3^2}V_{133} \quad \text{and} \quad V_{112} = -\frac{a_1^2}{4a_3^2}V_{233}. \quad (109)$$

Equations (107) and (108) thus become a pair of equations for V_{122} and V_{112}. The characteristic equation which follows can be readily written down; it determines the remaining four roots that belong to the even modes (in addition to a spurious root $-4\Omega^2$ to which we have made reference in § 42).

Considering next equations (91)–(93) governing the odd modes, we find that when $a_1 = a_2$ the equations reduce to the forms

$$\lambda^3 V_{123}+\lambda^2\Omega(V_{113}-V_{223})+\lambda(\delta S_{123}-2\Omega^2 V_{123})+$$
$$+2\Omega(\delta\mathfrak{W}_{23;2}-\delta\mathfrak{W}_{13;1}) = 0, \quad (110)$$

$$\lambda^3(V_{113}-V_{223})-4\lambda^2\Omega V_{123}+\lambda[2\Omega^2(V_{223}-V_{113})+\delta S_{113}-\delta S_{223}]+$$
$$+4\Omega(\delta\mathfrak{W}_{13;2}+\delta\mathfrak{W}_{23;1}) = 0, \quad (111)$$

and

$$\lambda^4(V_{113}+V_{223}-\tfrac{2}{3}V_{333})+\lambda^2[2\Omega^2(V_{113}+V_{223})+\delta S_{113}+\delta S_{223}]-$$
$$-8\Omega^2(\delta\mathfrak{W}_{13;1}+\delta\mathfrak{W}_{23;2}) = 0. \quad (112)$$

It will be observed that in equation (112) the term in λV_{123} is absent. The reason is that for spheroids $\delta\mathfrak{W}_{13;2} = \delta\mathfrak{W}_{23;1}$ (see equation (152) of Chapter 3).

Equations (110)–(112) must be further supplemented by the divergence condition

$$V_{113}+V_{223} = -\frac{a_1^2}{a_3^2}V_{333}. \quad (113)$$

Considering equations (110)–(113), we first observe that by equations (153) and (157) of Chapter 3,

$$-2\delta\mathfrak{W}_{13;1}+2\delta\mathfrak{W}_{23;2} = 2(B_{13}+a_1^2 B_{113})(V_{113}-V_{223}) \quad (114)$$

and $\quad \delta S_{113}-\delta S_{223} = [3(B_{11}+B_{13})+(5a_1^2+a_3^2)B_{113}](V_{113}-V_{223}). \quad (115)$

With these substitutions, we find (after some further reductions) that equations (110) and (111) become

$$\{\lambda^3+\lambda[3(B_{11}+B_{13})+(5a_1^2+a_3^2)B_{113}-2\Omega^2]\}(V_{113}-V_{223})-$$
$$-4\Omega[\lambda^2+2(B_{13}+a_1^2 B_{113})]V_{123} = 0 \quad (116)$$

and

$$\{\lambda^3+\lambda[2B_{11}+4B_{13}+(4a_1^2+2a_3^2)B_{113}-2\Omega^2]\}V_{123}+$$
$$+\Omega[\lambda^2+2(B_{13}+a_1^2 B_{113})](V_{113}-V_{223}) = 0. \quad (117)$$

These equations therefore involve only $(V_{113}-V_{223})$ and V_{123}. They can accordingly be considered independently of equations (112) and (113).

Before writing down the characteristic equation which follows from equations (116) and (117), we may note that the coefficients of $V_{113}-V_{223}$ in equation (116) and of V_{123} in equation (117) are the same; for, their difference, namely $B_{11}-B_{13}+(a_1^2-a_3^2)B_{113}$, clearly vanishes. The resulting characteristic equation is, therefore, of the form

$$\lambda^2[\lambda^2+3(B_{11}+B_{13})+(5a_1^2+a_3^2)B_{113}-2\Omega^2]^2+$$
$$+4\Omega^2[\lambda^2+2(B_{13}+a_1^2 B_{113})]^2 = 0. \quad (118)$$

Writing $-\sigma^2$ in place of λ^2, we can factorize equation (118) to give

$$\sigma^3+2\Omega\sigma^2 \quad [3(B_{11}+B_{13})+(5a_1^2+a_3^2)B_{113}-2\Omega^2]\sigma-$$
$$-4\Omega(B_{13}+a_1^2 B_{113}) = 0, \quad (119)$$

and a similar equation with $-\sigma$ in place of σ. It should be noted that in deriving equation (119) no use was made of the divergence condition (113).

It remains to consider equations (112) and (113). For consistency with equations (116) and (117) we must now put

$$V_{113} = V_{223} \quad \text{and} \quad V_{123} = 0. \quad (120)$$

The divergence condition (113) then gives

$$V_{113} = V_{223} = -\frac{a_1^2}{2a_3^2}V_{333}. \quad (121)$$

With these substitutions, we find, after some further reductions, that equation (112) leads to the characteristic equation

$$\left(\frac{2}{3}+\frac{a_1^2}{a_3^2}\right)\sigma^4+\left\{2[(a_1^2+2a_3^2)B_{133}-5a_3^2 B_{333}-2B_{33}]-\right.$$
$$-\frac{a_1^2}{a_3^2}[2\Omega^2+3(B_{11}+B_{13})+(7a_1^2+5a_3^2)B_{113}-6a_3^2 B_{133}]\Big\}\sigma^2-$$
$$-8\Omega^2\left[a_1^2 B_{133}-\frac{a_1^2}{a_3^2}(B_{13}+2a_1^2 B_{113})\right] = 0. \quad (122)$$

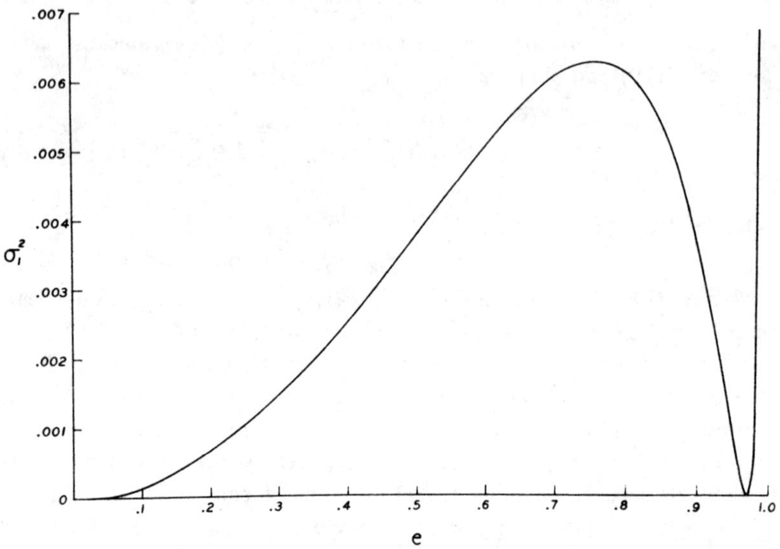

FIG. 11a. The square of the characteristic frequency (in the unit $\pi G \rho$) of oscillation of the Maclaurin spheroid belonging to the mode that becomes unstable at the point $(e = 0{\cdot}9694)$, where a pear-shaped sequence bifurcates.

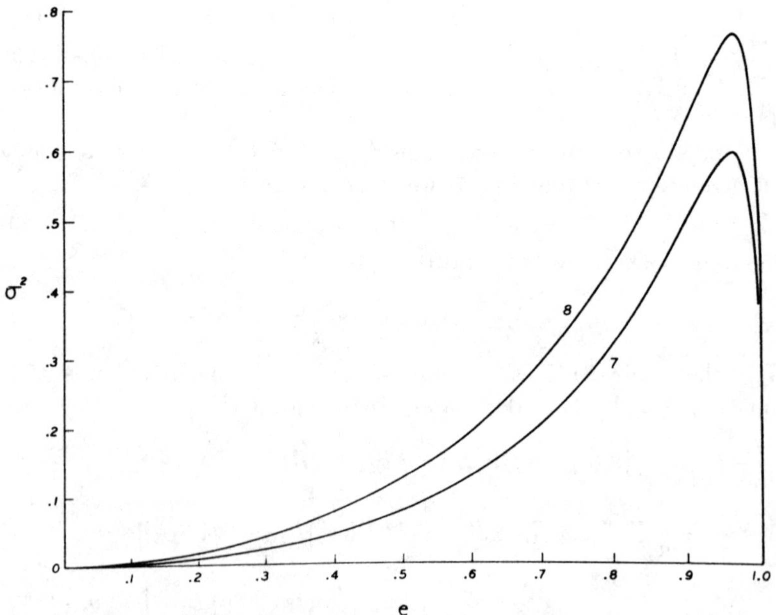

FIG. 11b. The squares of the characteristic frequencies (in the unit $\pi G \rho$) of two of the five odd modes of third-harmonic oscillation of the Maclaurin spheroid.

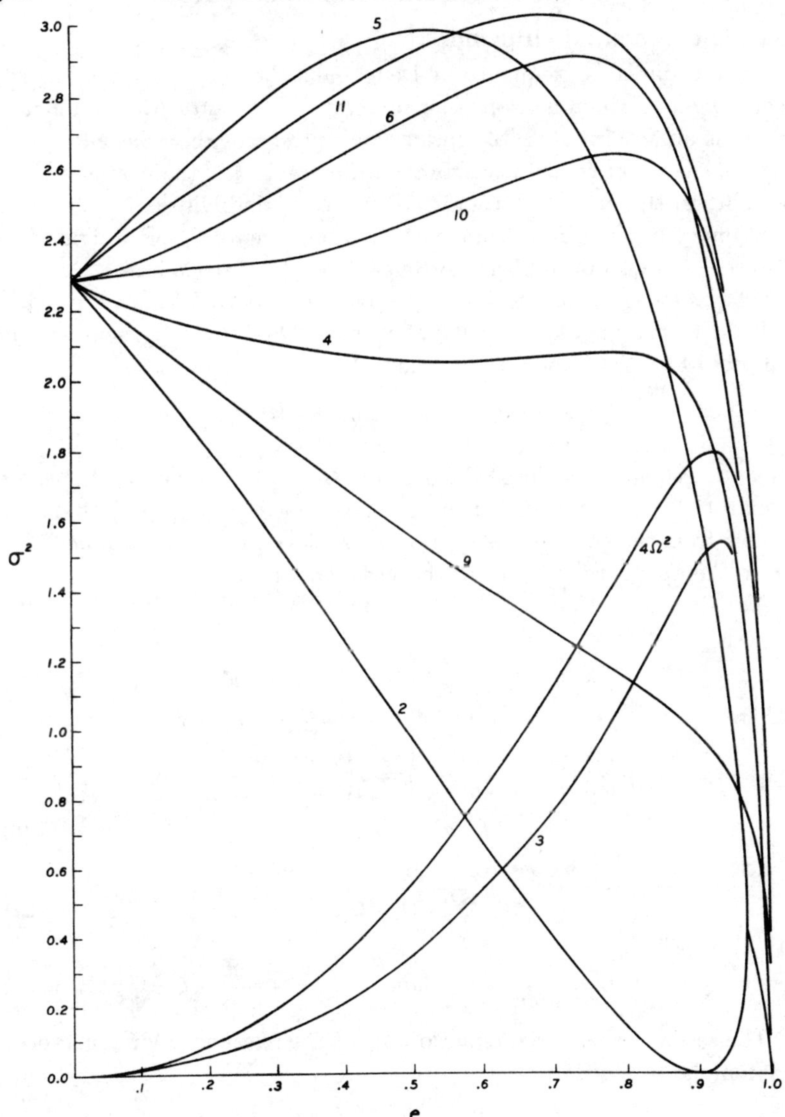

FIG. 11c. The squares of the characteristic frequencies (in the unit $\pi G\rho$) of the remaining eight modes of third-harmonic oscillation of the Maclaurin spheroid.

The characteristic frequencies of the Maclaurin spheroid belonging to the third harmonics have been evaluated; and the results are illustrated in Figs. 11a, 11b, and 11c.

44. The Dedekind ellipsoids

It is a direct consequence of Dedekind's theorem (Chapter 4, § 28) that for any state of motions that preserves a constant ellipsoidal figure, there is an adjoint state of motions that preserves the *same* ellipsoidal figure. As we have already shown in Chapter 1, § 5, the configurations adjoint to the Jacobi ellipsoids are the Dedekind ellipsoids; these are stationary in an inertial frame and maintain their ellipsoidal figures by internal motions of uniform vorticity about the least axis.

The vorticity ζ of the internal motions in a Dedekind ellipsoid is related to the angular velocity of rotation Ω of the congruent Jacobi ellipsoid by (cf. Chapter 1, equation (31))

$$\zeta^2 = \frac{(a_1^2+a_2^2)^2}{a_1^2 a_2^2}\Omega^2 = 2\frac{(a_1^2+a_2^2)^2}{a_1^2 a_2^2}B_{12}; \tag{123}$$

and the geometry of the Dedekind ellipsoid is determined by the same equation (4). While these results concerning the Dedekind ellipsoids follow from the general theory, it is instructive to derive them *ab initio* from the appropriate forms of the virial equations.

We first note that a state of uniform vortical motion about the x_3-axis specified by

$$u_1 = Q_1 x_2, \qquad u_2 = Q_2 x_1, \quad \text{and} \quad u_3 = 0, \tag{124}$$

where Q_1 and Q_2 are constants, will satisfy the kinematical requirement

$$u_j\frac{\partial}{\partial x_j}\left(\frac{x_1^2}{a_1^2}+\frac{x_2^2}{a_2^2}+\frac{x_3^2}{a_3^2}-1\right) = 2\left(\frac{u_1 x_1}{a_1^2}+\frac{u_2 x_2}{a_2^2}\right) = 0, \tag{125}$$

only if
$$Q_1/a_1^2 = -Q_2/a_2^2. \tag{126}$$

Since the vorticity associated with the motion (124) is given by

$$\zeta = Q_2-Q_1, \tag{127}$$

we may write

$$Q_1 = -\frac{a_1^2}{a_1^2+a_2^2}\zeta \quad \text{and} \quad Q_2 = +\frac{a_2^2}{a_1^2+a_2^2}\zeta. \tag{128}$$

The second-order virial equation under the stationary circumstances contemplated requires

$$2\mathfrak{T}_{ij}+\mathfrak{W}_{ij} = -\Pi\delta_{ij}. \tag{129}$$

For internal motions specified by equations (124), the tensors $\mathfrak{T}_{ij}$ and $\mathfrak{W}_{ij}$ are diagonal (in the frame considered) and, moreover,

$$\mathfrak{T}_{11} = \tfrac{1}{2}Q_1^2 I_{22}, \qquad \mathfrak{T}_{22} = \tfrac{1}{2}Q_2^2 I_{11}, \quad \text{and} \quad \mathfrak{T}_{33} = 0. \tag{130}$$

Equation (129) now gives

$$Q_1^2 a_2^2-2A_1 a_1^2 = Q_2^2 a_1^2-2A_2 a_2^2 = -2A_3 a_3^2. \tag{131}$$

Adding $2a_1^2 a_2^2 A_{12}$ to each of the three sides of the triangle of equations (131), we obtain

$$a_1^2\left(\frac{Q_1^2 a_2^2}{a_1^2}-2B_{12}\right) = a_2^2\left(\frac{Q_2^2 a_1^2}{a_2^2}-2B_{12}\right) = 2(a_1^2 a_2^2 A_{12}-a_3^2 A_3). \quad (132)$$

Since
$$\frac{Q_1^2 a_2^2}{a_1^2} = \frac{Q_2^2 a_1^2}{a_2^2} = \frac{a_1^2 a_2^2}{(a_1^2+a_2^2)^2}\,\zeta^2 = -Q_1 Q_2, \quad (133)$$

it follows from the equalities (132) that

$$-Q_1 Q_2 = \frac{a_1^2 a_2^2}{(a_1^2+a_2^2)^2}\,\zeta^2 = 2B_{12} \quad \text{and} \quad a_1^2 a_2^2 A_{12} = a_3^2 A_3. \quad (134)$$

We thus recover equation (123) and the geometrical condition (4).

45. The stability of the Dedekind ellipsoids and the point of bifurcation of a sequence of pear-shaped configurations

One result concerning the normal modes of oscillation of a Dedekind ellipsoid, which can be deduced directly from Dedekind's general theorem, is that its characteristic frequencies of oscillation (in the inertial frame), belonging to the second harmonics, must be the same as those for the congruent Jacobi ellipsoid (in the rotating frame). This equality in the frequencies arises from the fact that during an oscillation belonging to the second harmonics, the figures continue to be ellipsoids; and since the equilibrium configurations are adjoint they must remain adjoint during the oscillations.† However, this equality in the frequencies of oscillation, belonging to the second harmonics, cannot extend to the oscillations of the higher harmonics. In particular, the points where the pear-shaped configurations bifurcate from the Dedekind and the Jacobian sequences cannot be the same; and it is of interest to demonstrate explicitly how this difference is manifested.

For the purposes of determining a neutral point belonging to the third harmonics, we may use the first variation of the equilibrium relation,

$$2(\mathfrak{T}_{ij;k}+\mathfrak{T}_{ik;j})+\mathfrak{W}_{ik;j}+\mathfrak{W}_{ij;k} = -\Pi_k\delta_{ij}-\Pi_j\delta_{ik}, \quad (135)$$

which follows from the virial equations of the third order (cf. Chapter 2, equation (56)).

The fifteen relations that follow from equation (135), after the elimination of the Π_i's, are of three different types (cf. Chapter 2, equations (78), (79), and (83)):

$$2\mathfrak{W}_{ij;j}+4\mathfrak{T}_{ij;j} = 0, \quad (136)$$

$$\mathfrak{W}_{ij;k}+\mathfrak{W}_{ik;j}+2(\mathfrak{T}_{ij;k}+\mathfrak{T}_{ik;j}) = 0, \quad (137)$$

---

† For an explicit demonstration of these facts see the original Paper XXIV.

and
$$S_{ijj}+2R_{ijj} = 0, \tag{138}$$

where
$$R_{ijj} = -4\mathfrak{I}_{ij;j}-2\mathfrak{I}_{jj;i}+2\mathfrak{I}_{ii;i}, \tag{139}$$

($i \neq j \neq k$; and no summation over repeated indices in equations (136)–(139)), and S_{ijj} has the same meaning as in equation (84) of Chapter 2.

At a neutral point, the first variations of the relations (136)–(138) must vanish non-trivially; and as in the case of the Jacobi ellipsoid, it will suffice to consider only the relations that are odd in the index 1 and even in the indices 2 and 3. Thus, to locate a point of bifurcation where a sequence of pear-shaped configurations may branch off, we must consider (cf. equations (20)–(23))

$$\delta J_1 = -2\delta\mathfrak{W}_{12;2}-4\delta\mathfrak{I}_{12;2} = 0, \tag{140}$$

$$\delta J_2 = -2\delta\mathfrak{W}_{13;3}-4\delta\mathfrak{I}_{13;3} = 0, \tag{141}$$

$$\delta J_3 = \delta S_{122}+2\delta R_{122} = 0, \tag{142}$$

and
$$\delta J_4 = \delta S_{133}+2\delta R_{133} = 0. \tag{143}$$

By equations (153) and (157) of Chapter 3, $\delta\mathfrak{W}_{12;2}$, $\delta\mathfrak{W}_{13;3}$, δS_{122}, and δS_{133} are expressible as linear combinations of V_{111}, V_{122}, and V_{133}. It remains to express the $\delta\mathfrak{I}_{ij;k}$'s that occur in equations (140)–(143) in terms of the $V_{i;jk}$'s which are odd in the index 1.

In Chapter 2, § 15 (equations (141)–(149)) we have expressed the first variations of the various tensors that occur in the case when the internal motion u_i is a linear function of the coordinates of the form

$$u_i = Q_{ij}x_j, \tag{144}$$

where $\mathbf{Q}$ is some constant matrix. In particular, for quasi-static deformations, equation (148) of Chapter 2 gives

$$2\delta\mathfrak{I}_{ij;k} = Q_{il}Q_{jm}V_{k;lm}-Q_{jl}(Q_{km}V_{i;lm}+Q_{lm}V_{i;km})-$$
$$-Q_{il}(Q_{km}V_{j;lm}+Q_{lm}V_{j;km}). \tag{145}$$

For the case we are presently considering, the only non-vanishing elements of $\mathbf{Q}$ are

$$Q_{12} = Q_1 \quad \text{and} \quad Q_{21} = Q_2, \tag{146}$$

and we find from equation (145) that

$$2\delta\mathfrak{I}_{11;1} = -2Q_1 Q_2 V_{1;11}-Q_1^2 V_{1;22}, \tag{147}$$

$$2\delta\mathfrak{I}_{22;1} = -4Q_1 Q_2 V_{2;12}+Q_2^2 V_{1;11}, \qquad 2\delta\mathfrak{I}_{33;1} = 0, \tag{148}$$

$$2\delta\mathfrak{I}_{12;2} = -Q_1 Q_2(V_{1;22}+V_{2;21})-Q_2^2 V_{1;11}, \tag{149}$$

and
$$2\delta\mathfrak{I}_{13;3} = -Q_1 Q_2 V_{3;31}. \tag{150}$$

Combining the foregoing equations appropriately, we find

$$2\delta R_{122} = (2Q_2^2 - 4Q_1 Q_2)V_{1;11} + (4Q_1 Q_2 - 2Q_1^2)V_{1;22} + 12Q_1 Q_2 V_{2;12} \quad (151)$$

and

$$2\delta R_{133} = -4Q_1 Q_2 V_{1;11} - 2Q_1^2 V_{1;22} + 4Q_1 Q_2 V_{3;31}. \quad (152)$$

Equations (149)–(152), together with the known expressions for $\delta\mathfrak{W}_{12;2}$, $\delta\mathfrak{W}_{13;3}$, δS_{122}, and δS_{133}, enable us to express δJ_i as linear combinations of the *five* $V_{i;jk}$'s that are odd in the index 1 and even in the indices 2 and 3. Thus, we may write

$$\delta J_i = \langle i \,|\, 1; 11 \rangle V_{1;11} + \langle i \,|\, 1; 22 \rangle V_{1;22} + \langle i \,|\, 1; 33 \rangle V_{1;33} +$$
$$+ \langle i \,|\, 2; 12 \rangle V_{2;12} + \langle i \,|\, 3; 13 \rangle V_{3;13}, \quad (153)$$

where $\langle i \,|\, 1; 11 \rangle$, etc., are certain matrix elements which are known. Equation (153), supplemented by the solenoidal condition

$$\frac{3}{a_1^2} V_{1;11} + \frac{1}{a_2^2}(V_{1;22} + 2V_{2;12}) + \frac{1}{a_3^2}(V_{1;33} + 2V_{3;13}) = 0, \quad (154)$$

provides a system of five linear equations for the five $V_{i;jk}$'s that occur in these equations. The vanishing of the determinant of this system of equations is the necessary and sufficient condition for the occurrence of a neutral point. It is found that the neutral point occurs where

$$a_2/a_1 = 0\cdot441330, \qquad a_3/a_1 = 0\cdot350409, \qquad 2B_{12} = 0\cdot287813,$$

and

$$\zeta = 1\cdot45237. \quad (155)$$

These values should be contrasted with those given in equation (28) for the corresponding point of bifurcation along the Jacobian sequence.

BIBLIOGRAPHICAL NOTES

The present chapter brings together the results on the Jacobi and the Dedekind ellipsoids as well as those on the Maclaurin spheroids pertaining to their third-harmonic oscillations.

§ 39. The argument in this section, relating to the origin of Jacobi's formula $\Omega^2 = 2B_{12}$, is briefly noted in Paper XXIII.

§ 40. The major parts of the discussion in this section are derived from Papers IX and XVI. The direct determination of the point of bifurcation in § (a) has, however, not been published before. It would appear that the basic idea underlying the determination is, in its essence, contained in a method due to Jeans:

> J. H. Jeans, *Problems of Cosmogony and Stellar Dynamics* (Cambridge, England, Cambridge University Press, 1919), pp. 78–85; also *Astronomy and Cosmogony* (Cambridge, England, Cambridge University Press, 1929), pp. 216–19.

In contrast to Jeans, whose treatment required the elaborate machinery of ellipsoidal harmonics, we had no occasion even to mention the topic. The comparison of Jeans' treatment of this problem (not to mention Darwin's) and the one given in the text provides an illustration of how the systematic use of Ferrers' potentials enables the restoration to the subject of its essentially elementary character.

The concept of *linear independence of displacements modulo the ellipsoid*, introduced in this section, is exploited in Papers XXX and XXXII for the purposes of determining the effects of general relativity, in the post-Newtonian approximation, on the Maclaurin and the Jacobian figures.

FIG. 8, illustrating the critical Jacobi ellipsoid and its deformation into a pear-shaped figure, is taken from the following paper by Darwin:

> G. H. DARWIN, "On the pear-shaped figure of equilibrium of a rotating mass of liquid," *Phil. Trans. R. Soc. (London)*, **198** (1901), 301–31 ; see also *Scientific Papers*, **3** (Cambridge, England, 1910), 314.

§ 41. The principal results in this section are taken from Paper XVI (§ 5).

§§ 42 and 43. These sections provide a condensed presentation of the analysis set out in greater detail in Papers XIII and XIV. The reader may find a reference to these papers helpful in following the account in the text.

As we have mentioned in Chapter 1, Cartan was the first to show rigorously that the Jacobi ellipsoid becomes dynamically unstable at the point of bifurcation of the pear-shaped configurations. For an extension of Cartan's method to the solution of the characteristic-value problem see:

> S. YABUSHITA, "Initial motions of a Jacobi ellipsoid away from its unstable form," *Astrophys. J.*, **141** (1965), 232–39.

§§ 44 and 45. The analysis in these sections is, in the main, derived from Paper XXIV. The equality of the characteristic frequencies of oscillation, belonging to the second harmonics, of congruent Dedekind and Jacobi ellipsoids is established in Paper XXIV, directly, without appealing to Dedekind's theorem.

In Paper XXXV, the neutral points along the Maclaurin and the Jacobian sequences, belonging to the *fourth* harmonics, are located with the aid of the virial equations of the fourth order. It is shown, for example, that along the Maclaurin sequence, there are three neutral points at

$$e = 0{\cdot}93275 \quad \text{where} \quad \Omega^2 = 2(B_{11}+a_1^2 B_{111}+a_1^4 B_{1111}) = 0{\cdot}44923,$$

$$e = 0{\cdot}98097 \quad \text{where} \quad \Omega^2 = 2B_{11}+7a_1^2(B_{111}-a_3^2 B_{1113}) = 0{\cdot}37290,$$

and $\qquad e = 0{\cdot}98531 \quad \text{where} \quad \Omega^2 = 0{\cdot}34619.$

Along the Jacobian sequence, on the other hand, there is only one neutral point at

$$a_2/a_1 = 0{\cdot}2972 \quad \text{and} \quad a_3/a_1 = 0{\cdot}2575.$$

7

THE RIEMANN ELLIPSOIDS

46. Introduction

THE central topic of this chapter is Riemann's solution to the problem of the stationary ellipsoidal figures that are admissible under Dirichlet's general formulation. As we have seen in Chapter 4, a homogeneous fluid mass that maintains, at all times, under its own gravitation, an ellipsoidal figure allows the following description. The coordinate frame, in which the principal axes of the ellipsoid are at rest, is rotating with an angular velocity $\Omega(t)$ with respect to an inertial frame; and in this rotating frame, the ellipsoid has internal motions having a uniform vorticity $\zeta(t)$. Riemann's enumeration of the admissible figures is concerned with the case when Ω and ζ are independent of time and are constants. In addition to considering these stationary figures of Riemann, we shall be concerned also with their stability.

While Riemann's paper, devoted to Dirichlet's problem, bears the stamp of his superb craftsmanship, there are some very surprising lapses and some definitely erroneous conclusions. One may perhaps speculate that Riemann's oversights in this paper are, in some way, related to the circumstances of his life in 1860 and to his personal relationship to Dirichlet, to whose chair in Göttingen he had succeeded only the year before.

47. Riemann's theorem

Under the stationary conditions contemplated, we may suppose that the principal axes of the ellipsoid are at rest in a frame of reference rotating with a constant angular velocity Ω; and that in this rotating frame, there are internal motions having a uniform constant vorticity ζ. And we seek the conditions, under the postulated circumstances, that a homogeneous ellipsoid with semi-axes a_1, a_2, and a_3 will be a figure of equilibrium.

Let the orientation of the coordinate axes of the rotating frame be so chosen that they lie along the principal axes of the ellipsoid; and let the components of Ω and ζ in this frame be Ω_1, Ω_2, and Ω_3 and ζ_1, ζ_2, and ζ_3. The kinematical requirement, that the motion $(\mathbf{u})$, associated with $\boldsymbol{\zeta}$, preserves the ellipsoidal boundary, leads to the following expressions for its components:

$$u_1 = -\frac{a_1^2}{a_1^2+a_2^2}\,\zeta_3\,x_2 + \frac{a_1^2}{a_1^2+a_3^2}\,\zeta_2\,x_3,$$

$$u_2 = -\frac{a_2^2}{a_2^2+a_3^2}\,\zeta_1\,x_3 + \frac{a_2^2}{a_2^2+a_1^2}\,\zeta_3\,x_1, \qquad (1)$$

$$u_3 = -\frac{a_3^2}{a_3^2+a_1^2}\,\zeta_2\,x_1 + \frac{a_3^2}{a_3^2+a_2^2}\,\zeta_1\,x_2.$$

To obtain the conditions that the ellipsoid will also be in gravitational equilibrium, we shall make use of the second-order virial equations in the form given in equation (64) of Chapter 2. We have

$$2\mathfrak{T}_{ij}+\mathfrak{W}_{ij}+\Omega^2 I_{ij}-\Omega_i\Omega_k I_{kj}+2\epsilon_{ilm}\Omega_m\int_V \rho u_l x_j\,d\mathbf{x} = -\delta_{ij}\,\Pi. \qquad (2)$$

Consider first the non-diagonal components of equation (2). The $(2, 3)$- and the $(3, 2)$-components of the equation give

$$2\mathfrak{T}_{23}-\Omega_2\Omega_3 I_{33}-2\Omega_3\int_V \rho u_1 x_3\,d\mathbf{x} = 0 \qquad (3)$$

and

$$2\mathfrak{T}_{32}-\Omega_3\Omega_2 I_{22}+2\Omega_2\int_V \rho u_1 x_2\,d\mathbf{x} = 0, \qquad (4)$$

since in the chosen coordinate system, the tensor I_{ij} and $\mathfrak{W}_{ij}$ are diagonal and, moreover, for $\mathbf{u}$ given by equation (1),

$$\int_V \rho u_i x_j\,d\mathbf{x} = 0 \quad \text{if} \quad i = j. \qquad (5)$$

Adding and subtracting equations (3) and (4), we get

$$4\mathfrak{T}_{23}-\Omega_2\Omega_3(I_{22}+I_{33})+2\int_V \rho u_1(\Omega_2 x_2-\Omega_3 x_3)\,d\mathbf{x} = 0 \qquad (6)$$

and

$$\Omega_2\Omega_3(I_{22}-I_{33})-2\int_V \rho u_1(\Omega_2 x_2+\Omega_3 x_3)\,d\mathbf{x} = 0. \qquad (7)$$

For the motion specified in equation (1)

$$2\mathfrak{T}_{23} = -\frac{a_2^2 a_3^2}{(a_1^2+a_2^2)(a_1^2+a_3^2)}\,\zeta_2\,\zeta_3\,I_{11}, \qquad (8)$$

and

$$\int_V \rho u_1 x_2\,d\mathbf{x} = -\frac{a_1^2}{a_1^2+a_2^2}\,\zeta_3\,I_{22} \quad \text{and} \quad \int_V \rho u_1 x_3\,d\mathbf{x} = +\frac{a_1^2}{a_1^2+a_3^2}\,\zeta_2\,I_{33}. \qquad (9)$$

Inserting these relations in equations (6) and (7) and substituting for

I_{ij} its value in terms of the mass and the semi-axes of the ellipsoid, we find after some rearrangements

$$a_2^2+a_3^2+\frac{2a_1^2a_3^2}{a_1^2+a_3^2}\frac{\zeta_2}{\Omega_2}+\frac{2a_1^2a_2^2}{a_1^2+a_2^2}\frac{\zeta_3}{\Omega_3}+\frac{2a_1^2a_2^2a_3^2}{(a_1^2+a_3^2)(a_1^2+a_2^2)}\frac{\zeta_2}{\Omega_2}\frac{\zeta_3}{\Omega_3}=0 \qquad (10)$$

and

$$a_3^2+\frac{2a_1^2a_3^2}{a_1^2+a_3^2}\frac{\zeta_2}{\Omega_2}=a_2^2+\frac{2a_1^2a_2^2}{a_1^2+a_2^2}\frac{\zeta_3}{\Omega_3}, \qquad (11)$$

where in writing these equations in these forms, *we have explicitly supposed that* Ω_2 *and* Ω_3 *are different from zero.*

Now letting

$$\beta=-\frac{a_3^2}{a_1^2+a_3^2}\frac{\zeta_2}{\Omega_2} \quad \text{and} \quad \gamma=-\frac{a_2^2}{a_1^2+a_2^2}\frac{\zeta_3}{\Omega_3}, \qquad (12)$$

we can rewrite equations (10) and (11) in the forms

$$\beta+\gamma-\beta\gamma=\frac{a_3^2+a_2^2}{2a_1^2} \quad \text{and} \quad \beta-\gamma=\frac{a_3^2-a_2^2}{2a_1^2}. \qquad (13)$$

These equations provide for β and γ the equations

$$\beta^2-\frac{4a_1^2+a_3^2-a_2^2}{2a_1^2}\beta+\frac{a_3^2}{a_1^2}=0 \qquad (14)$$

and

$$\gamma^2-\frac{4a_1^2+a_2^2-a_3^2}{2a_1^2}\gamma+\frac{a_2^2}{a_1^2}=0. \qquad (15)$$

The roots of these equations are

$$\beta=\frac{1}{4a_1^2}\{4a_1^2-a_2^2+a_3^2\pm\sqrt{[4a_1^2-(a_2+a_3)^2][4a_1^2-(a_2-a_3)^2]}\} \qquad (16)$$

and

$$\gamma=\frac{1}{4a_1^2}\{4a_1^2-a_3^2+a_2^2\pm\sqrt{[4a_1^2-(a_2+a_3)^2][4a_1^2-(a_2-a_3)^2]}\}, \qquad (17)$$

where the signs in front of the radicals, in the two expressions, go together. It will appear subsequently that the two roots for β and γ, given by equations (16) and (17), correspond to the fact that, consistent with Dedekind's theorem, two states of internal motions are compatible with the same external figure.

From equations (12), (16), and (17) it follows that, if Ω_2 and Ω_3 are different from zero, the ratios ζ_2/Ω_2 and ζ_3/Ω_3 become determinate. In particular, equation (14) expressed in terms of ζ_2/Ω_2 is

$$\left(\frac{\zeta_2}{\Omega_2}\right)^2+(4a_1^2-a_2^2+a_3^2)\frac{a_1^2+a_3^2}{2a_1^2a_3^2}\left(\frac{\zeta_2}{\Omega_2}\right)+\frac{(a_1^2+a_3^2)^2}{a_1^2a_3^2}=0. \qquad (18)$$

On the other hand, if Ω_1 is also different from zero, then the $(1,2)$- and the $(2, 1)$-components of equation (2) would have similarly led to the equation

$$\left(\frac{\zeta_2}{\Omega_2}\right)^2 + (4a_3^2 - a_2^2 + a_1^2)\frac{a_1^2 + a_3^2}{2a_1^2 a_3^2}\left(\frac{\zeta_2}{\Omega_2}\right) + \frac{(a_1^2 + a_3^2)^2}{a_1^2 a_3^2} = 0. \qquad (19)$$

Equations (18) and (19) are clearly incompatible unless $a_1 = a_3$; and a similar consideration of the equations governing ζ_3/Ω_3 would have required $a_1 = a_2$. It therefore follows that *non-trivial solutions are obtained only if no more than two of the three pairs of components* (ζ_1, Ω_1), (ζ_2, Ω_2), *and* (ζ_3, Ω_3) *are different from zero.* This is Riemann's fundamental theorem.

An alternative way of stating Riemann's theorem is that *equilibrium demands that either ζ and Ω are parallel, in which case they are both directed along one of the principal axes of the ellipsoid, or ζ and Ω are not parallel, in which case they lie in a principal plane of the ellipsoid.*

It is important to observe, even at this stage, that the two cases distinguished by Riemann's theorem lead to essentially different types of configurations: the configurations that arise when only one pair of components, (ζ_3, Ω_3) say, is different from zero are, in no sense, "special cases" of the configurations that arise when two pairs of components, (ζ_2, Ω_2) and (ζ_3, Ω_3) say, are different from zero. We shall find that in the former case, the equilibrium figures, which we shall call the *S-type ellipsoids*, can be arranged in linear sequences (special examples of which are the Jacobian and the Dedekind sequences) while in the latter case they are of three distinct types with entirely different structures and properties.

48. The equilibrium figures in the case ζ and Ω are parallel: the S-type ellipsoids

In this and in the following two sections, we shall consider the equilibrium and the stability of the Riemann ellipsoids in which the directions of ζ and Ω are parallel and lie along one of the principal axes. There is clearly no loss of generality in supposing that Ω and ζ are along the x_3-direction and that

$$a_1 \geqslant a_2. \qquad (20)$$

The components of the internal motion having an assigned vorticity ζ about the x_3-direction can be written in the form (cf. equations (124)–(128) of Chapter 6)

$$u_1 = Q_1 x_2, \qquad u_2 = Q_2 x_1, \qquad u_3 = 0, \qquad (21)$$

where $\qquad Q_1 = -\dfrac{a_1^2}{a_1^2 + a_2^2}\zeta \quad$ and $\quad Q_2 = +\dfrac{a_2^2}{a_1^2 + a_2^2}\zeta. \qquad (22)$

Some elementary relations which follow from these definitions and which we shall find useful are

$$a_1^2 Q_2 = -a_2^2 Q_1, \qquad a_1^2 Q_2^2 = -a_2^2 Q_1 Q_2, \quad \text{and} \quad a_2^2 Q_1^2 = -a_1^2 Q_1 Q_2,$$

(23)

where
$$Q_1 Q_2 = -\frac{a_1^2 a_2^2}{(a_1^2+a_2^2)^2}\, \zeta^2.$$

(24)

In the present context, the non-diagonal components of equation (2), considered in § 47, are trivially satisfied, while the diagonal components give

$$2\mathfrak{T}_{11}+\Omega^2 I_{11}+\mathfrak{W}_{11}+2\Omega \int_V \rho u_2 x_1 \, d\mathbf{x}$$
$$= 2\mathfrak{T}_{22}+\Omega^2 I_{22}+\mathfrak{W}_{22}-2\Omega \int_V \rho u_1 x_2 \, d\mathbf{x} = \mathfrak{W}_{33}. \quad (25)$$

For the internal motion specified by equation (21), equations (25) give

$$a_2^2 Q_1^2+a_1^2(\Omega^2+2Q_2\Omega)-2A_1 a_1^2$$
$$= a_1^2 Q_2^2+a_2^2(\Omega^2-2Q_1\Omega)-2A_2 a_2^2 = -2A_3 a_3^2. \quad (26)$$

In view of the relations (23), we can rewrite the equalities (26) in the form

$$a_1^2(\Omega^2-Q_1 Q_2)-2A_1 a_1^2 = a_2^2(\Omega^2-Q_1 Q_2)-2A_2 a_2^2$$
$$= -2A_3 a_3^2 - \frac{2a_1^2 a_2^2}{a_1^2+a_2^2}\, \zeta\Omega. \quad (27)$$

Adding $2a_1^2 a_2^2 A_{12}$ to each of the three sides of this triangle of equalities, we obtain

$$a_1^2(\Omega^2-Q_1 Q_2-2B_{12}) = a_2^2(\Omega^2-Q_1 Q_2-2B_{12})$$
$$= 2\left(a_1^2 a_2^2 A_{12}-a_3^2 A_3 - \frac{a_1^2 a_2^2}{a_1^2+a_2^2}\, \zeta\Omega\right). \quad (28)$$

It follows from these equalities that

$$\Omega^2-Q_1 Q_2 = 2B_{12}$$

(29)

and
$$\frac{a_1^2 a_2^2}{a_1^2+a_2^2}\, \zeta\Omega = a_1^2 a_2^2 A_{12}-a_3^2 A_3.$$

(30)

We shall now define a *Riemann sequence* as one along which

$$f = \zeta/\Omega$$

(31)

is a constant. By this definition, *the Jacobian and the Dedekind sequences are Riemann sequences for $f = 0$ and $f = \pm\infty$, respectively.*

In terms of f, we have the relations

$$Q_1 = -\frac{a_1^2\Omega}{a_1^2+a_2^2}\, f, \qquad Q_2 = +\frac{a_2^2\Omega}{a_1^2+a_2^2}\, f, \quad \text{and} \quad Q_1 Q_2 = -\frac{a_1^2 a_2^2}{(a_1^2+a_2^2)^2}\, \Omega^2 f^2;$$

(32)

and equations (29) and (30) become

$$\left[1+\frac{a_1^2 a_2^2}{(a_1^2+a_2^2)^2}f^2\right]\Omega^2 = 2B_{12} \tag{33}$$

and

$$\frac{a_1^2 a_2^2}{a_1^2+a_2^2}f\Omega^2 = a_1^2 a_2^2 A_{12}-a_3^2 A_3. \tag{34}$$

Eliminating Ω^2 between equations (33) and (34), we obtain

$$\frac{a_1^2 a_2^2}{(a_1^2+a_2^2)^2}f^2+\frac{2a_1^2 a_2^2 B_{12}}{a_3^2 A_3-a_1^2 a_2^2 A_{12}}\frac{f}{a_1^2+a_2^2}+1 = 0. \tag{35}$$

For a given f equation (35) determines the ratios of the axes of the ellipsoids that are compatible with equilibrium; and the value of Ω^2, that is to be associated with a particular solution of equation (35), then follows from equation (33).

It will be recalled that the a_1-axis was *defined* as the longer of the two axes in the plane transverse to the common direction of ζ and Ω (*chosen along the x_3-axis*). We shall now show that *this choice ensures that a_1 is also the longest of the three axes*.

First, we observe that the equalities (27) can be written as the pair of equations

$$\Omega^2-Q_1 Q_2+2Q_2\Omega = 2\frac{a_1^2-a_3^2}{a_1^2}B_{13} \tag{36}$$

and

$$\Omega^2-Q_1 Q_2-2Q_1\Omega = 2\frac{a_2^2-a_3^2}{a_2^2}B_{23}. \tag{37}$$

Inserting for Q_1 and Q_2 their values given in equations (32), we find that the left-hand side of equation (36) can be expressed in the form

$$\Omega^2\left[\left(1+\frac{a_2^2}{a_1^2+a_2^2}f\right)^2+\frac{a_2^2(a_1^2-a_2^2)}{(a_1^2+a_2^2)^2}f^2\right]. \tag{38}$$

This expression is positive-definite since, by definition, $a_1 \geqslant a_2$; it now follows from equation (36) that

$$a_1 \geqslant a_3. \tag{39}$$

(a) The adjoint Riemann ellipsoids and Dedekind's theorem

Let

$$x = \frac{a_1 a_2}{a_1^2+a_2^2}f. \tag{40}$$

The relations (32) and (33) now take the forms

$$\Omega^2 = \frac{2B_{12}}{1+x^2}, \qquad -Q_1 Q_2 = \frac{2x^2}{1+x^2}B_{12}, \tag{41}$$

$$Q_1\Omega = -2B_{12}\frac{a_1}{a_2}\frac{x}{1+x^2}, \qquad Q_2\Omega = +2B_{12}\frac{a_2}{a_1}\frac{x}{1+x^2}, \tag{42}$$

while equation (35) becomes

$$x^2 + \frac{2a_1 a_2 B_{12}}{a_3^2 A_3 - a_1^2 a_2^2 A_{12}}\, x + 1 = 0. \tag{43}$$

From equation (43) it is apparent that *if x is a root, then so is $1/x$.* Therefore, a Riemann ellipsoid for a given f is equally a figure of equilibrium for

$$f^\dagger = \frac{(a_1^2 + a_2^2)^2}{a_1^2 a_2^2}\, \frac{1}{f}. \tag{44}$$

From Dedekind's general theorem, we can conclude that the two configurations associated with f and $f^\dagger$ must be adjoints of one another in the sense defined in Chapter 4, § 28. It is instructive to verify that this is the case.

If Ω and $\Omega^\dagger$ denote the angular velocities associated with the solutions x and $1/x$, then according to the first of the relations (41)

$$(\Omega^\dagger)^2 = 2 B_{12} \frac{x^2}{1+x^2} = \Omega^2 x^2, \tag{45}$$

so that we may write

$$\Omega^\dagger = \Omega x = \frac{a_1 a_2}{a_1^2 + a_2^2} f\Omega = \frac{a_1 a_2}{a_1^2 + a_2^2} \zeta. \tag{46}$$

(There is clearly no loss of any essential generality by having chosen the positive sign in equation (46).)

To show that the configurations belonging to x and $1/x$ are adjoints of one another, we must consider the fluid motion $\mathbf{u}^{(0)}$ in the inertial frame. Since

$$\mathbf{u}^{(0)} = \mathbf{u} + \boldsymbol{\Omega} \times \mathbf{x}, \tag{47}$$

it follows from equations (21) and (22) that

$$u_1^{(0)} = u_1 - \Omega x_2 = -\Omega\left(1 + \frac{a_1^2}{a_1^2 + a_2^2} f\right) x_2,$$

$$u_2^{(0)} = u_2 + \Omega x_1 = +\Omega\left(1 + \frac{a_2^2}{a_1^2 + a_2^2} f\right) x_1,$$

and
$$u_3^{(0)} = 0. \tag{48}$$

Alternatively, making use of equations (40) and (46), we can write for the non-vanishing components of $\mathbf{u}^{(0)}$ the expressions

$$u_1^{(0)} = -\Omega\left(1 + \frac{a_1}{a_2} x\right) x_2 = -(a_2\Omega + a_1\Omega^\dagger)\frac{x_2}{a_2}$$

and
$$u_2^{(0)} = +\Omega\left(1 + \frac{a_2}{a_1} x\right) x_1 = +(a_1\Omega + a_2\Omega^\dagger)\frac{x_1}{a_1}. \tag{49}$$

Therefore, replacing Ω by $\Omega^\dagger$ (and $\Omega^\dagger$ by Ω) on the right-hand sides of equations (49) results only in transposing, apart from a change of sign, the coefficients of x_2/a_2 and x_1/a_1; and this result of the transposition is the content of Dedekind's theorem relative to adjoint configurations in the present context.

Since the constants appropriate to adjoint configurations are obtained by replacing x by $1/x$ in the relevant formulas, it follows from equations (41) and (42) that *the values of Ω^2 and $-Q_1 Q_2$ are interchanged when one passes from a configuration to its adjoint, while the values of $Q_1\Omega$ and $Q_2\Omega$ are unchanged.* Since, by equation (43),

$$x+\frac{1}{x} = -\frac{2a_1 a_2 B_{12}}{a_3^2 A_3 - a_1^2 a_2^2 A_{12}}, \tag{50}$$

we can rewrite the expressions (42) for $Q_1\Omega$ and $Q_2\Omega$ in the manifestly "invariant forms"

$$Q_1\Omega = +\frac{1}{a_2^2}(a_3^2 A_3 - a_1^2 a_2^2 A_{12}) \quad \text{and} \quad Q_2\Omega = -\frac{1}{a_1^2}(a_3^2 A_3 - a_1^2 a_2^2 A_{12}). \tag{51}$$

A further fact which follows from equation (47) may be noted. The vorticity $\zeta^{(0)}$, in the inertial frame, is related to the vorticity ζ, in the rotating frame, by

$$\zeta^{(0)} = \zeta + 2\Omega = (2+f)\Omega. \tag{52}$$

From this last equation it follows that *the Riemann sequence for $f = -2$ is irrotational* (i.e. $\zeta^{(0)} = 0$).

We shall conclude this section with an evaluation of the angular momentum L about the x_3-axis in the inertial frame. By definition

$$L = \int_V \rho[x_1 u_2^{(0)} - x_2 u_1^{(0)}]\,d\mathbf{x}; \tag{53}$$

or, making use of the expressions for $u_1^{(0)}$ and $u_2^{(0)}$ given in equations (48), we find

$$L = \tfrac{1}{5}M(a_1^2+a_2^2)\Omega\left[1+\frac{2a_1^2 a_2^2}{(a_1^2+a_2^2)^2}f\right]. \tag{54}$$

The angular momentum $L^\dagger$ of the adjoint configuration is given by

$$L^\dagger = \tfrac{1}{5}M(a_1^2+a_2^2)\Omega^\dagger\left[1+\frac{2a_1^2 a_2^2}{(a_1^2+a_2^2)^2}f^\dagger\right], \tag{55}$$

or, by virtue of equations (44) and (46),

$$L^\dagger = \tfrac{1}{5}Ma_1 a_2 \zeta(1+2/f). \tag{56}$$

We observe that $L^\dagger$ vanishes for $f = -2$. Therefore, *the configurations adjoint to the irrotational ellipsoids have zero angular momentum.*

(b) *The stable Maclaurin spheroids as the first members of Riemann sequences*

We shall now show that a stable Maclaurin spheroid, when viewed from a rotating frame of reference in which an even mode of oscillation belonging to the second harmonics is neutralized (in the sense considered in Chapter 5, § 36), is a limiting form of a Riemann ellipsoid.

As we have already remarked in Chapter 5, § 36, a Maclaurin spheroid rotating with the angular velocity Ω_{Mc}, when viewed from a frame of reference rotating with an angular velocity Ω, will appear to have stationary internal motions with the uniform vorticity

$$\zeta = 2(\Omega_{\mathrm{Mc}} - \Omega). \tag{57}$$

At the same time, it follows from equation (24), that when $a_1 = a_2$ (as it is the case for a Maclaurin spheroid)

$$Q_1 Q_2 = -\tfrac{1}{4}\zeta^2 = -(\Omega_{\mathrm{Mc}} - \Omega)^2. \tag{58}$$

But in order that the Maclaurin spheroid may be considered as a Riemann ellipsoid, equation (29) must be satisfied. Hence, the angular velocity Ω, which should be ascribed to it, if it is to be considered as a limiting form of a Riemann ellipsoid, should satisfy the equation

$$\Omega^2 + (\Omega_{\mathrm{Mc}} - \Omega)^2 = 2B_{11}, \tag{59}$$

or

$$\Omega = \tfrac{1}{2}[\Omega_{\mathrm{Mc}} \pm \sqrt{(4B_{11} - \Omega_{Mc}^2)}]. \tag{60}$$

The corresponding value of ζ is

$$\zeta = \Omega_{\mathrm{Mc}} \mp \sqrt{(4B_{11} - \Omega_{\mathrm{Mc}}^2)}. \tag{61}$$

Accordingly, a Maclaurin spheroid, rotating with an angular velocity Ω_{Mc}, may be considered as the first member of the Riemann sequences for

$$f = 2\frac{\Omega_{\mathrm{Mc}} - \sqrt{(4B_{11} - \Omega_{\mathrm{Mc}}^2)}}{\Omega_{\mathrm{Mc}} + \sqrt{(4B_{11} - \Omega_{\mathrm{Mc}}^2)}} \quad \text{and} \quad f^\dagger = 2\frac{\Omega_{\mathrm{Mc}} + \sqrt{(4B_{11} - \Omega_{\mathrm{Mc}}^2)}}{\Omega_{\mathrm{Mc}} - \sqrt{(4B_{11} - \Omega_{\mathrm{Mc}}^2)}}. \tag{62}$$

Alternatively, *these two Riemann sequences may be said to bifurcate from the Maclaurin sequence at* Ω_{Mc}.

Comparison of equation (60) with equation (106) of Chapter 5 shows that the bifurcation of the two Riemann sequences from the point Ω_{Mc} on the Maclaurin sequence is consistent with the fact that the two even modes of oscillation of the Maclaurin spheroid are neutralized when viewed from frames of reference rotating with the angular velocities given by equation (60).

There are several noteworthy features connected with this association of the Riemann sequences with the Maclaurin sequence.

(i) *Only the stable members of the Maclaurin sequence can be considered as limiting forms of Riemann ellipsoids.* The Maclaurin spheroids become unstable for $\Omega_{\mathrm{Mc}}^2 > 4B_{11}$; and when this is the case Ω and ζ given by equations (60) and (61) become complex; and this is incompatible with their meanings as real quantities.

(ii) The definitions of f and $f^\dagger$ are consistent with the relation (44), since it now requires

$$f^\dagger = 4/f; \tag{63}$$

and this relation is satisfied.

(iii) When $\Omega_{\mathrm{Mc}} = 0$,

$$f = f^\dagger = -2, \qquad \Omega^2 = B_{11}, \quad \text{and} \quad \zeta^2 = 4B_{11}, \tag{64}$$

where

$$B_{11} = 4/15 \tag{65}$$

is the value appropriate to a sphere. In other words, *the irrotational Riemann sequence for $f = -2$ bifurcates from the non-rotating sphere.* However, the sphere considered as the first member of the irrotational sequence is viewed from a frame rotating with the angular velocity $\Omega = \sqrt{B_{11}}$.

(iv) When $\Omega_{\mathrm{Mc}}^2 = 2B_{11}$,

$$f = 0, \qquad \Omega = \Omega_{\mathrm{Mc}}, \quad \text{and} \quad \zeta = 0$$

and

$$f^\dagger = \pm\infty, \quad \Omega^\dagger = 0, \quad \text{and} \quad \zeta = 2\Omega_{\mathrm{Mc}}. \tag{66}$$

The Maclaurin spheroid at this point is thus the first member of both the Jacobian and the Dedekind sequences. When considered as the first member of the Jacobian sequence, it is viewed from a frame in which it appears as at rest; but when considered as the first member of the Dedekind sequence, it is viewed from an inertial frame in which it is ascribed a vortical motion appropriate to its rotation. These results are in accord with what we already know about these two sequences.

(v) When $\Omega_{\mathrm{Mc}}^2 = 4B_{11}$,

$$f = f^\dagger = 2, \qquad \Omega = \tfrac{1}{2}\Omega_{\mathrm{Mc}}, \quad \text{and} \quad \zeta = \Omega_{\mathrm{Mc}}. \tag{67}$$

(vi) The Maclaurin spheroids, with angular velocities in the range $0 \leqslant \Omega_{\mathrm{Mc}}^2 < 2B_{11}$, are the first members of Riemann sequences for f in the range $-2 \leqslant f < 0$ and $-2 \geqslant f > -\infty$; and the Maclaurin spheroids, with angular velocities in the range $2B_{11} < \Omega_{\mathrm{Mc}}^2 \leqslant 4B_{11}$, are the first members of Riemann sequences for f in the range $0 < f \leqslant 2$ and $+\infty > f \geqslant 2$. Since every value of f has been accounted for, we conclude that there is no Riemann sequence which does not bifurcate from the Maclaurin sequence.

The variation of f and $f^\dagger$ along the Maclaurin sequence is shown in Fig. 12.

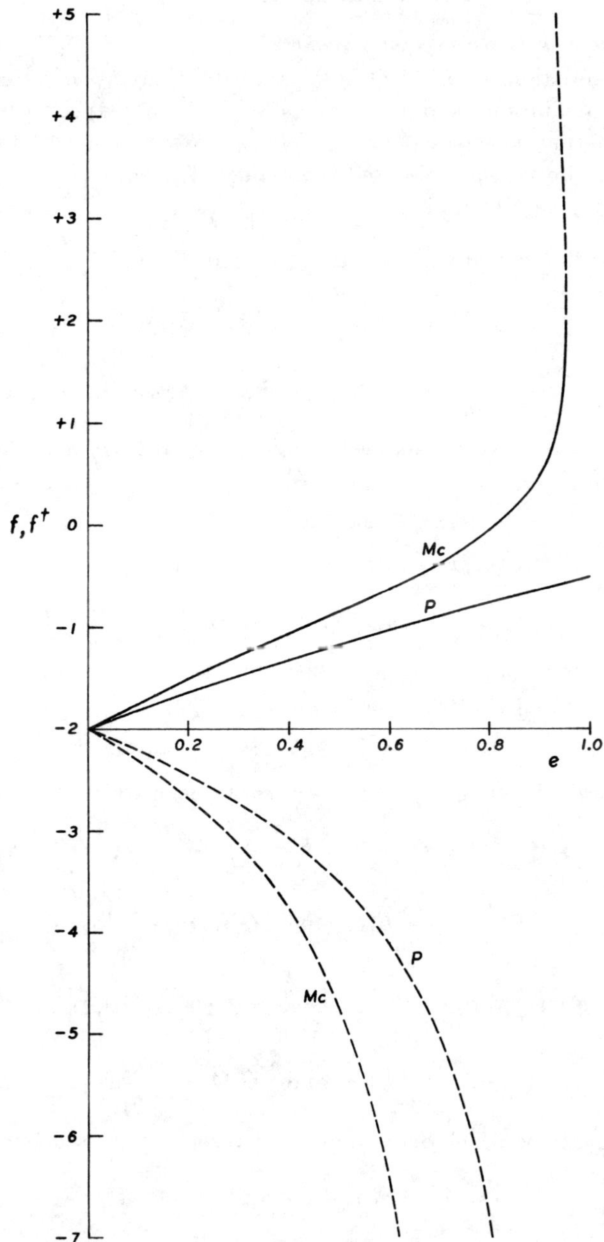

Fig. 12. The variation of f (*full-line curves*) and $f^\dagger$ (*dashed curves*) along the Maclaurin sequence (labeled Mc) and the sequence of the prolate spheroids (labeled P).

(c) *The bounding self-adjoint sequences*

From equations (40)–(43) it is apparent that any set of values a_1, a_2 ($< a_1$ by definition), and a_3 ($< a_1$, as we have proved) consistent with these equations provides a solution. The only restriction is that equation (43) for x allows real roots; and the reality of x requires

$$a_1 a_2 B_{12} \geqslant |a_1^2 a_2^2 A_{12} - a_3^2 A_3|. \tag{68}$$

This condition is equivalent to the two inequalities

$$\frac{a_3^2}{a_1 a_2} A_3 \leqslant a_1 a_2 A_{12} + B_{12} \qquad \text{(case i)} \tag{69}$$

and

$$\frac{a_3^2}{a_1 a_2} A_3 \geqslant a_1 a_2 A_{12} - B_{12} \qquad \text{(case ii).} \tag{70}$$

By using the relations among the index symbols, we find that these inequalities have the alternative forms

$$\frac{a_1 + a_2}{a_1 a_2} a_3^2 A_3 \leqslant a_2 A_2 + a_1 A_1 \qquad \text{(case i)} \tag{71}$$

and

$$\frac{a_1 - a_2}{a_1 a_2} a_3^2 A_3 \geqslant a_2 A_2 - a_1 A_1 \qquad \text{(case ii).} \tag{72}$$

When the inequalities (69) and (70) degenerate into equalities, equation (43) allows two equal roots:

$$x = -1 \text{ in case i} \quad \text{and} \quad x = +1 \text{ in case ii;} \tag{73}$$

and we are led to two *self-adjoint sequences* along which

$$x = -1, \qquad f = f^\dagger = -\frac{a_1^2 + a_2^2}{a_1 a_2}, \qquad \Omega^2 = -Q_1 Q_2 = B_{12},$$

$$Q_1 \Omega = +\frac{a_1}{a_2} B_{12}, \quad \text{and} \quad Q_2 \Omega = -\frac{a_2}{a_1} B_{12}; \tag{74}$$

and

$$x = +1, \qquad f = f^\dagger = +\frac{a_1^2 + a_2^2}{a_1 a_2}, \qquad \Omega^2 = -Q_1 Q_2 = B_{12},$$

$$Q_1 \Omega = -\frac{a_1}{a_2} B_{12}, \quad \text{and} \quad Q_2 \Omega = +\frac{a_2}{a_1} B_{12}. \tag{75}$$

The equations which determine the two self-adjoint sequences are

$$\frac{a_3^2}{a_1 a_2} A_3 = a_1 a_2 A_{12} + B_{12} \qquad (x = -1) \tag{76}$$

and

$$\frac{a_3^2}{a_1 a_2} A_3 = a_1 a_2 A_{12} - B_{12} \qquad (x = +1). \tag{77}$$

Notice that when $a_1 = a_2$, these equations become

$$\frac{a_3^2}{a_1^2} A_3 = a_1^2 A_{11} + B_{11} = A_1 \quad (x = -1) \tag{78}$$

and

$$\frac{a_3^2}{a_1^2} A_3 = a_1^2 A_{11} - B_{11} = A_1 - 2B_{11} \quad (x = +1); \tag{79}$$

and these equations imply that the self-adjoint sequences $x = -1$ and $x = +1$ intersect the Maclaurin sequence at the points where

$$\Omega^2_{\mathrm{Mc}} = 0 \quad \text{and} \quad \Omega^2_{\mathrm{Mc}} = 4B_{11}, \tag{80}$$

respectively. In other words, the sphere is the first member of the self-adjoint sequence $x = -1$, while the Maclaurin spheroid on the verge of dynamical instability is the first member of the self-adjoint sequence $x = +1$.

We conclude that *the domain of occupancy of the Riemann ellipsoids, in which the directions of Ω and ζ are parallel, is bounded by the stable part of the Maclaurin sequence and the self-adjoint sequences $x = -1$ and $x = +1$.*

In Table VI the properties of the Riemann ellipsoids along the two self-adjoint sequences are listed.

(d) *The irrotational sequence, $\zeta^{(0)} = 0$ and $f = -2$*

Among the various Riemann sequences, the one of greatest interest, besides the Jacobi and the Dedekind sequences, is the irrotational sequence for $f = -2$. As we have already remarked, this sequence bifurcates from the sphere; and the configurations adjoint to those along this sequence have zero angular momentum.

By setting $f = -2$ in equations (32), (33), and (35), we obtain

$$Q_1 = +\frac{2a_1^2\Omega}{a_1^2 + a_2^2}, \qquad Q_2 = -\frac{2a_2^2\Omega}{a_1^2 + a_2^2}, \tag{81}$$

$$\Omega^2 = \frac{2(a_1^2 + a_2^2)^2}{a_1^4 + 6a_1^2 a_2^2 + a_2^4} B_{12}, \tag{82}$$

and

$$(a_3^2 A_3 - a_1^2 a_2^2 A_{12})(a_1^4 + 6a_1^2 a_2^2 + a_2^4) = 4a_1^2 a_2^2 (a_1^2 + a_2^2) B_{12}. \tag{83}$$

An important feature of these irrotational ellipsoids is that they are all *prolate*, i.e.

$$a_1 \geqslant a_3 \geqslant a_2. \tag{84}$$

This result follows from inserting in equation (37) the values of Q_1 and Q_2 given in equations (81). We find

$$\frac{(a_1^2 - a_2^2)(3a_1^2 + a_2^2)}{(a_1^2 + a_2^2)^2} \Omega^2 = 2\frac{a_3^2 - a_2^2}{a_2^2} B_{23}. \tag{85}$$

TABLE VI

The properties of the Riemann ellipsoids along the self-adjoint sequences

a_2	a_3	Ω^2	f	Q_1	Q_2	a_3	Ω^2	f	Q_1	Q_2
			The sequence $x = -1$					The sequence $x = +1$		
0	0	0	$-\infty$		0	0	0	$+\infty$		0
0·08	0·10361	0·028055	$-12\cdot58000$	2·09371	$-0\cdot013400$	0·057817	0·016884	12·58000	$-1\cdot62423$	0·010395
0·12	0·16154	0·050998	$-8\cdot45333$	1·88190	$-0\cdot027099$	0·079299	0·027893	8·45333	$-1\cdot39178$	0·020042
0·16	0·21970	0·075026	$-6\cdot41000$	1·71193	$-0\cdot043825$	0·098178	0·038445	6·41000	$-1\cdot22546$	0·031372
0·20	0·27691	0·098533	$-5\cdot20000$	1·56950	$-0\cdot062780$	0·11513	0·048172	5·20000	$-1\cdot09740$	0·043896
0·24	0·33250	0·12067	$-4\cdot40667$	1·44740	$-0\cdot083370$	0·13056	0·056956	4·40667	$-0\cdot99440$	0·057277
0·28	0·38609	0·14103	$-3\cdot85143$	1·34121	$-0\cdot105151$	0·14476	0·064793	3·85143	$-0\cdot90909$	0·071272
0·32	0·43746	0·15946	$-3\cdot44500$	1·24787	$-0\cdot12778$	0·15794	0·071724	3·44500	$-0\cdot83692$	0·085700
0·36	0·48651	0·17594	$-3\cdot13778$	1·16514	$-0\cdot15100$	0·17025	0·077817	3·13778	$-0\cdot77488$	0·100424
0·40	0·53322	0·19055	$-2\cdot90000$	1·09129	$-0\cdot17461$	0·18181	0·083143	2·90000	$-0\cdot72086$	0·11534
0·44	0·57760	0·20339	$-2\cdot71273$	1·02498	$-0\cdot19844$	0·19270	0·087777	2·71273	$-0\cdot67335$	0·13036
0·48	0·61971	0·21461	$-2\cdot56333$	0·96513	$-0\cdot22237$	0·20302	0·091789	2·56333	$-0\cdot63118$	0·14542
0·52	0·65961	0·22434	$-2\cdot44308$	0·91086	$-0\cdot24630$	0·21281	0·095245	2·44308	$-0\cdot59350$	0·16048
0·56	0·69740	0·23272	$-2\cdot34571$	0·86145	$-0\cdot27015$	0·22214	0·098204	2·34571	$-0\cdot55960$	0·17549
0·60	0·73316	0·23988	$-2\cdot26667$	0·81629	$-0\cdot29386$	0·23105	0·100721	2·26667	$-0\cdot52894$	0·19042
0·64	0·76700	0·24594	$-2\cdot20250$	0·77488	$-0\cdot31739$	0·23958	0·102843	2·20250	$-0\cdot50108$	0·20524
0·68	0·79900	0·25101	$-2\cdot15059$	0·73678	$-0\cdot34069$	0·24775	0·104616	2·15059	$-0\cdot47565$	0·21994
0·72	0·82927	0·25520	$-2\cdot10889$	0·70163	$-0\cdot36373$	0·25560	0·106076	2·10889	$-0\cdot45235$	0·23450
0·76	0·85789	0·25861	$-2\cdot07579$	0·66913	$-0\cdot38649$	0·26316	0·107260	2·07579	$-0\cdot43093$	0·24890
0·80	0·88497	0·26131	$-2\cdot05000$	0·63898	$-0\cdot40895$	0·27044	0·108198	2·05000	$-0\cdot41117$	0·26315
0·84	0·91058	0·26338	$-2\cdot03048$	0·61096	$-0\cdot43109$	0·27746	0·108916	2·03048	$-0\cdot39289$	0·27722
0·88	0·93481	0·26490	$-2\cdot01636$	0·58487	$-0\cdot45292$	0·28425	0·109441	2·01636	$-0\cdot37593$	0·29112
0·92	0·95774	0·26592	$-2\cdot00696$	0·56052	$-0\cdot47442$	0·29081	0·109793	2·00696	$-0\cdot36016$	0·30484
0·96	0·97945	0·26654	$-2\cdot00167$	0·53778	$-0\cdot49562$	0·29714	0·109988	2·00167	$-0\cdot34546$	0·31838
1·00	1·00000	0·26667	$-2\cdot00000$	0·51640	$-0\cdot51640$	0·30333	0·110055	2·00000	$-0\cdot33175$	0·33175

Since the left-hand side of this equation is positive-definite, we must have

$$a_3 \geqslant a_2; \tag{86}$$

and combining this result with the inequality (39), valid for all Riemann sequences, we obtain (84).

In Table VII we list the properties of these irrotational ellipsoids. In some ways, the most striking feature of this irrotational sequence is the

TABLE VII

The properties of the irrotational ellipsoids†

ϕ	0°	30°	45°	55°	57°	59°	60°
θ		48°·828	54°·2405	59°·822	61°·190	62°·656	63°·4267
a_2/a_1	1·00000	0·86603	0·70711	0·57358	0·54464	0·51504	0·50000
a_3/a_1	1·00000	0·92647	0·81899	0·70608	0·67822	0·64829	0·63252
Ω^2	0·26667	0·26715	0·26875	0·26882	0·26811	0·26686	0·26598
$Q_1 \, Q_2$	−0·26667	−0·26170	−0·23889	−0·20029	−0·18921	−0·17688	−0·17023
Ω	0·51640	0·51687	0·51841	0·51848	0·51779	0·51659	0·51573
Q_1	0·51640	0·59070	0·69121	0·78026	0·79867	0·81657	0·82517
Q_2	−0·51640	−0·44303	−0·34561	−0·25670	−0·23691	−0·21661	−0·20629
$f^\dagger$	−2·00000	−2·04167	−2·25000	−2·68430	−2·83391	−3·01755	−3·12500
$\Omega^\dagger$	0·51640	0·51150	0·48870	0·44754	0·43499	0·42057	0·41259
$Q_1^\dagger$	0·51640	0·59682	0·73314	0·90394	0·95071	1·00301	1·03147
$Q_2^\dagger$	−0·51640	−0·44762	−0·36657	−0·29739	−0·28201	−0·26606	−0·25787

ϕ	70°	71°	72°	75°	78°	80°	85°
θ	72°·618	73°·6775	74°·756	78°·074	81°·4035	83°·535	87°·991
a_2/a_1	0·34202	0·32557	0·30902	0·25882	0·20791	0·17365	0·08716
a_3/a_1	0·44248	0·42025	0·39752	0·32685	0·25417	0·20603	0·09389
Ω^2	0·23768	0·23164	0·22475	0·19825	0·16228	0·13332	0·05285
$Q_1 \, Q_2$	−0·08914	−0·08029	−0·07153	−0·04666	−0·02578	−0·01515	−0·00158
Ω	0·48753	0·48129	0·47407	0·44525	0·40283	0·36513	0·22988
Q_1	0·87294	0·87031	0·86550	0·83450	0·77228	0·70888	0·45630
Q_2	−0·10211	−0·09225	−0·08265	−0·05591	−0·03338	−0·02138	−0·00347
$f^\dagger$	−5·33280	−5·77022	−6·28381	−8·49760	−12·58839	−17·59680	−66·82684
$\Omega^\dagger$	0·29856	0·28335	0·26745	0·21601	0·16057	0·12310	0·03977
$Q_1^\dagger$	1·42543	1·47832	1·53414	1·72030	1·93752	2·10269	2·63759
$Q_2^\dagger$	−0·16674	−0·15669	−0·14650	−0·11524	−0·08375	−0·06340	−0·02004

$$\dagger \; \cos\phi = a_2/a_1; \; \sin\theta = \sqrt{[(a_1^2-a_3^2)/(a_1^2-a_2^2)]}.$$

extreme sensitivity of the figure to very small changes in Ω^2: thus Ω^2 changes by less than 1 per cent from its value 4/15 (in the unit $\pi G\rho$) for figures comprised in the range $1 \geqslant a_2/a_1 \geqslant 0.5$ and $1 \geqslant a_3/a_1 \geqslant 0.6325$ (see Fig. 13).

(e) *Prolate spheroids among the S-type ellipsoids*

We have seen that the irrotational sequence is entirely *prolate* (i.e. $a_1 \geqslant a_3 \geqslant a_2$); but the Jacobian sequence, which is a Riemann sequence for $f = 0$, is entirely *oblate* (i.e. $a_1 \geqslant a_2 \geqslant a_3$). Moreover, since the Riemann sequences for $f \neq -2$ *all* have as their first members Maclaurin spheroids with $\Omega^2_{\mathrm{Mc}} > 0$, it is clear that all these sequences begin as oblate objects. Indeed, it follows from remark (vi) in § (b) that all Riemann ellipsoids for positive values of f must be oblate since they are represented by points below the Jacobian–Dedekind locus in the $(a_2/a_1, a_3/a_1)$-plane. However, Riemann sequences for at least some negative values of f must *end* as prolate objects. And the question arises

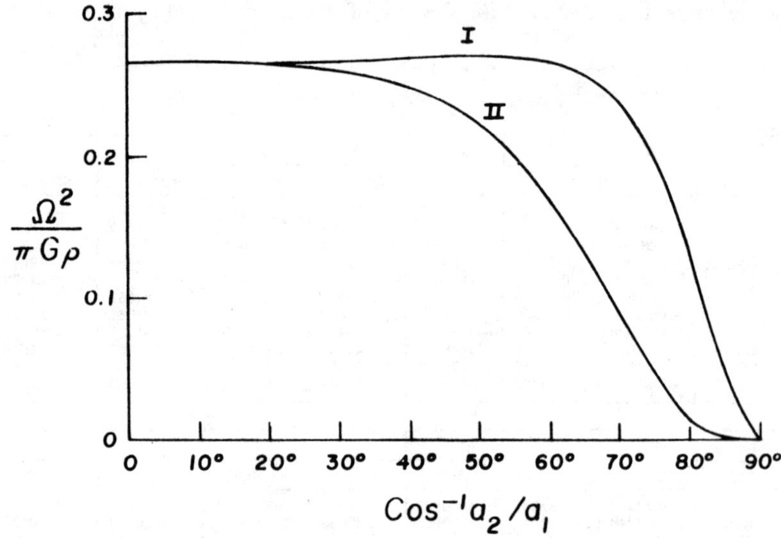

FIG. 13. The variation of $\Omega^2/\pi G\rho$ along the irrotational sequence (curve labeled I) and the adjoint configurations of zero angular momentum (curve labeled II).

as to which among the Riemann sequences for negative values of f include prolate ellipsoids and in particular prolate spheroids. This question can be answered with the aid of equation (43).

When $a_2 = a_3$,

$$a_3^2 A_3 - a_1^2 a_2^2 A_{12} = a_2^2 A_2 - a_1^2 a_2^2 A_{12} = a_2^2 B_{12}, \tag{87}$$

and equation (43) becomes

$$x^2 + 2\frac{a_1}{a_2}x + 1 = 0, \tag{88}$$

or

$$x = -\frac{a_1}{a_2} \pm \left(\frac{a_1^2}{a_2^2} - 1\right)^{\frac{1}{2}}. \tag{89}$$

In terms of the eccentricity

$$e = (1 - a_2^2/a_1^2)^{\frac{1}{2}}, \tag{90}$$

the solution for x is

$$x = -\left(\frac{1 \mp e}{1 \pm e}\right)^{\frac{1}{2}}. \tag{91}$$

The corresponding expression for f is

$$f = -\frac{2-e^2}{1\pm e}. \tag{92}$$

This solution for f allows all values less than $-1/2$ (see Fig. 12 in which the variation of f with e is also included). Therefore, along all Riemann sequences for $f < -1/2$, a prolate spheroid occurs; therefore, in the $(a_2/a_1, a_3/a_1)$-plane, the curves representing these Riemann sequences

TABLE VIII

The physical parameters of the prolate spheroids included among the Riemann ellipsoids

e	f	Ω^2	Q_1	Q_2	$f^\dagger$	$(\Omega^\dagger)^2$	$Q_1{}^\dagger$	$Q_2{}^\dagger$
0	$-2{\cdot}00000$	$0{\cdot}26667$	$0{\cdot}51640$	$-0{\cdot}51640$	$-2{\cdot}00000$	$0{\cdot}26667$	$0{\cdot}51640$	$-0{\cdot}51640$
0·05	$-1{\cdot}90238$	$0{\cdot}28018$	$0{\cdot}50411$	$-0{\cdot}50285$	$-2{\cdot}10263$	$0{\cdot}25350$	$0{\cdot}52999$	$-0{\cdot}52866$
0·10	$-1{\cdot}80909$	$0{\cdot}29289$	$0{\cdot}49199$	$-0{\cdot}48707$	$-2{\cdot}21111$	$0{\cdot}23963$	$0{\cdot}54391$	$-0{\cdot}53847$
0·15	$-1{\cdot}71957$	$0{\cdot}30566$	$0{\cdot}48075$	$-0{\cdot}46994$	$-2{\cdot}32647$	$0{\cdot}22592$	$0{\cdot}55919$	$-0{\cdot}54661$
0·20	$-1{\cdot}63333$	$0{\cdot}31810$	$0{\cdot}47000$	$-0{\cdot}45120$	$-2{\cdot}45000$	$0{\cdot}21207$	$0{\cdot}57563$	$-0{\cdot}55261$
0·25	$-1{\cdot}55000$	$0{\cdot}33016$	$0{\cdot}45968$	$-0{\cdot}43095$	$-2{\cdot}58333$	$0{\cdot}19810$	$0{\cdot}59344$	$-0{\cdot}55635$
0·30	$-1{\cdot}46923$	$0{\cdot}34177$	$0{\cdot}44970$	$-0{\cdot}40923$	$-2{\cdot}72857$	$0{\cdot}18403$	$0{\cdot}61284$	$-0{\cdot}55769$
0·35	$-1{\cdot}39074$	$0{\cdot}35285$	$0{\cdot}44001$	$-0{\cdot}38611$	$-2{\cdot}88846$	$0{\cdot}16989$	$0{\cdot}63412$	$-0{\cdot}55644$
0·40	$-1{\cdot}31429$	$0{\cdot}36324$	$0{\cdot}43050$	$-0{\cdot}36162$	$-3{\cdot}06667$	$0{\cdot}15568$	$0{\cdot}65760$	$-0{\cdot}55238$
0·45	$-1{\cdot}23966$	$0{\cdot}37281$	$0{\cdot}42109$	$-0{\cdot}33582$	$-3{\cdot}26818$	$0{\cdot}14141$	$0{\cdot}68372$	$-0{\cdot}54526$
0·50	$-1{\cdot}16667$	$0{\cdot}38131$	$0{\cdot}41107$	$-0{\cdot}30875$	$-3{\cdot}50000$	$0{\cdot}12710$	$0{\cdot}71303$	$-0{\cdot}53477$
0·55	$-1{\cdot}09516$	$0{\cdot}38846$	$0{\cdot}40211$	$-0{\cdot}28047$	$-3{\cdot}77222$	$0{\cdot}11278$	$0{\cdot}74628$	$-0{\cdot}52053$
0·60	$-1{\cdot}02500$	$0{\cdot}39386$	$0{\cdot}39224$	$-0{\cdot}25103$	$-4{\cdot}10000$	$0{\cdot}09846$	$0{\cdot}78448$	$-0{\cdot}50206$
0·65	$-0{\cdot}95606$	$0{\cdot}39692$	$0{\cdot}38183$	$-0{\cdot}22051$	$-4{\cdot}50714$	$0{\cdot}08420$	$0{\cdot}82904$	$-0{\cdot}47877$
0·70	$-0{\cdot}88824$	$0{\cdot}39682$	$0{\cdot}37055$	$-0{\cdot}18898$	$-5{\cdot}03333$	$0{\cdot}07003$	$0{\cdot}88209$	$-0{\cdot}44986$
0·75	$-0{\cdot}82143$	$0{\cdot}39225$	$0{\cdot}35789$	$-0{\cdot}15658$	$-5{\cdot}75000$	$0{\cdot}05604$	$0{\cdot}94688$	$-0{\cdot}41426$
0·80	$-0{\cdot}75556$	$0{\cdot}38112$	$0{\cdot}34297$	$-0{\cdot}12347$	$-6{\cdot}80000$	$0{\cdot}04235$	$1{\cdot}02892$	$-0{\cdot}37041$
0·82	$-0{\cdot}72945$	$0{\cdot}37409$	$0{\cdot}33606$	$-0{\cdot}11009$	$-7{\cdot}37556$	$0{\cdot}03700$	$1{\cdot}06860$	$-0{\cdot}35007$
0·84	$-0{\cdot}70348$	$0{\cdot}36508$	$0{\cdot}32838$	$-0{\cdot}09668$	$-8{\cdot}09000$	$0{\cdot}03175$	$1{\cdot}11360$	$-0{\cdot}32784$
0·86	$-0{\cdot}67763$	$0{\cdot}35364$	$0{\cdot}31972$	$-0{\cdot}08325$	$-9{\cdot}00286$	$0{\cdot}02662$	$1{\cdot}16536$	$-0{\cdot}30346$
0·88	$-0{\cdot}65191$	$0{\cdot}33908$	$0{\cdot}30974$	$-0{\cdot}06988$	$-10{\cdot}21333$	$0{\cdot}02164$	$1{\cdot}22597$	$-0{\cdot}27658$
0·90	$-0{\cdot}62632$	$0{\cdot}32045$	$0{\cdot}29794$	$-0{\cdot}05661$	$-11{\cdot}90000$	$0{\cdot}01687$	$1{\cdot}29868$	$-0{\cdot}24675$
0·92	$-0{\cdot}60083$	$0{\cdot}29628$	$0{\cdot}28350$	$-0{\cdot}04354$	$-14{\cdot}42000$	$0{\cdot}01234$	$1{\cdot}38884$	$-0{\cdot}21333$
0·94	$-0{\cdot}57546$	$0{\cdot}26412$	$0{\cdot}26491$	$-0{\cdot}03084$	$-18{\cdot}60667$	$0{\cdot}00817$	$1{\cdot}50636$	$-0{\cdot}17534$
0·96	$-0{\cdot}55020$	$0{\cdot}21944$	$0{\cdot}23900$	$-0{\cdot}01874$	$-26{\cdot}96000$	$0{\cdot}00448$	$1{\cdot}67300$	$-0{\cdot}13116$
0·98	$-0{\cdot}52505$	$0{\cdot}15146$	$0{\cdot}19656$	$-0{\cdot}00778$	$-51{\cdot}98000$	$0{\cdot}00153$	$1{\cdot}95570$	$-0{\cdot}07745$

intersect the line $a_2 = a_3$. We conclude then that *the Riemann sequences for $f < -1/2$ start as oblate objects and end as prolate objects, while the Riemann sequences for $f > -1/2$ consist entirely of oblate objects.*

Finally, we may note that by equations (41) and (45), the values of Ω^2 that are to be associated with a prolate spheroid of eccentricity e are

$$\Omega^2 = (1\pm e)B_{12}(e). \tag{93}$$

In Table VIII the properties of these prolate spheroids are listed; and in Fig. 14 the variation of Ω^2 along the sequence is illustrated.

The principal results of this section are exhibited in Fig. 15. In this figure the domain of occupancy of the S-type Riemann ellipsoids is exhibited; and some typical Riemann sequences, including the Jacobian–

Dedekind sequence, are delineated. (The figure also includes the results of the stability analysis of §§ 49 and 50.)

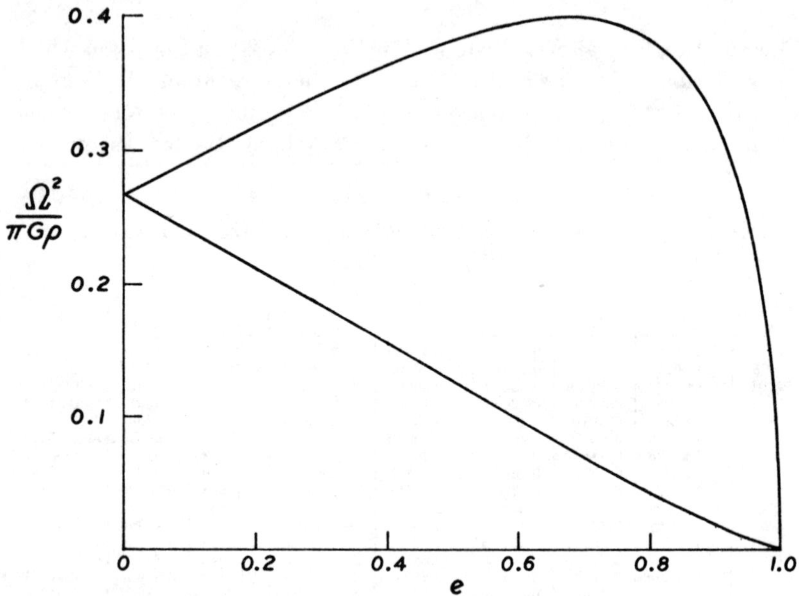

FIG. 14. The variation of $\Omega^2/\pi G\rho$ along the sequence of prolate spheroids for the two adjoint configurations associated with each figure.

49. The stability of the S-type ellipsoids with respect to second-harmonic oscillations

We turn now to the question of the stability of the S-type Riemann ellipsoids with respect to oscillations belonging to the second harmonics. For this purpose, we consider, as in Chapters 5 and 6, the linearized version of the second-order virial equation. Since internal motions, linear in the coordinates, are now present we must use equation (152) of Chapter 2 with the various terms in this equation having the values given in equations (143), (144), and (145) of the same chapter. If in considering these equations, we suppose that the Lagrangian displacement $\boldsymbol{\xi}(\mathbf{x}, t)$ is of the form

$$\boldsymbol{\xi}(\mathbf{x}, t) = e^{\lambda t}\boldsymbol{\xi}(\mathbf{x}), \tag{94}$$

where λ is a characteristic-value parameter to be determined, the virial equation gives

$$\lambda^2 V_{i;j} - 2\lambda Q_{jl} V_{i;l} - 2\lambda\Omega\epsilon_{il3} V_{l;j} - 2\Omega\epsilon_{il3}(Q_{lk} V_{j;k} - Q_{jk} V_{l;k}) + Q_{jl}^2 V_{i;l} + Q_{il}^2 V_{j;l}$$
$$= \Omega^2(V_{ij} - \delta_{i3} V_{3j}) + \delta\mathfrak{W}_{ij} + \delta_{ij}\delta\Pi. \tag{95}$$

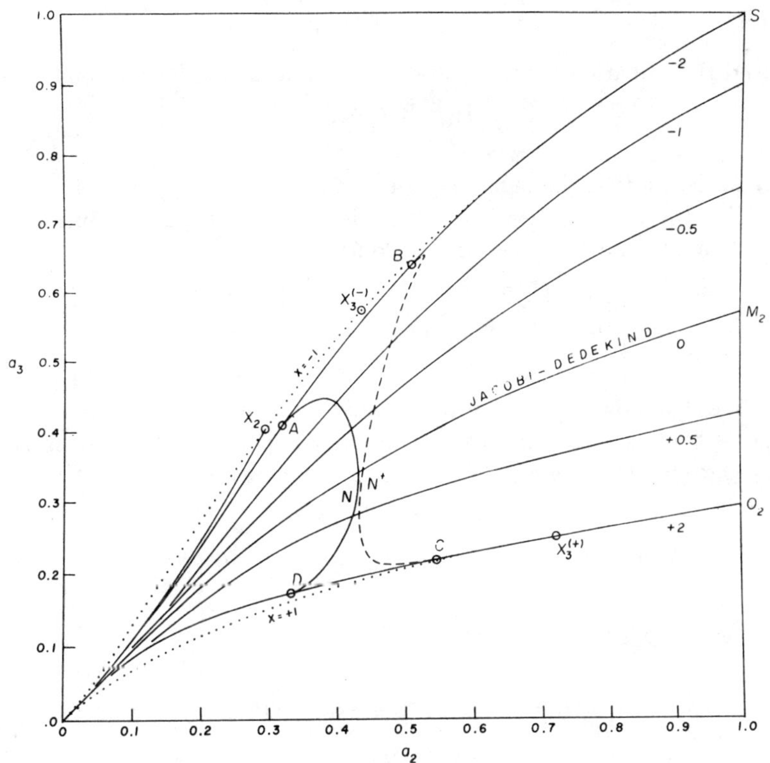

FIG. 15. The delineation of the Riemann sequences in the (a_2, a_3)-plane (a_1 has been set equal to 1). The stable part of the Maclaurin sequence is represented by the segment $O_2 S$ of the line $a_2 = 1$. At O_2 the Maclaurin spheroid becomes unstable by overstable oscillations and at M_2 the Jacobian and the Dedekind sequences bifurcate (labeled by 0).

The different Riemann sequences are labeled by the values of f to which they belong; these sequences are bounded by the two self-adjoint sequences (the dotted curves labeled $x = -1$ and $x = +1$). The sequences belonging to f in the range $-2 \leqslant f \leqslant +2$ form a non-intersecting family of continuous curves which join points on the line $O_2 S$ to the origin. The sequences belonging to $f < -2$ and $f > +2$ are represented by curves (not shown) which consist of two parts: a part which joins a point on the line SM_2 (or $M_2 O_2$) to a point on the self-adjoint sequence for $x = -1$ (or $x = +1$) and a part which joins the point on the self-adjoint sequence to the origin. Along the self-adjoint sequence $x = -1$, instability by a mode of oscillation belonging to the second harmonics sets in at the point indicated by X_2, and the locus of points at which instability by this mode sets in is the curve which joins X_2 to the origin. The curve labeled AND is the locus of neutral points, belonging to the third harmonics, along the Riemann sequences for $-2 \leqslant f \leqslant +2$; and the curve labeled $BN^\dagger C$ is the corresponding locus for configurations adjoint to the Riemann ellipsoids represented in the domain included between the same sequences $f = -2$ and $f = +2$. The continuation of the loci AND and $BN^\dagger C$ into the domains included between the sequences $x = -1$ and $f = -2$ (and, similarly, between the sequences $x = +1$ and $f = +2$) are represented by curves (not shown) joining the points A and B to $X_3^{(-)}$ on the sequence $x = -1$ (and, similarly, by curves joining the points D and C to the point $X_3^{(+)}$ on the sequence $x = +1$); $X_3^{(-)}$ and $X_3^{(+)}$ are the neutral points, belonging to the third harmonics, along the self-adjoint sequences $x = -1$ and $x = +1$, respectively.

And this equation must be supplemented, as usual, by the condition

$$\frac{V_{11}}{a_1^2}+\frac{V_{22}}{a_2^2}+\frac{V_{33}}{a_3^2} = 0 \tag{96}$$

required by the solenoidal character of $\boldsymbol{\xi}$.

For the S-type Riemann ellipsoids we are presently considering, matrices $\mathbf{Q}$ and $\mathbf{Q}^2$ have the simple forms

$$\mathbf{Q} = \begin{vmatrix} 0 & Q_1 & 0 \\ Q_2 & 0 & 0 \\ 0 & 0 & 0 \end{vmatrix} \quad \text{and} \quad \mathbf{Q}^2 = \begin{vmatrix} Q_1 Q_2 & 0 & 0 \\ 0 & Q_1 Q_2 & 0 \\ 0 & 0 & 0 \end{vmatrix}. \tag{97}$$

We shall now write down the explicit forms which the different components of equation (95) take in view of the special form of the matrices $\mathbf{Q}$ and $\mathbf{Q}^2$. The five equations even in the index 3 are

$$\tfrac{1}{2}\lambda^2 V_{33} = \delta\mathfrak{W}_{33}+\delta\Pi, \tag{98}$$

$$(\tfrac{1}{2}\lambda^2+Q_1 Q_2-\Omega^2)V_{11}-2\lambda Q_1 V_{1;2}-2\lambda\Omega V_{2;1}-\Omega(Q_2 V_{11}-Q_1 V_{22})$$
$$= \delta\mathfrak{W}_{11}+\delta\Pi, \tag{99}$$

$$(\tfrac{1}{2}\lambda^2+Q_1 Q_2-\Omega^2)V_{22}-2\lambda Q_2 V_{2;1}+2\lambda\Omega V_{1;2}+\Omega(Q_1 V_{22}-Q_2 V_{11})$$
$$= \delta\mathfrak{W}_{22}+\delta\Pi, \tag{100}$$

$$\lambda^2 V_{1;2}-\lambda Q_2 V_{11}+Q_1 Q_2 V_{12}-\lambda\Omega V_{22} = \delta\mathfrak{W}_{12}+\Omega^2 V_{12} = -(2B_{12}-\Omega^2)V_{12}, \tag{101}$$

$$\lambda^2 V_{2;1}-\lambda Q_1 V_{22}+Q_1 Q_2 V_{12}+\lambda\Omega V_{11} = \delta\mathfrak{W}_{12}+\Omega^2 V_{12} = -(2B_{12}-\Omega^2)V_{12}, \tag{102}$$

where in equations (101) and (102) we have substituted for $\delta\mathfrak{W}_{12}$ its known expression. Similarly, the four equations odd in the index 3 are

$$\lambda^2 V_{1;3}-2\lambda\Omega V_{2;3}+Q_1 Q_2 V_{3;1}-2\Omega Q_2 V_{3;1} = \delta\mathfrak{W}_{13}+\Omega^2 V_{13} = -(2B_{13}-\Omega^2)V_{13}, \tag{103}$$

$$\lambda^2 V_{2;3}+2\lambda\Omega V_{1;3}+Q_1 Q_2 V_{3;2}+2\Omega Q_1 V_{3;2} = \delta\mathfrak{W}_{23}+\Omega^2 V_{23} = -(2B_{23}-\Omega^2)V_{23}, \tag{104}$$

$$\lambda^2 V_{3;1}-2\lambda Q_1 V_{3;2}+Q_1 Q_2 V_{3;1} = \delta\mathfrak{W}_{31} = -2B_{13}V_{13}, \tag{105}$$

$$\lambda^2 V_{3;2}-2\lambda Q_2 V_{3;1}+Q_1 Q_2 V_{3;2} = \delta\mathfrak{W}_{32} = -2B_{23}V_{23}. \tag{106}$$

(a) *The characteristic equation for the even modes of oscillation*

Considering first equations (98)–(102) governing the even modes, we observe that in view of the relation (29), equations (101) and (102) become

$$\lambda^2 V_{1;2}-\lambda Q_2 V_{11}-\lambda\Omega V_{22} = 0 \tag{107}$$

and

$$\lambda^2 V_{2\,1}-\lambda Q_1 V_{22}+\lambda\Omega V_{11} = 0. \tag{108}$$

Excluding the possibility that λ may be zero—a possibility to which we shall return presently—we may conclude from equations (107) and (108) that

$$\lambda V_{1;2} = Q_2 V_{11} + \Omega V_{22} \tag{109}$$

and

$$\lambda V_{2;1} = Q_1 V_{22} - \Omega V_{11}. \tag{110}$$

Eliminating $V_{1;2}$ and $V_{2;1}$ from equations (99) and (100), we obtain

$$(\tfrac{1}{2}\lambda^2 + \Omega^2 - Q_1 Q_2 - \Omega Q_2)V_{11} - 3\Omega Q_1 V_{22} = \delta\mathfrak{W}_{11} + \delta\Pi \tag{111}$$

and

$$(\tfrac{1}{2}\lambda^2 + \Omega^2 - Q_1 Q_2 + \Omega Q_1)V_{22} + 3\Omega Q_2 V_{11} = \delta\mathfrak{W}_{22} + \delta\Pi. \tag{112}$$

Next, eliminating $\delta\Pi$ from equations (111) and (112) with the aid of equation (98), we obtain the pair of equations

$$(\tfrac{1}{2}\lambda^2 + 2B_{12} - \Omega Q_2)V_{11} - 3\Omega Q_1 V_{22} - \tfrac{1}{2}\lambda^2 V_{33} = \delta\mathfrak{W}_{11} - \delta\mathfrak{W}_{33}$$
$$= -(3B_{11} - B_{13})V_{11} + (B_{23} - B_{12})V_{22} + (3B_{33} - B_{13})V_{33} \tag{113}$$

and

$$(\tfrac{1}{2}\lambda^2 + 2B_{12} + \Omega Q_1)V_{22} + 3\Omega Q_2 V_{11} - \tfrac{1}{2}\lambda^2 V_{33} = \delta\mathfrak{W}_{22} - \delta\mathfrak{W}_{33}$$
$$= -(3B_{22} - B_{23})V_{22} + (B_{13} - B_{12})V_{11} + (3B_{33} - B_{23})V_{33}, \tag{114}$$

where we have made use of the relation (29) and further substituted for $\delta\mathfrak{W}_{11} - \delta\mathfrak{W}_{33}$ and $\delta\mathfrak{W}_{22} - \delta\mathfrak{W}_{33}$ in accordance with equation (149) of Chapter 3.

Equations (113) and (114) together with equation (96) lead to the following characteristic equation for λ^2:

$$\begin{vmatrix} \tfrac{1}{2}\lambda^2 + 2B_{12} - \Omega Q_2 + 3B_{11} - B_{13} & B_{12} - B_{23} - 3\Omega Q_1 & -\tfrac{1}{2}\lambda^2 - 3B_{33} + B_{13} \\ B_{12} - B_{13} + 3\Omega Q_2 & \tfrac{1}{2}\lambda^2 + 2B_{12} + \Omega Q_1 + 3B_{22} - B_{23} & -\tfrac{1}{2}\lambda^2 - 3B_{33} + B_{23} \\ \dfrac{1}{a_1^2} & \dfrac{1}{a_2^2} & \dfrac{1}{a_3^2} \end{vmatrix} = 0. \tag{115}$$

A somewhat simpler form of equation (115) is

$$\begin{vmatrix} \tfrac{1}{2}\lambda^2 + 2B_{12} - \Omega Q_2 + 3B_{11} - B_{13} & B_{12} - B_{23} - 3\Omega Q_1 & 3(B_{12} + B_{11} - B_{33}) - B_{23} - \Omega(Q_2 + 3Q_1) \\ B_{12} - B_{13} + 3\Omega Q_2 & \tfrac{1}{2}\lambda^2 + 2B_{12} + \Omega Q_1 + 3B_{22} - B_{23} & 3(B_{12} + B_{22} - B_{33}) - B_{13} + \Omega(Q_1 + 3Q_2) \\ \dfrac{1}{a_1^2} & \dfrac{1}{a_2^2} & \dfrac{1}{a_1^2} + \dfrac{1}{a_2^2} + \dfrac{1}{a_3^2} \end{vmatrix} = 0. \tag{116}$$

Returning to equations (107) and (108) and the possibility of a non-trivial root $\lambda^2 = 0$, we observe that we do indeed have such a root belonging to a proper solution associated with the possibility

$$V_{11} = V_{22} = V_{33} = 0 \quad \text{and} \quad V_{12} \neq 0. \tag{117}$$

In other words, all Riemann ellipsoids do allow a non-trivial neutral mode. We have verified this result in the special case of the Jacobi ellipsoid in Chapter 6 (see equation (72)).

(b) *The characteristic equation for the odd modes of oscillation*

Turning next to the equations (103)–(106) governing the odd modes of oscillation, we first rewrite them in the forms

$$(\lambda^2+2B_{13}-\Omega^2)V_{1;3}+(2B_{13}-\Omega^2+Q_1Q_2-2\Omega Q_2)V_{3;1}-2\lambda\Omega V_{2;3} = 0, \tag{118}$$

$$(\lambda^2+2B_{23}-\Omega^2)V_{2;3}+(2B_{23}-\Omega^2+Q_1Q_2+2\Omega Q_1)V_{3;2}+2\lambda\Omega V_{1;3} = 0, \tag{119}$$

$$(\lambda^2+2B_{13}+Q_1Q_2)V_{3;1}+2B_{13}V_{1;3}-2\lambda Q_1 V_{3;2} = 0, \tag{120}$$

$$(\lambda^2+2B_{23}+Q_1Q_2)V_{3;2}+2B_{23}V_{2;3}-2\lambda Q_2 V_{3;1} = 0. \tag{121}$$

On the other hand, according to equations (36) and (37)

$$2B_{13}-(\Omega^2-Q_1Q_2+2\Omega Q_2) = 2\frac{a_3^2}{a_1^2}B_{13} \tag{122}$$

and

$$2B_{23}-(\Omega^2-Q_1Q_2-2\Omega Q_1) = 2\frac{a_3^2}{a_2^2}B_{23}. \tag{123}$$

With this simplification of the coefficients of $V_{3;1}$ and $V_{3;2}$ in equations (118) and (119), equations (118)–(121) lead to the following characteristic equation:

$$\begin{vmatrix} \lambda^2+2B_{13}-\Omega^2 & 2a_3^2 B_{13}/a_1^2 & -2\lambda\Omega & 0 \\ 2B_{13} & \lambda^2+2B_{13}+Q_1Q_2 & 0 & -2\lambda Q_1 \\ 2\lambda\Omega & 0 & \lambda^2+2B_{23}-\Omega^2 & 2a_3^2 B_{23}/a_2^2 \\ 0 & -2\lambda Q_2 & 2B_{23} & \lambda^2+2B_{23}+Q_1Q_2 \end{vmatrix} = 0. \tag{124}$$

By a judicious manipulation of this determinant, it can be shown that equation (124) allows the roots

$$\lambda^2 = -\Omega^2 \quad \text{and} \quad \lambda^2 = Q_1 Q_2; \tag{125}$$

and factoring out $(\lambda^2+\Omega^2)(\lambda^2-Q_1Q_2)$, we find that the characteristic equation reduces to

$$\lambda^4+(4B_{13}+4B_{23}+2B_{12})\lambda^2+(4B_{13}-\Omega Q_1)(4B_{23}+\Omega Q_2) = 0. \tag{126}$$

We observe that the characteristic equations (116) and (126), apart from constants that depend only on the semi-axes of the ellipsoid, involve Ω, Q_1, and Q_2 only in the combinations ΩQ_1 and ΩQ_2; and these combinations are the same for a configuration and its adjoint (as is manifest from equations (51)). Hence the roots of equations (116) and (126) are the same for adjoint configurations; and, moreover, the roots $-\Omega^2$ and $Q_1 Q_2$, which the characteristic equation (124) also allows, are simply interchanged when we pass from a configuration to its adjoint. Thus, *the characteristic frequencies of oscillation, belonging to the second harmonics, of adjoint configurations are the same.* We could have deduced this equality of the characteristic frequencies from Dedekind's general theorem, as we did in the special case of the Jacobi and the Dedekind ellipsoids in Chapter 6, § 45. The present analysis provides a direct verification for the S-type ellipsoids, in general.

An additional fact of some interest is that for an irrotational ellipsoid one of the roots of equation (126) is, again, $Q_1 Q_2$, so that in this case, the four characteristic roots belonging to the odd modes of oscillation are

$$\lambda^2 - -\Omega^2, \quad \lambda^2 = Q_1 Q_2 \text{ (double root)}, \quad \text{and} \quad \lambda^2 = -(4B_{13}+4B_{23}+\Omega^2).$$

$$(127)$$

(c) The locus of marginal stability

From an examination of equation (116) it appears that the characteristic frequencies determined by this equation are always real. Therefore, instability via an even mode of oscillation does not arise. This fact is in accord with the circumstance that the Riemann sequences bifurcate from the Maclaurin sequence by the neutralization of an even mode. The occurrence of $\lambda^2 = 0$ as a non-trivial root, which was noted in § (a), is related to this same circumstance.

Turning next to equation (126) governing the odd modes, we observe that instability by one of these modes can arise. Indeed, a locus of marginally stable configurations will be defined by the equation

$$(4B_{13}-\Omega Q_1)(4B_{23}+\Omega Q_2) = 0.$$

$$(128)$$

We shall show presently that

$$4B_{23}+\Omega Q_2 > 0,$$

$$(129)$$

so that the locus of marginal stability in the $(a_2/a_1, a_3/a_1)$-plane will be determined by the equation

$$4B_{13} = \Omega Q_1;$$

$$(130)$$

or, making use of the expression for ΩQ_1 given in equations (51), we have

$$4a_2^2 B_{13} = a_3^2 A_3 - a_1^2 a_2^2 A_{12}.$$

$$(131)$$

An alternative form of this equation which we shall find useful is

$$(4a_2^2+a_1^2-a_3^2)B_{13}-a_1^2 B_{12} = 0. \tag{132}$$

It remains to prove the inequality (129). In a manner similar to the reduction of equation (130) to equation (132), we find

$$4B_{23}+\Omega Q_2 = \frac{1}{a_1^2}[(4a_1^2+a_2^2-a_3^2)B_{23}-a_2^2 B_{12}], \tag{133}$$

or, substituting for the index symbols their integral expressions, we obtain

$$4B_{23}+\Omega Q_2 = \frac{a_2 a_3}{a_1} \int_0^\infty \left(\frac{4a_1^2+a_2^2-a_3^2}{a_3^2+u}-\frac{a_2^2}{a_1^2+u}\right)\frac{u\,du}{(a_2^2+u)\Delta}$$

$$= \frac{a_2 a_3}{a_1} \int_0^\infty [(a_1^2+a_2^2)(a_1^2-a_3^2)+3a_1^4+(4a_1^2-a_3^2)u]\frac{u\,du}{\Delta^3}$$

$$> 0, \tag{134}$$

since the integrand, in view of $a_1 \geqslant a_3$, is positive-definite.

Returning to the locus of marginal stability given by equation (131), we shall now show that it intersects the self-adjoint sequence $x = -1$ defined by equation (76). At the intersection—if one is assumed to exist—both equations (76) and (131) must be satisfied simultaneously; and this requirement gives

$$4a_2 B_{13} = a_1 B_{12}. \tag{135}$$

Inserting this last relation in equation (132) (which is another form of the same locus (131)) we obtain

$$4a_2^2-4a_1 a_2+a_1^2-a_3^2 = 0, \tag{136}$$

or
$$[2a_2-(a_1+a_3)][2a_2-(a_1-a_3)] = 0. \tag{137}$$

An intersection can therefore occur where

$$2a_2 = a_1+a_3 \quad \text{or} \quad 2a_2 = a_1-a_3. \tag{138}$$

Actually, the intersection occurs where

$$2a_2 = a_1-a_3. \tag{139}$$

(We shall see in §§ 51 (b) and (d), that the two loci (138) define self-adjoint sequences bounding the ellipsoids of types I and III if the roles of a_1 and a_2 are interchanged in the convention with respect to the coordinate axes adopted in § 48.)

In Table IX the locus of marginally stable configurations defined by equation (132) is specified numerically. The locus is further delineated in Fig. 15.

The foregoing analysis shows that instability via any of the modes of oscillation belonging to the second harmonics does not arise along Riemann sequences for $f \geqslant -2$. (The stability of the irrotational ellipsoids, along the sequence $f = -2$, for the odd modes of oscillation is

TABLE IX

The properties of the marginally stable ellipsoids along the curve of bifurcation adjoining the domains of occupancy of the ellipsoids of types S and III†

$(a_2/a_1)_{III}$	$(a_3/a_1)_{III}$	$(a_2/a_1)_S$	$(a_3/a_1)_S$	Ω	Q_1	Q_2
3·3746	1·3746	0·29633	0·40734			
3·4	1·3718	0·29412	0·40348	0·40404	0·106768	−1·2342
				0·36301	0·118835	−1·3737
3·6	1·3511	0·27778	0·37530	0·42480	0·085852	−1·1126
				0·30907	0·118001	−1·5293
4·0	1·3143	0·25000	0·32857	0·42439	0·062645	−1·0023
				0·25058	0·106098	−1·6976
4·4	1·2828	0·22727	0·29154	0·41323	0·048004	−0·9294
				0·21122	0·093916	−1·8182
5·0	1·2436	0·20000	0·24873	0·39173	0·033872	−0·8468
				0·16936	0·078346	1·0586
5·6	1·2121	0·17857	0·21645	0·36969	0·024927	−0·7817
				0·13959	0·066016	−2·0703
6·4	1·1789	0·15625	0·18420	0·34234	0·017372	−0·7115
				0·11118	0·053490	−2·1910
7·2	1·1530	0·13889	0·16014	0·31806	0·012622	−0·6543
				0·09088	0·044175	−2·2901
8·0	1·1325	0·12500	0·14156	0·29676	0·009477	−0·6065
				0·07582	0·037095	−2·3741
9·0	1·1123	0·11111	0·12359	0·27376	0·006870	−0·5564
				0·06183	0·030418	−2·4638
10·0	1·0966	0·10000	0·10966	0·25410	0·005146	−0·5146
				0·05146	0·025410	−2·5410
11·0	1·0841	0·09091	0·09855	0·23715	0·003960	−0·4791
				0·04356	0·021559	−2·6087
12·0	1·0739	0·08333	0·08949	0·22241	0·003115	−0·4485
				0·03738	0·018534	−2·6690
13·0	1·0655	0·07692	0·08196	0·20948	0·002496	−0·4219
				0·03245	0·016114	−2·7233
14·0	1·0586	0·07143	0·07561	0·19805	0·002033	−0·3984
				0·02846	0·014147	−2·7727

† With the convention regarding the axes adopted in Fig. 16, the entries in the first two columns refer to the ellipsoids of type III, while the entries in the next two columns refer to the ellipsoids of type S. The two entries in the last three columns, for each value of $(a_2/a_1,\ a_3/a_1)$, refer to the adjoint ellipsoids.

manifest from the explicit expressions (127) for the characteristic roots λ^2.) But instability via an odd mode of oscillation does arise along all sequences for $f < -2$; and the domain of instability is confined to a small segment (see Fig. 15).

50. The loci of neutral points belonging to the third harmonics in the domain of occupancy of the S-type ellipsoids

We have seen in § 49 that the Riemann ellipsoids for $f \geqslant -2$ are stable with respect to all modes of oscillation belonging to the second harmonics. In analogy with the Jacobian and the Dedekind sequences, we may expect that instability along the sequences for $f \geqslant -2$ will be manifested, first, by a mode of oscillation belonging to the third harmonics. And, moreover, the resulting points of onset of instability will not be the same for adjoint configurations.

In this section, we shall show how the integral properties provided by the third-order virial equations enable us to determine the loci of points, in the domain of occupancy of these ellipsoids, that separate the regions of stability from the regions of instability.

Under stationary conditions, the third-order virial theorem gives (cf. Chapter 2, equation (74))

$$2(\mathfrak{T}_{ij;k}+\mathfrak{T}_{ik;j})+\Omega^2(I_{ijk}-\delta_{i3}I_{3jk})+\mathfrak{W}_{ik;j}+\mathfrak{W}_{ij;k}+$$
$$+2\Omega\epsilon_{il3}\int_V \rho u_l x_j x_k \, d\mathbf{x} = -\delta_{ij}\Pi_k-\delta_{ik}\Pi_j, \qquad (140)$$

where we have specialized the equation for the case when Ω, as at present, is in the x_3-direction.

At a neutral point, belonging to the third harmonics, the first variation of the equations resulting from equation (140) must be satisfied non-trivially. The relevant variational equation is

$$2(\delta\mathfrak{T}_{ij;k}+\delta\mathfrak{T}_{ik;j})+\Omega^2(V_{ijk}-\delta_{i3}V_{3jk})+\delta\mathfrak{W}_{ik;j}+\delta\mathfrak{W}_{ij;k}+$$
$$+2\Omega\epsilon_{il3}\delta\int_V \rho u_l x_j x_k \, d\mathbf{x} = -\delta_{ij}\delta\Pi_k-\delta_{ik}\delta\Pi_j. \qquad (141)$$

Since the internal motion that is now present is linear in the coordinates, the first variations of $\mathfrak{T}_{ij;k}$ and of the Coriolis term that occur in equation (141) are expressible, even as the variations of $\mathfrak{W}_{ij;k}$, as linear combinations of the $V_{ij;k}$'s. Thus, equations (148) and (149) of Chapter 2, under the present quasi-static conditions, give

$$2\delta\mathfrak{T}_{ij;k} = Q_{il}Q_{jm}V_{k;lm}-Q_{jl}(Q_{km}V_{i;lm}+Q_{lm}V_{i;km})-Q_{il}(Q_{km}V_{j;lm}+Q_{lm}V_{j;km})$$
$$(142)$$

and

$$\delta\int_V \rho u_l x_j x_k \, d\mathbf{x} = Q_{lm}(V_{j;km}+V_{k;jm})-Q_{jm}V_{l;km}-Q_{km}V_{l;jm}. \qquad (143)$$

Equations (142) and (143) together with the known expressions for the $\delta\mathfrak{W}_{ij;k}$'s given in Chapter 3 (equations (152)–(155)) enable us to write the right-hand side of equation (141) as a linear combination of the $V_{ij;k}$'s.

For our present purposes of isolating the neutral point, it will suffice to consider the five equations which are odd in the index 1 and even in the indices 2 and 3. These equations are

$$4\delta\mathfrak{T}_{11;1}+2\delta\mathfrak{W}_{11;1}+\Omega^2 V_{111}+4\Omega(Q_2 V_{1;11}-Q_1 V_{2;12}) = -2\delta\Pi_1, \quad (144)$$

$$2\delta\mathfrak{T}_{12;2}+2\delta\mathfrak{T}_{22;1}+\delta\mathfrak{W}_{12;2}+\delta\mathfrak{W}_{22;1}+\Omega^2 V_{122}+2\Omega(Q_2 V_{1;11}-Q_1 V_{2;12}) = -\delta\Pi_1, \quad (145)$$

$$2\delta\mathfrak{T}_{13;3}+2\delta\mathfrak{T}_{33;1}+\delta\mathfrak{W}_{13;3}+\delta\mathfrak{W}_{33;1} = -\delta\Pi_1, \quad (146)$$

$$4\delta\mathfrak{T}_{12;2}+2\delta\mathfrak{W}_{12;2}+\Omega^2 V_{122} = 0, \quad (147)$$

$$4\delta\mathfrak{T}_{13;3}+2\delta\mathfrak{W}_{13;3}+\Omega^2 V_{133}+4\Omega Q_2 V_{3;31} = 0, \quad (148)$$

where in evaluating the Coriolis term in accordance with equation (143), we have used the present form of **Q** given in equation (97). Eliminating $\delta\Pi_1$ appropriately from the foregoing equations, we obtain, in addition to equations (147) and (148), the pair of equations

$$2\delta R_{122}+\delta S_{122}+\Omega^2(V_{111}-3V_{122}) = 0 \quad (149)$$

and

$$2\delta R_{133}+\delta S_{133}+\Omega^2(V_{111}-V_{133})+4\Omega Q_2 V_{1;11}-4\Omega Q_1 V_{2;12}-4\Omega Q_2 V_{3;13} = 0, \quad (150)$$

where
$$\delta S_{ijj} = -4\delta\mathfrak{W}_{ij;j}-2\delta\mathfrak{W}_{jj;i}+2\delta\mathfrak{W}_{ii;i} \quad (151)$$

and
$$\delta R_{ijj} = -4\delta\mathfrak{T}_{ij;j}-2\delta\mathfrak{T}_{jj;i}+2\delta\mathfrak{T}_{ii;i} \quad (152)$$

(no summation over the repeated indices in equations (151) and (152)).

Expressions for $\delta\mathfrak{W}_{12;2}$, $\delta\mathfrak{W}_{13;3}$, δS_{122}, and δS_{133}, as linear combinations of V_{111}, V_{122}, and V_{133}, can be obtained from equations (153) and (157) of Chapter 3; and the expressions for $\delta\mathfrak{T}_{12;2}$, $\delta\mathfrak{T}_{13;3}$, δR_{122}, and δR_{133}, appropriate in the present context, have been given in equations (149)–(152) of Chapter 6. Therefore, equations (147)–(150) together with the solenoidal condition,

$$\frac{3}{a_1^2} V_{1;11}+\frac{1}{a_2^2}(V_{1;22}+2V_{2;12})+\frac{1}{a_3^2}(V_{1;33}+2V_{3;13}) = 0, \quad (153)$$

provide a system of five homogeneous linear equations for the five quantities $V_{1;11}$, $V_{1;22}$, $V_{1;33}$, $V_{2;12}$, and $V_{3;13}$. The condition for the occurrence of a neutral point is that the determinant of the system vanishes. The neutral points along the Jacobian and the Dedekind sequences have already been determined by making use of this condition (Chapter 5, §§ 40 and 45). The neutral points along other Riemann sequences have been determined similarly; and by considering the adjoint configurations along the same sequences, the places where the neutral points for these configurations occur have also been determined. In the same way, the two neutral points along the sequence of the prolate spheroids, as well as the two self-adjoint sequences, were also determined. The results of these calculations are summarized in Table X. And in Fig. 15 the loci separating the regions of stability from the regions of instability, in the domain of occupancy of these Riemann ellipsoids, are drawn.

51. The Riemann ellipsoids in which the directions of $\boldsymbol{\Omega}$ and $\boldsymbol{\zeta}$ are not parallel; the ellipsoids of types I, II, and III

By Riemann's theorem, if $\boldsymbol{\Omega}$ and $\boldsymbol{\zeta}$ are not parallel to one of the principal axes of the ellipsoid, they must lie in a principal plane of the ellipsoid. We shall suppose then that Ω_2 and Ω_3 are different from zero while Ω_1 and ζ_1 are zero. *We thus distinguish the a_1-axis as the one along which neither $\boldsymbol{\Omega}$ nor $\boldsymbol{\zeta}$ has a component.* Under these circumstances, the internal motion $\mathbf{u}$ has the components (cf. equation (1))

$$u_1 = -\frac{a_1^2}{a_1^2+a_2^2}\,\zeta_3 x_2 + \frac{a_1^2}{a_1^2+a_3^2}\,\zeta_2 x_3 = \Omega_3\gamma\frac{a_1^2}{a_2^2}x_2 - \Omega_2\beta\frac{a_1^2}{a_3^2}x_3,$$

$$u_2 = +\frac{a_2^2}{a_2^2+a_1^2}\,\zeta_3 x_1 = -\Omega_3\gamma x_1,$$

$$u_3 = -\frac{a_3^2}{a_3^2+a_1^2}\,\zeta_2 x_1 = +\Omega_2\beta x_1, \tag{154}$$

where β and γ are defined in equation (12). The ratios ζ_2/Ω_2 and ζ_3/Ω_3 are determined by the solutions for β and γ given in equations (16) and (17).

It remains to determine the values of Ω_2 and Ω_3 that are to be associated with the ellipsoid. The necessary additional conditions follow from the diagonal components of equation (2). When $\Omega_1 = \zeta_1 = 0$, equation (2) gives

$$2\mathfrak{T}_{11}+(\Omega_2^2+\Omega_3^2)I_{11}+\mathfrak{W}_{11}+2\int_V \rho x_1(\Omega_3 u_2-\Omega_2 u_3)\,d\mathbf{x} = -\Pi, \tag{155}$$

$$2\mathfrak{T}_{22}+\Omega_3^2 I_{22}+\mathfrak{W}_{22}-2\Omega_3\int_V \rho u_1 x_2\,d\mathbf{x} = -\Pi, \tag{156}$$

and

$$2\mathfrak{T}_{33}+\Omega_2^2 I_{33}+\mathfrak{W}_{33}+2\Omega_2\int_V \rho u_1 x_3\,d\mathbf{x} = -\Pi. \tag{157}$$

On evaluating the components of the kinetic-energy tensor and the

Table X

The neutral points along the Riemann sequences belonging to the third harmonics

	Neutral points along the Riemann sequences for									
	$f = -2$	$f = -1\cdot8$	$f = -1\cdot5$	$f = -1\cdot2$	$f = -1\cdot0$	$f = 0$†	$f = 0\cdot5$	$f = 1\cdot5$	$f = 1\cdot9$	$f = 2\cdot0$
a_2/a_1	0·32098	0·33855	0·36492	0·38859	0·40152	0·43223	0·42253	0·36898	0·33948	0·33202
a_3/a_1	0·41397	0·42886	0·44579	0·45130	0·44657	0·34507	0·28004	0·19716	0·17787	0·17379
Ω^2	0·22981	0·24582	0·27018	0·29207	0·30315	0·28403	0·23315	0·14336	0·12036	0·11556
$-Q_1 Q_2$	0·07784	0·07348	0·06304	0·04794	0·03624	0	0·00749	0·03402	0·04026	0·04134
Q_1	0·86922	0·80067	0·68805	0·56344	0·47415	0	−0·20486	−0·49988	−0·59104	−0·61239
Q_2	−0·08955	−0·09177	−0·09162	−0·08508	−0·07644	0	0·03657	0·06806	0·06812	0·06751

† The entries in this column refer to the Jacobian sequence.

	Neutral points for the adjoint configurations along the Riemann sequences for									
	$f = -2$	$f = -1\cdot8$	$f = -1\cdot5$	$f = -1\cdot2$	$f = -1\cdot0$	$f = 0$†	$f = 0\cdot5$	$f = 1\cdot5$	$f = 1\cdot9$	$f = 2\cdot0$
$f^\dagger$	−3·02418	−3·22397	−4·02098	−5·28598	−6·53927	$\pm\infty$	15·02537	4·65671	3·21389	2·85043
a_2/a_1	0·51407	0·53308	0·51530	0·49391	0·48185	0·44133	0·43336	0·45760	0·50989	0·54174
a_3/a_1	0·64729	0·65904	0·62111	0·56900	0·53188	0·35041	0·28412	0·21660	0·21528	0·22030
$(\Omega^\dagger)^2$	0·17645	0·16089	0·11744	0·07546	0·05169	0	0·00785	0·04738	0·07159	0·08068
$-Q_1^\dagger Q_2^\dagger$	0·26681	0·28817	0·31480	0·33242	0·33800	0·28781	0·23599	0·14708	0·12110	0·11499
$Q_1^\dagger$	1·00481	1·00700	1·08882	1·16733	1·20657	−1·21560	−1·12099	−0·83810	−0·68251	−0·62594
$Q_2^\dagger$	−0·26553	−0·28616	−0·28912	−0·28477	−0·28014	0·23677	0·21052	0·17549	0·17744	0·18370

† The entries in this column refer to the Dedekind sequence.

	Neutral points along the sequence of the prolate spheroids		Neutral points along the self adjoint sequences for	
	$f = -0\cdot61563$	$f = -10\cdot65076$	$x = -1$	$x = +1$
$f = f^\dagger$	..	..	−2·71414	2·10714
a_2/a_1	0·41815	0·46386	0·43966	0·72189
a_3/a_1	0·41815	0·46386	0·57722	0·25597
Ω^2	0·31111	0·02021	0·20328	0·10614
$-Q_1 Q_2$	0·01494	0·33406	0·20328	0·10614
Q_1	0·29228	1·24601	1·02550	−0·45130
Q_2	−0·05110	−0·26810	−0·19823	0·23518

moments of the velocities that occur, we find

$$\left[\Omega_3^2\left(1-2\gamma+\frac{a_1^2}{a_2^2}\gamma^2\right)+\Omega_2^2\left(1-2\beta+\frac{a_1^2}{a_3^2}\beta^2\right)\right]I_{11}+\mathfrak{W}_{11} = -\Pi, \tag{158}$$

$$\Omega_3^2\left(\gamma^2-2\gamma+\frac{a_2^2}{a_1^2}\right)I_{11}+\mathfrak{W}_{22} = -\Pi, \tag{159}$$

and

$$\Omega_2^2\left(\beta^2-2\beta+\frac{a_3^2}{a_1^2}\right)I_{11}+\mathfrak{W}_{33} = -\Pi. \tag{160}$$

From equations (14) and (15) governing β and γ we obtain the relations

$$\beta^2-2\beta+\frac{a_3^2}{a_1^2} = \frac{a_3^2-a_2^2}{2a_1^2}\beta, \qquad \gamma^2-2\gamma+\frac{a_2^2}{a_1^2} = \frac{a_2^2-a_3^2}{2a_1^2}\gamma, \tag{161}$$

$$1-2\beta+\frac{a_1^2}{a_3^2}\beta^2 = \frac{4a_1^2-a_2^2-3a_3^2}{2a_3^2}\beta, \tag{162}$$

and

$$1-2\gamma+\frac{a_1^2}{a_2^2}\gamma^2 = \frac{4a_1^2-a_3^2-3a_2^2}{2a_2^2}\gamma. \tag{163}$$

Inserting these relations in equations (158)–(160) and making use of the known expressions for the potential-energy tensors, we find

$$(\Omega_2^2\beta+\Omega_3^2\gamma)-\tfrac{1}{2}(4a_1^2-a_2^2-a_3^2)\left(\frac{\Omega_2^2\beta}{a_3^2}+\frac{\Omega_3^2\gamma}{a_2^2}\right)+2A_1 = \frac{5\Pi}{Ma_1^2}, \tag{164}$$

$$\frac{a_3^2-a_2^2}{2a_2^2}\Omega_3^2\gamma+2A_2 = \frac{5\Pi}{Ma_2^2}, \tag{165}$$

and

$$\frac{a_2^2-a_3^2}{2a_3^2}\Omega_2^2\beta+2A_3 = \frac{5\Pi}{Ma_3^2}, \tag{166}$$

where M denotes the mass of the ellipsoid. By combining equations (165) and (166) suitably, we obtain the pair of equations

$$\beta\Omega_2^2+\gamma\Omega_3^2 = 4\frac{A_3a_3^2-A_2a_2^2}{a_3^2-a_2^2} = 4B_{23} \tag{167}$$

and

$$\frac{1}{2}\left(\frac{\Omega_2^2\beta}{a_3^2}+\frac{\Omega_3^2\gamma}{a_2^2}\right)+2A_{23} = \frac{5\Pi}{Ma_2^2a_3^2}. \tag{168}$$

From equations (164), (167), and (168), we now find

$$\frac{5\Pi}{2Ma_1^2a_2^2a_3^2} = \frac{2B_{23}+(4a_1^2-a_2^2-a_3^2)A_{23}+A_1}{4a_1^4-a_1^2(a_2^2+a_3^2)+a_2^2a_3^2}. \tag{169}$$

Inserting this relation in equations (165) and (166) and simplifying, we obtain the formulas:

$$\Omega_2^2\beta = \frac{4a_3^2}{a_2^2-a_3^2}\frac{a_1^2(3a_2^2-4a_1^2+a_3^2)B_{23}+a_2^2(A_1a_1^2-A_3a_3^2)}{4a_1^4-a_1^2(a_2^2+a_3^2)+a_2^2a_3^2} \tag{170}$$

and
$$\Omega_3^2 \gamma = \frac{4a_2^2}{a_3^2 - a_2^2} \, \frac{a_1^2(3a_3^2 - 4a_1^2 + a_2^2)B_{23} + a_3^2(A_1 a_1^2 - A_2 a_2^2)}{4a_1^4 - a_1^2(a_2^2 + a_3^2) + a_2^2 a_3^2}. \tag{171}$$

Alternative forms of the foregoing formulas which we shall find useful are

$$\Omega_2^2 \beta = \frac{4a_3^2(a_2^2 - a_1^2)}{(a_2^2 - a_3^2)} \, \frac{(4a_1^2 + a_2^2 - a_3^2)B_{23} - a_2^2 B_{12}}{4a_1^4 - a_1^2(a_2^2 + a_3^2) + a_2^2 a_3^2} \tag{172}$$

and
$$\Omega_3^2 \gamma = \frac{4a_2^2(a_3^2 - a_1^2)}{(a_3^2 - a_2^2)} \, \frac{(4a_1^2 + a_3^2 - a_2^2)B_{23} - a_3^2 B_{13}}{4a_1^4 - a_1^2(a_2^2 + a_3^2) + a_2^2 a_3^2}. \tag{173}$$

Equations (172) and (173), together with equations (16) and (17), determine the angular velocities and the vorticities that are to be associated with an ellipsoid with semi-axes a_1, a_2, and a_3.

Finally, substituting for the index symbols that appear in equations (169), (172), and (173) their integral expressions (cf. Chapter 3, § 21) we find

$$\Omega_2^2 \beta = 4a_1 a_2 a_3 \frac{a_0^2(a_2^2 - a_1^2)}{(a_2^2 - a_3^2)D} \int_0^\infty [(4a_1^2 - a_3^2)u + a_1^2(4a_1^2 + a_2^2 - a_3^2) - a_2^2 a_3^2] \frac{u \, du}{\Delta^3},$$
$$\tag{174}$$

$$\Omega_3^2 \gamma = 4a_1 a_2 a_3 \frac{a_2^2(a_3^2 - a_1^2)}{(a_3^2 - a_2^2)D} \int_0^\infty [(4a_1^2 - a_2^2)u + a_1^2(4a_1^2 + a_3^2 - a_2^2) - a_2^2 a_3^2] \frac{u \, du}{\Delta^3},$$
$$\tag{175}$$

and
$$\frac{5\Pi}{2a_1^2 a_2^2 a_3^2 M} = \frac{1}{D} \int_0^\infty (3u^2 + 6a_1^2 u + D) \frac{du}{\Delta^3}, \tag{176}$$

where
$$D = 4a_1^4 - a_1^2(a_2^2 + a_3^2) + a_2^2 a_3^2. \tag{177}$$

Equations (174)–(176), in essentially these forms, were given by Riemann though the manner of his derivation was very different.

(a) *The domain of occupancy in the $(a_2/a_1, a_3/a_1)$-plane*

It is clear that any set of values (a_1, a_2, a_3), that is consistent with equations (16), (17), (169), (172), and (173) (or, equivalently, equations (16), (17), and (174)–(177)) and leads to realizable values for the various physical parameters, provides an admissible solution. We shall now show how the physical requirements, that β, γ, Ω_2, and Ω_3 are real and that $\Pi \geqslant 0$, limit the domain of occupancy of these ellipsoidal figures of equilibrium in the $(a_2/a_1, a_3/a_1)$-plane.

First, we observe that since all the equations are symmetric in the indices 2 and 3, the domain of occupancy in the $(a_2/a_1, a_3/a_1)$-plane must be symmetrically disposed about the $45°$ line $a_2 = a_3$. Therefore, without loss of generality, we may restrict ourselves to the part of the plane

$$a_2 \geqslant a_3. \tag{178}$$

Next, we observe that the reality of β and γ requires that

$$either \ \ 2a_1 \geqslant (a_2+a_3) \ \ or \ \ 2a_1 \leqslant |a_2-a_3| = a_2-a_3; \tag{179}$$

and these two cases must be considered separately.

(i) *Case I*: $2a_1 \geqslant (a_2+a_3)$ *and* $a_2 \geqslant a_3$. Under the restrictions of this case

$$4a_1^2 \pm (a_3^2-a_2^2) \geqslant (a_2+a_3)^2 \pm (a_3^2-a_2^2) \geqslant 0. \tag{180}$$

In view of this inequality it follows from equations (14) and (15) that

$$\beta > 0 \ \ \text{and} \ \ \gamma > 0. \tag{181}$$

The reality of Ω_2 and Ω_3 now requires that the quantities on the right-hand sides of equations (174) and (175) are positive. Clearly,

$$D \geqslant a_1^2(a_2+a_3)^2 - a_1^2(a_2^2+a_3^2) + a_2^2 a_3^2 > 0. \tag{182}$$

Also, making use of the inequality (180), we have

$$a_1^2[4a_1^2 \pm (a_2^2-a_3^2)] - a_2^2 a_3^2 \geqslant \tfrac{1}{4}(a_2+a_3)^2[(a_2+a_3)^2 \pm (a_2^2-a_3^2)] - a_2^2 a_3^2. \tag{183}$$

The right-hand side of this inequality is

$$\tfrac{1}{2}a_2[(a_2+a_3)^3 - 2a_2 a_3^2] \ \ \text{or} \ \ \tfrac{1}{2}a_3[(a_3+a_2)^3 - 2a_3 a_2^2]; \tag{184}$$

and in either case it is positive. Since $4a_1^2 - a_3^2$ and $4a_1^2 - a_2^2$ are also positive, the integrands on the right-hand sides of equations (174) and (175) are positive-definite; the integrals are, therefore, positive. As D and β have already been shown to be positive, it follows that the reality of Ω_2 requires

$$a_2 \geqslant a_1. \tag{185}$$

Hence we are, in this case, limited to the domain

$$2a_1 \geqslant (a_2+a_3) \ \ \text{and} \ \ a_2 \geqslant a_1 \geqslant a_3. \tag{186}$$

Under these circumstances the reality of Ω_3 is also assured. Moreover, since $D > 0$, the quantity on the right-hand side of equation (176) is manifestly positive-definite and assures that $\Pi > 0$. Therefore, all ellipsoids represented in the triangle $SMcR_I$ in Fig. 16 are allowed figures of equilibrium: *we shall call them Riemann ellipsoids of type I.*

(ii) *The case* $2a_1 < (a_2-a_3)$ *and* $a_2 \geqslant a_3$. In this case

$$4a_1^2 < a_2^2 - a_3^2, \tag{187}$$

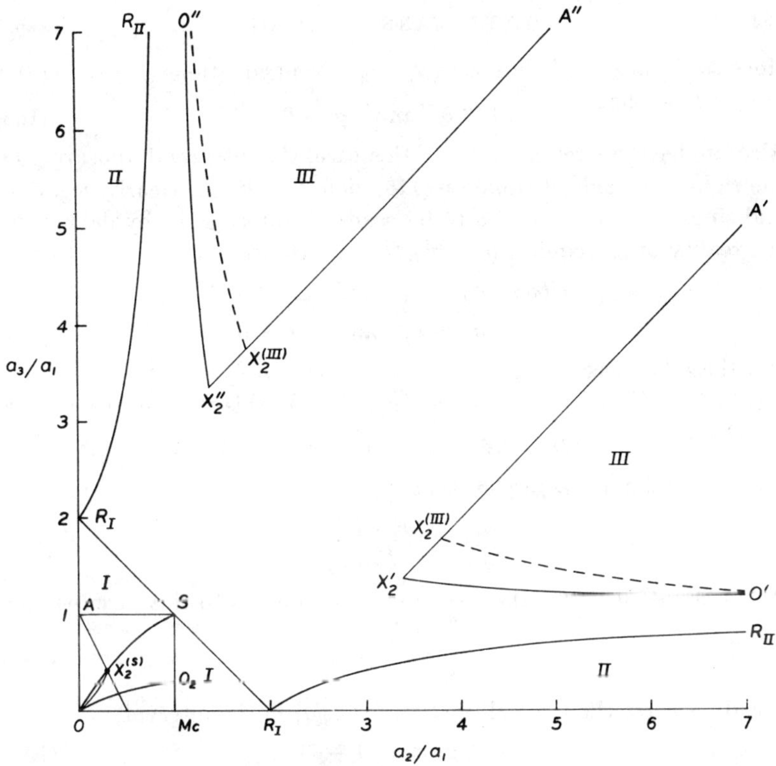

FIG. 16. The domain of occupancy of the Riemann ellipsoids in the $(a_2/a_1, a_3/a_1)$-plane. The stable part of the Maclaurin sequence is represented by the segment $O_2 S$ on the line $a_2 = a_1$. At O_2 the Maclaurin spheroid becomes unstable by overstable oscillations.

The Riemann ellipsoids of type S (for which the directions of rotation and vorticity coincide with the a_3-axis) are included between the self-adjoint sequences represented by SO and $O_2 O$. Along the arc $X_2^{(S)}O$ the Riemann ellipsoids of type S become unstable by an odd mode of oscillation belonging to the second harmonics.

The Riemann ellipsoids, in which the directions of rotation and vorticity do not coincide but lie in the (a_2, a_3)-plane, are of three types—I, II, and III—with the domains of occupancy shown. Ellipsoids of type I adjoin the Maclaurin sequence and are bounded on one side (SR_I) by a self-adjoint sequence. Along the locus $R_I R_{II}$, which limits the domain of occupancy of the ellipsoids of type II, the pressure is zero. And along the loci $X_2'O'$ and $X_2''O''$, limiting the domain of occupancy of ellipsoids of type III, the directions of $\mathbf{\Omega}$ and $\mathbf{\zeta}$ coincide with one of the principal axes (the a_3-axis in the case $a_2 > a_3$ and the a_2-axis in the case $a_2 < a_3$). The locus $X_2'O'$ (for the case $a_2 > a_3$) is transformed into $X_2^{(S)}O$ if the roles of a_1 and a_2 are interchanged; and simultaneously the domain of occupancy $A'X_2'O'$, similarly, becomes transformed into the domain $AX_2^{(S)}O$. The dotted curve $X_2^{(III)}O'$ defines the locus of configurations, among the ellipsoids of type III, that are marginally overstable by a mode of oscillation belonging to the second harmonics.

since $2a_1$ is necessarily less than a_2+a_3. From equations (14) and (15) it now follows that

$$\beta < 0 \quad\text{and}\quad \gamma > 0. \tag{188}$$

Also, under the circumstances of this case, the integrand appearing on the right-hand side of equation (175), defining $\Omega_3^2\,\gamma$, is clearly negative; and since γ has been shown to be positive and $a_2 \geqslant a_3$ (by definition), the reality of Ω_3 requires $(a_3^2-a_1^2)/D \geqslant 0$. Hence

$$either \quad a_3 < a_1 \quad\text{and}\quad D < 0,$$
$$or \quad a_3 > a_1 \quad\text{and}\quad D > 0; \tag{189}$$

and these two cases must be considered separately.

(iii) *Case II*: $2a_1 \leqslant (a_2-a_3)$ *and* $a_3 \leqslant a_1$. In this case we must require (cf. (189))

$$D = 4a_1^4 - a_1^2(a_2^2+a_3^2) + a_2^2 a_3^2 \leqslant 0. \tag{190}$$

This restriction on D implies that

$$\frac{a_3}{a_1} \leqslant \left(\frac{a_2^2-4a_1^2}{a_2^2-a_1^2}\right)^{\frac{1}{2}}. \tag{191}$$

It can be verified that the restriction appropriate to this case, namely

$$\frac{a_3}{a_1} \leqslant \frac{a_2}{a_1} - 2, \tag{192}$$

already assures the inequality (191) for a_2/a_1 in the interval

$$2 \leqslant a_2/a_1 \leqslant 1+\sqrt{3}; \tag{193}$$

but outside of this interval, we must require that the inequality (191) is satisfied. With a_2/a_1 so restricted, the reality of Ω_3 is assured.

Turning next to equation (172) defining $\Omega_2^2\,\beta$ and rewriting it in the manner

$$\Omega_2^2\beta = 4a_1 a_2 a_3 \frac{a_3^2(a_2^2-a_1^2)}{(a_2^2-a_3^2)D} \int_0^\infty \left(\frac{4a_1^2+a_2^2-a_3^2}{a_3^2+u} - \frac{a_2^2}{a_1^2+u}\right)\frac{u\,du}{(a_2^2+u)\Delta}, \tag{194}$$

we observe that the integrand is positive, since

$$\frac{4a_1^2-a_3^2}{a_3^2+u} + a_2^2\left(\frac{1}{a_3^2+u} - \frac{1}{a_1^2+u}\right) > 0. \tag{195}$$

The integral on the right-hand side of equation (194) is therefore positive and since, further, $a_2 \geqslant a_1 \geqslant a_3$, $D < 0$, and $\beta < 0$, the reality of Ω_2 is also assured. On the other hand, since $D < 0$, the positive definiteness of Π is not manifest from equation (176). Indeed, the requirement (cf. the alternative expression (169) for Π)

$$2B_{23} + (4a_1^2-a_2^2-a_3^2)A_{23} + A_1 \leqslant 0 \tag{196}$$

provides a limit on a_3/a_1 for an assigned a_2/a_1 $(\geqslant 2)$ which, it will appear,

TABLE XI

The properties of the ellipsoids of type II along the limiting locus $\Pi = 0$†

a_2/a_1	a_3/a_1	Ω_2	ζ_2	Ω_3	ζ_3
2·0	0	0	0	0	0
2·4	0·2140	0·9441	1·2345	0·3612	−1·1870
		0·2526	4·6133	0·4214	−1·0175
2·8	0·3590	1·1403	0·7219	0·3498	−1·4920
		0·2296	3·5853	0·4726	−1·1042
3·2	0·4625	1·2360	0·5103	0·3102	−1·6820
		0·1944	3·2441	0·4789	−1·0895
3·6	0·5397	1·2912	0·3908	0·2702	−1·8234
		0·1633	3·0893	0·4702	−1·0476
4·0	0·5994	1·3260	0·3130	0·2351	−1·9378
		0·1380	3·0071	0·4559	−0·9991
4·4	0·6468	1·3492	0·2581	0·2055	−2·0346
		0·1177	2·9586	0·4397	−0·9509
4·8	0·6854	1·3654	0·2173	0·1808	−2·1191
		0·1014	2·9279	0·4231	−0·9054
5·2	0·7174	1·3770	0·1860	0·1601	−2·1941
		0·0881	2·9074	0·4069	−0·8632
5·6	0·7443	1·3856	0·1611	0·1427	−2·2617
		0·0772	2·8929	0·3914	−0·8245
6·0	0·7672	1·3920	0·1411	0·1279	−2·3232
		0·0682	2·8825	0·3767	−0·7890
7·0	0·8117	1·4024	0·1051	0·0007	2·1565
		0·0514	2·8660	0·3439	−0·7123
8·0	0·8439	1·4081	0·0814	0·0800	−2·5677
		0·0401	2·8569	0·3160	−0·6496
9·0	0·8681	1·4115	0·0650	0·0656	−2·6628
		0·0322	2·8512	0·2923	−0·5976
10·0	0·8868	1·4135	0·0531	0·0548	−2·7456
		0·0263	2·8474	0·2718	−0·5537
11·0	0·9016	1·4145	0·0442	0·0465	−2·8187
		0·0220	2·8447	0·2541	−0·5161
12·0	0·9135	1·4155	0·0373	0·0400	−2·8840
		0·0186	2·8426	0·2387	−0·4835

† The two entries in the last four columns, for each value of $(a_2/a_1,\, a_3/a_1)$, refer to the adjoint ellipsoids.

is more stringent than either (191) or (192). A more convenient form of the condition (196) is

$$A_{23}(a_2^2 + a_3^2 - 2a_1^2) \geqslant 1. \tag{197}$$

The locus in the $(a_2/a_1, a_3/a_1)$-plane along which the inequality (197) becomes an equality is specified in Table XI. And the curve labelled $R_I R_{II}$ in Fig. 16 defines this locus.

We shall call the ellipsoids, represented in the domain limited by the a_2-axis and the inequality (197), *the Riemann ellipsoids of type II*.

(iv) *Case III*: $2a_1 \leqslant (a_2 - a_3)$ *and* $a_2 \geqslant a_3 \geqslant a_1$. In this case a_1 is the least axis and

$$D = 3a_1^4 + (a_2^2 - a_1^2)(a_3^2 - a_1^2) > 0, \tag{198}$$

as required by (189). And since $D > 0$, Π is necessarily positive. The reality of Ω_3 has already been assured. It remains to insure the reality of Ω_2. Since β is negative (cf. (188)), the reality requires (cf. equation (172))

$$(4a_1^2 + a_2^2 - a_3^2)B_{23} - a_2^2 B_{12} \leqslant 0. \tag{199}$$

And this inequality requires that for a given a_2 ($\geqslant 3 \cdot 3746$, as we shall see presently), a_3 ($\leqslant a_2$) exceed a certain lower limit. The limit is determined by the condition

$$(4a_1^2 + a_2^2 - a_3^2)B_{23} - a_2^2 B_{12} = 0. \tag{200}$$

We shall call the ellipsoids limited by the locus (200) and the line $2a_1 = a_2 - a_3$ the *Riemann ellipsoids of type III*.

(b) *The ellipsoids of type III as branching off from the ellipsoids of type S along a curve of bifurcation*

It is clear, from the way the locus (200) was defined, that along this locus

$$\Omega_2 = \zeta_2 = 0. \tag{201}$$

Accordingly, for these ellipsoids, the only non-vanishing components of $\boldsymbol{\Omega}$ and $\boldsymbol{\zeta}$ are Ω_3 and ζ_3 along the x_3-axis. These ellipsoids are, therefore, of the type S considered in §§ 48–50. But to be in agreement with the convention adopted in § 48, namely, that a_1 is the longest axis, we must interchange the present roles of the indices 1 and 2 since, on our present convention, a_2 is the longest axis of the ellipsoids of type III. With the indices 1 and 2 interchanged, the loci limiting the domain of these ellipsoids become (see Figs. 15 and 16)

$$a_2 = 0, \qquad 2a_2 = a_1 - a_3, \tag{202}$$

and
$$(4a_2^2 + a_1^2 - a_3^2)B_{13} - a_1^2 B_{12} = 0. \tag{203}$$

The locus (203) is seen to be identical with the locus of marginal stability of the S-type ellipsoids determined in § 49 (equation (132)). Moreover, in view of the relation (139) that obtains at the point of intersection of the locus (132) with the bounding self-adjoint sequence $x = -1$, it follows that the curve along which the Riemann ellipsoids of type S become marginally unstable is the curve along which the Riemann ellipsoids of type III branch off. We may, therefore, say that *the ellipsoids of type III bifurcate from the ellipsoids of type S along a locus on which an exchange of stability must occur*.

(c) *The Maclaurin spheroids as limiting forms of the Riemann ellipsoids of type I*

We have seen in § 48 (b) how the stable members of the Maclaurin sequence can be considered as limiting forms of the Riemann ellipsoids of type S. We shall now show how the *entire* Maclaurin sequence can be considered as limiting forms of the Riemann ellipsoids of type I.

Let $a_2/a_1 \to 1$ while a_3 remains finite. From equations (16) and (17), we find that in this limit

$$\beta = \frac{1}{4a_1^2}[3a_1^2+a_3^2\pm\sqrt{(9a_1^2-a_3^2)(a_1^2-a_3^2)}]$$

and $\qquad \gamma = \frac{1}{4a_1^2}[5a_1^2-a_3^2\pm\sqrt{(9a_1^2-a_3^2)(a_1^2-a_3^2)}] \quad (a_1 = a_2).$ (204)

At the same time it follows from equations (172) and (173) that

$$\Omega_2^2\beta = 0 \quad \text{and} \quad \Omega_3^2\gamma = 4B_{13}.$$ (205)

From equations (12) we may now conclude that

$$\zeta_2 = 0 \quad \text{and} \quad \zeta_3 = -2\gamma\Omega_3.$$ (206)

Hence, on the line $a_1 = a_2$, the Riemann ellipsoids of type I become spheroids and are attributed the parameters

$$\Omega_3^2 = 4B_{13}/\gamma, \quad \zeta_3 = -2\gamma\Omega_3, \quad \text{and} \quad \zeta_2 = \Omega_2 = 0 \quad (a_2 \to a_1),$$ (207)

while ζ_2/Ω_2 tends to a finite value.

The relations (207) give

$$(\Omega_3+\tfrac{1}{2}\zeta_3)^2 = \Omega_3^2(1-\gamma)^2 = 4B_{13}\frac{(1-\gamma)^2}{\gamma}.$$ (208)

Now, when $a_1 \to a_2$, equation (15) governing γ is

$$\gamma^2-\left[2+\frac{1}{2}\left(1-\frac{a_3^2}{a_1^2}\right)\right]\gamma+1 = 0,$$ (209)

or $\qquad (\gamma-1)^2 = \tfrac{1}{2}\gamma\left(1-\frac{a_3^2}{a_1^2}\right).$ (210)

In view of this last relation, equation (208) becomes

$$(\Omega_3+\tfrac{1}{2}\zeta_3)^2 = 2B_{13}(1-a_3^2/a_1^2) = \Omega_{\text{Mc}}^2,$$ (211)

or $\qquad\qquad \Omega_3+\tfrac{1}{2}\zeta_3 = \Omega_{\text{Mc}}.$ (212)

But equation (212) is exactly the relation that must be satisfied if what we are viewing is in fact a Maclaurin spheroid rotating uniformly with an angular velocity Ω_{Mc} (in the inertial frame). The Riemann ellipsoids of type I, therefore, degenerate into Maclaurin spheroids when $a_2 \to a_1$

(from the right); but they are viewed from a frame of reference in which they are attributed internal motions with the vorticity ζ_3.

From equations (207) and (211) giving Ω_3 and Ω_{Mc}, we derive

$$\Omega_3^2 - \Omega_3 \Omega_{\mathrm{Mc}} = 4B_{13}\left\{\frac{1}{\gamma} - \left[\frac{1}{2\gamma}\left(1 - \frac{a_3^2}{a_1^2}\right)\right]^{\frac{1}{2}}\right\}, \tag{213}$$

or, making use of equation (210), we have

$$\Omega_3^2 - \Omega_3 \Omega_{\mathrm{Mc}} - 4B_{13} = 0, \tag{214}$$

or

$$\Omega_3 = \tfrac{1}{2}[\Omega_{\mathrm{Mc}} \pm \sqrt{(16B_{13} + \Omega_{\mathrm{Mc}}^2)}]. \tag{215}$$

Comparison of equation (215) with equation (105) of Chapter 5, shows that the branching of the ellipsoids of type I from the Maclaurin sequence is consistent with the fact that the two odd modes of oscillation of the Maclaurin spheroid are neutralized when viewed from frames of reference rotating with the angular velocities given by equation (215).

Finally, we may note that had we defined a_1 as the longest axis (instead of as the axis along which neither $\boldsymbol{\Omega}$ nor $\boldsymbol{\zeta}$ has a component) then the hypotenuse $2a_1 = a_2 + a_3$, of the right-angled triangle ($SMc\,R_I$ in Fig. 16), which now limits the domain of the ellipsoids of type I, would have been the straight line $2a_2 = a_1 + a_3$ (cf. equation (138) and the remarks following equation (139)).

(d) The adjoint ellipsoids and Dedekind's theorem

For any pair of values $(a_2/a_1,\ a_3/a_1)$ which represents a point in the permissible domain of occupancy of the Riemann ellipsoids of types I, II, and III, there are two states of motion, compatible with equilibrium, corresponding to the two roots for β and γ given by equations (16) and (17)—except along the lines,

$$a_2 + a_3 = 2a_1 \quad \text{and} \quad a_2 - a_3 = 2a_1, \tag{216}$$

where the two roots coincide. It is clear that the two physically distinct configurations that one, generally, obtains in this way must be adjoints of one another in the sense of Dedekind's theorem. Similarly, the ellipsoids of type I, along the line $a_2 + a_3 = 2a_1$, and the ellipsoids of type III, along the line $a_2 - a_3 = 2a_1$, must be *self-adjoint*. These facts can be verified directly from the relevant equations; but we shall not do so.

As we have remarked already, had we defined a_1 as the longest axis, the ellipsoids of type I would be limited by the a_2-axis and the lines $a_2 = a_1$ and $a_1 + a_3 = 2a_2$; and the ellipsoids of type III would be limited by the a_3-axis, the locus of marginal stability of the S-type ellipsoids, and the line $a_1 - a_3 = 2a_2$.

(e) The disklike ellipsoids on the a_2-axis

As $a_3 \to 0$, the ellipsoids of types I and III become disklike and their asymptotic properties are of some interest.

It can be readily shown that in the limit

$$\epsilon = a_3/a_1 \to 0 \tag{217}$$

the index symbols A_i have the behavior

$$A_1 = \alpha_1 \epsilon, \quad A_2 = \alpha_2 \epsilon, \quad \text{and} \quad A_3 = 2, \tag{218}$$

where α_1 and α_2 are certain constants expressible in terms of the *complete* elliptic integrals (cf. Chapter 3, § 17, equations (31)–(35))

$$E(\theta) = \int_0^{\pi/2} d\phi \, (1 - \sin^2\theta \sin^2\phi)^{\frac{1}{2}} \quad \text{and} \quad F(\theta) = \int_0^{\pi/2} d\phi \, (1 - \sin^2\theta \sin^2\phi)^{-\frac{1}{2}}, \tag{219}$$

with the argument
$$\theta = \sec^{-1}(a_2/a_1). \tag{220}$$

Thus

$$\alpha_1 = \frac{2}{\sin^2\theta}[E(\theta) - F(\theta)\cos^2\theta] \quad \text{and} \quad \alpha_2 = \frac{2}{\tan^2\theta}[F(\theta) - E(\theta)]. \tag{221}$$

The corresponding forms of the two-index symbols are

$$B_{11} = \beta_{11}\epsilon, \quad B_{22} = \beta_{22}\epsilon, \quad B_{33} = \tfrac{4}{3},$$

$$B_{12} = \beta_{12}\epsilon, \quad B_{23} = \beta_{23}\epsilon, \quad \text{and} \quad B_{31} = \beta_{31}\epsilon, \tag{222}$$

where
$$\beta_{11} = \frac{(\alpha_1 - \alpha_2)a_2^2}{3(a_2^2 - a_1^2)}, \quad \beta_{22} = \frac{(\alpha_1 - \alpha_2)a_1^2}{3(a_2^2 - a_1^2)},$$

$$\beta_{12} = \frac{\alpha_2 a_2^2 - \alpha_1 a_1^2}{a_2^2 - a_1^2}, \quad \beta_{23} = \alpha_2, \quad \text{and} \quad \beta_{31} = \alpha_1. \tag{223}$$

From equations (16) and (17) we find that in this same limit

$$\beta = \frac{4a_1^2 - a_2^2}{2a_1^2}, \quad \gamma = 2, \quad \beta^\dagger = \frac{2a_1^2}{4a_1^2 - a_2^2}\epsilon^2, \quad \text{and} \quad \gamma^\dagger = \frac{a_2^2}{2a_1^2}, \tag{224}$$

where $\beta^\dagger$ and $\gamma^\dagger$ refer to the adjoint configurations.

Inserting the foregoing asymptotic forms of the various quantities in equations (172) and (173), we find

$$\Omega_2 = \omega_2 \epsilon^{\frac{1}{2}}, \quad \Omega_3 = \omega_3 \epsilon^{\frac{1}{2}}, \quad \Omega_2^\dagger = \omega_2^\dagger \epsilon^{\frac{1}{2}}, \quad \text{and} \quad \Omega_3^\dagger = \omega_3^\dagger \epsilon^{\frac{1}{2}}, \tag{225}$$

where
$$\omega_2 = \frac{2a_1^2}{a_2(4a_1^2 - a_2^2)}[2(3\alpha_2 + \alpha_1)a_2^2 - 8\alpha_2 a_1^2]^{\frac{1}{2}}, \quad \omega_3 = (2\alpha_2)^{\frac{1}{2}},$$

$$\omega_2^\dagger = \omega_2\beta = \frac{1}{a_2}[2(3\alpha_2 + \alpha_1)a_2^2 - 8\alpha_2 a_1^2]^{\frac{1}{2}}, \quad \text{and} \quad \omega_3^\dagger = \frac{a_1}{a_2}(8\alpha_2)^{\frac{1}{2}}. \tag{226}$$

The corresponding asymptotic forms of the components of the vorticity are

$$\zeta_2 = z_2\,\epsilon^{-\frac{1}{2}}, \quad \zeta_3 = z_3\,\epsilon^{\frac{1}{2}}, \quad \zeta_2^\dagger = z_2^\dagger\,\epsilon^{\frac{1}{2}}, \quad \text{and} \quad \zeta_3^\dagger = z_3^\dagger\,\epsilon^{\frac{1}{2}}, \quad (227)$$

where

$$z_2 = -\frac{4a_1^2 - a_2^2}{2a_1^2}\,\omega_2, \qquad z_3 = -2\frac{a_1^2 + a_2^2}{a_2^2}\,\omega_3,$$

$$z_2^\dagger = -\frac{2a_1^2}{4a_1^2 - a_2^2}\,\omega_2^\dagger, \quad \text{and} \quad z_3^\dagger = -\frac{a_1^2 + a_2^2}{2a_1^2}\,\omega^\dagger. \quad (228)$$

The properties of the disklike ellipsoids of type I, determined with the aid of the foregoing formulas, are listed in Table XII. Examples of ellipsoids of type II are not included since they are found to be unstable (see equation (234) below).

TABLE XII

The asymptotic properties of the disklike ellipsoids of type I

$\theta = \sec^{-1}(a_2/a_1)$	42°	43°	44°	45°	50°	58°
a_2/a_1	1·34563	1·36733	1·39016	1·41421	1·55572	1·88708
ω_2	+2·05429	+2·13629	+2·22576	+2·32434	+3·05912	+11·27045
ω_3	+1·46610	+1·45084	+1·43516	+1·41906	+1·33207	+1·16871
z_2	−2·24870	−2·27560	−2·30081	−2·32434	−2·41628	−2·47347
z_3	−4·55156	−4·45371	−4·35556	−4·25717	−3·76491	−2·99380
$\omega_2^\dagger$	+2·24870	+2·27560	+2·30081	+2·32434	+2·41628	+2·47347
$\omega_3^\dagger$	+2·17905	+2·12215	+2·06473	+2·00685	+1·71248	+1·23864
$z_2^\dagger$	−2·05429	−2·13629	−2·22576	−2·32434	−3·05912	−11·27045
$z_3^\dagger$	−3·06236	−3·04484	−3·02747	−3·01027	−2·92858	−2·82477

52. The stability of the ellipsoids of types I, II, and III with respect to second-harmonic oscillations

The stability of the Riemann ellipsoids of types I, II, and III with respect to oscillations belonging to the second harmonics can be determined, as in the other cases, with the aid of the linearized version of the second-order virial equations. The basic equation is, in fact, the same as the one used in § 49 in the context of the S-type ellipsoids. However, since the direction of $\mathbf{\Omega}$ does not now coincide with that of the a_3-axis, equation (95) must be replaced by the more general equation

$$\lambda^2 V_{i;j} - 2\lambda Q_{jl}V_{i;l} - 2\lambda\epsilon_{ilm}\Omega_m V_{l;j} - 2\epsilon_{ilm}\Omega_m(Q_{lk}V_{j;k} - Q_{jk}V_{l;k}) + Q_{jl}^2 V_{i;l} + Q_{il}^2 V_{j;l}$$
$$= \Omega^2 V_{ij} - \Omega_i\Omega_k V_{kj} + \delta\mathfrak{W}_{ij} + \delta_{ij}\delta\Pi. \quad (229)$$

In the present context the matrix $\mathbf{Q}$ has the form (cf. equation (154))

$$\mathbf{Q} = \begin{vmatrix} 0 & Q_{12} & Q_{13} \\ Q_{21} & 0 & 0 \\ Q_{31} & 0 & 0 \end{vmatrix}, \quad (230)$$

where
$$Q_{12} = +\Omega_3 \gamma \frac{a_1^2}{a_2^2}, \qquad Q_{21} = -\Omega_3 \gamma,$$

$$Q_{13} = -\Omega_2 \beta \frac{a_1^2}{a_3^2}, \quad \text{and} \quad Q_{31} = +\Omega_2 \beta. \tag{231}$$

This form for the matrix $\mathbf{Q}$ does not allow us, as hitherto, to group the nine equations, which follow from equation (229), into a set of "even" and a set of "odd" equations: the entire set of nine equations must be considered together. However, $\delta\Pi$ must be eliminated and the resulting eight equations must be supplemented by the solenoidal condition

$$\frac{V_{1;1}}{a_1^2} + \frac{V_{2;2}}{a_2^2} + \frac{V_{3;3}}{a_3^2} = 0. \tag{232}$$

The nine equations so obtained are linear and homogeneous in the nine $V_{i;j}$'s; and they lead to a secular determinant of order nine.

We shall not write down the explicit form of the secular determinant or indicate the extent to which reduction is possible. (The interested reader may wish to consult Paper XXVIII which contains some details.) But we shall state some general theorems that can be established.

(i) The secular matrix can be reduced to a form in which it is manifest that the characteristic frequencies of oscillation of an ellipsoid and its adjoint are the same—as is, indeed, required by Dedekind's theorem.

(ii) The characteristic equation allows the two characteristic frequencies $|\Omega|$ and $|\Omega^\dagger|$.

(iii) The characteristic equation also allows two non-trivial neutral modes of oscillation.

The characteristic equation that follows from equations (229) and (232) has been solved for some hundred ellipsoids in their domains of occupancy; and by interpolation among the roots so obtained the loci of marginal stability (rather, *marginal overstability*, as it happens) have been determined. The results of the calculations are summarized in Table XIII; and the loci of the marginally overstable configurations are delineated in Figs. 16 and 17.

It will be observed that among the ellipsoids of type I there are two disconnected domains of stability with respect to the oscillations considered. The existence of the stable domain adjoining the *stable* Maclaurin spheroids along SO_2 is, of course, to be expected. But the existence of the second domain, bounded by the segment $D_2 R_I$ of the a_2-axis, is unexpected. The point $a_2/a_1 = 1\cdot3707$ (see Table XIIIb, last column), limiting the stable disklike ellipsoids of type I, was determined by a

Table XIII

The properties of the marginally overstable Riemann ellipsoids (the angular velocities and the vorticities are expressed in the unit $(\pi G\rho)^{\frac{1}{2}}$)

(a) Type I (along the locus $O_2\,X_2^{(I)}$)

a_2/a_1 a_3/a_1	1·0000 0·3033	1·0526 0·3712	1·1111 0·4230	1·1765 0·4560	1·2500 0·4703	1·3333 0·4676	1·4286 0·4474	1·5385 0·4053	1·6722 0·3278
Ω_2	0	+0·1283	+0·2153	+0·2942	+0·3639	+0·4269	+0·4877	+0·5550	+0·7107
Ω_3	+0·7073	+0·7176	+0·7098	+0·6901	+0·6633	+0·6329	+0·5999	+0·5635	+0·5142
ζ_2	0	−1·5014	−1·8984	−2·1276	−2·2794	−2·3842	−2·4626	−2·5307	−2·4011
ζ_3	−2·7417	−2·5977	−2·3978	−2·1787	−1·9637	−1·7621	−1·5752	−1·3984	−1·1673
$\Omega_2^{\dagger}$	0	+0·4898	+0·6812	+0·8032	+0·8778	+0·9150	+0·9178	+0·8807	+0·7107
$\Omega_3^{\dagger}$	+1·3708	+1·2972	+1·1922	+1·0751	+0·9579	+0·8458	+0·7400	+0·6390	+0·5142
$\zeta_2^{\dagger}$	0	−0·3931	−0·6000	−0·7794	−0·9450	−1·1125	−1·3082	−1·5937	−2·4011
$\zeta_3^{\dagger}$	−1·4147	−1·4371	−1·4275	−1·3984	−1·3599	−1·3186	−1·2768	−1·2330	−1·1673

(b) Type I (along the locus $D_2\,Q$)

a_2/a_1 a_3/a_1	1·1582 0·1411	1·1846 0·1238	1·2124 0·1057	1·2418 0·0866	1·2727 0·0666	1·3050 0·0455	1·3707 ϵ
Ω_2	+0·0618	+0·0558	+0·0480	+0·0389	+0·0286	+0·0176	$+2\cdot1492\epsilon^{3/2}$
Ω_3	+0·5209	+0·4903	+0·4554	+0·4146	+0·3658	+0·3044	$+1\cdot4485\epsilon^{1/2}$
ζ_2	−4·1927	−4·7796	−5·4901	−6·4045	−7·6880	−9·7572	$-2\cdot2795\epsilon^{-1/2}$
ζ_3	−1·8047	−1·6695	−1·5236	−1·3629	−1·1810	−0·9654	$-4\cdot4390\epsilon^{1/2}$
$\Omega_2^{\dagger}$	+0·5802	+0·5829	+0·5737	+0·5506	+0·5098	+0·4436	$+2\cdot2795\epsilon^{1/2}$
$\Omega_3^{\dagger}$	+0·8927	+0·8229	+0·7479	+0·6658	+0·5737	+0·4661	$+2\cdot1135\epsilon^{1/2}$
$\zeta_2^{\dagger}$	−0·4469	−0·4573	−0·4598	−0·4523	−0·4308	−0·3873	$-2\cdot1492\epsilon^{1/2}$
$\zeta_3^{\dagger}$	−1·0530	−0·9947	−0·9277	−0·8488	−0·7529	−0·6305	$-3\cdot0422\epsilon^{1/2}$

(c) Type I (along the locus QR_I)

a_2/a_1 a_3/a_1	1·2907 0·1573	1·4954 0·1563	1·6417 0·1431	1·7679 0·1233	1·8651 0·0976
Ω_2	+0·0979	+0·1427	+0·1769	+0·2211	+0·2856
Ω_3	+0·5082	+0·4633	+0·4219	+0·3784	+0·3310
ζ_2	−4·6947	−5·1812	−5·5580	−6·0132	−6·7416
ζ_3	−1·6093	−1·3221	−1·1381	−0·9802	−0·8341
$\Omega_2^{\dagger}$	+0·7206	+0·7908	+0·7788	+0·7280	+0·6299
$\Omega_3^{\dagger}$	+0·7791	+0·6109	+0·5056	+0·4200	+0·3471
$\zeta_2^{\dagger}$	−0·6376	−0·9355	−1·2612	−1·8137	−2·8546
$\zeta_3^{\dagger}$	−1·0498	−1·0027	−0·9496	−0·8829	−0·7942

(d) Type III (along the locus $X_2^{(III)}O'$)

a_2/a_1 a_3/a_1	4·0000 1·7210	4·4141 1·6000	4·9777 1·4933	5·3909 1·4000
Ω_2	+0·4966	+0·5410	+0·5567	+0·5497
Ω_3	+0·3281	+0·2553	+0·1968	+0·1657
ζ_2	+0·5283	+0·3190	+0·1992	+0·1430
ζ_3	−1·9954	−2·1577	−2·2582	−2·2936
$\Omega_2^{\dagger}$	+0·2198	+0·1440	+0·0914	+0·0676
$\Omega_3^{\dagger}$	+0·4787	+0·4641	+0·4359	+0·4113
$\zeta_2^{\dagger}$	+1·0946	+1·2088	+1·1951	+1·1610
$\zeta_3^{\dagger}$	−1·4219	−1·1826	−1·0190	−0·9242

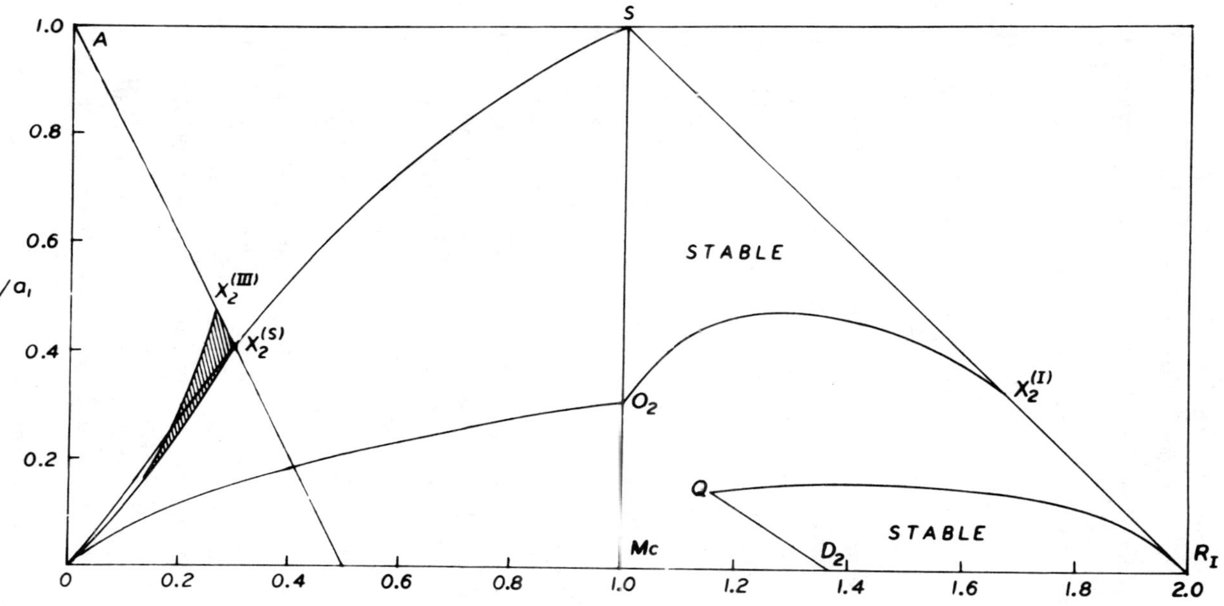

Fig. 17. The loci of marginally stable configurations in the $(a_2/a_1, a_3/a_1)$-plane. The ellipsoids of type S are bounded by two self-adjoint sequences (SO and $O_2 O$) and the stable part of the Maclaurin sequence represented by SO_2. Along the arc $X_2^{(S)}O$ the ellipsoids of type S become unstable by a mode of oscillation belonging to the second harmonics; and along this same arc the stability passes to the ellipsoids of type III whose domain of occupancy is $AX_2^{(S)}O$. The shaded region included between $X_2^{(III)}O$ and $X_2^{(S)}O$ represents the domain of stability for ellipsoids of type III with respect to oscillations belonging to the second harmonics.

The ellipsoids of type I occupy the triangle $SMc\,R_I$; and the region of the stable members is included in the two domains marked "stable." The domain $SO_2\,X_2^{(I)}$ of stable ellipsoids adjoining the stable Maclaurin spheroids is to be expected; but the domain $D_2\,QR_I$ including disklike ellipsoids along $D_2\,R_I$ is unexpected.

separate asymptotic calculation which shows that, in the limit $a_3/a_1 \to 0$, the characteristic frequencies have the behavior

$$\sigma \to \sigma_0 (a_3/a_1)^{\frac{1}{2}} \quad (\sigma_0 \text{ a constant depending on } a_2/a_1). \qquad (233)$$

The calculations further show that *all ellipsoids of type II are unstable.* In particular, the disklike ellipsoids of type II allow a proper mode which becomes unstable with the time-dependence

$$\exp\left[\left(6\,\frac{a_2^2 - 4a_1^2}{a_2^2} \cdot \frac{a_3}{a_1}\,\alpha_2\right)^{\frac{1}{2}} t\right] \quad \left(\frac{a_3}{a_1} \to 0\right), \qquad (234)$$

where α_2 is the constant defined in equations (220) and (221). (For the disklike ellipsoids of type I, this mode is stable, since for these objects $a_2 < 2a_1$.)

Finally, turning to the ellipsoids of type III, we find that there is a fringe of stable configurations (stable, that is, with respect to the oscillations considered) bordering the boundary $X_2' O'$ of the domain of occupancy (see Fig. 16). As we have shown in § 51, by interchanging the roles of the indices 1 and 2, the locus $X_2' O'$ is transformed into the locus $X_2^{(S)} O$ of the marginally stable ellipsoids of type S (see Fig. 17). One should expect that, under these circumstances, stability passes from the ellipsoids of type S to the ellipsoids of type III along their common curve of bifurcation; and this is exactly what happens. However, since the ellipsoids of type S become unstable by a mode of oscillation belonging to the third harmonics *prior* to the onset of instability by an odd mode of oscillation belonging to the second harmonics, it is very likely that *ellipsoids of type III are all unstable to a third-harmonic oscillation.*

It would appear then that only among the ellipsoids of types I and S do stable ones occur.

53. A class of finite-amplitude oscillations of the Maclaurin spheroid

The general equations governing Dirichlet's problem, that we have derived in Chapter 4, can be used to obtain a class of finite-amplitude oscillations of the Maclaurin spheroid. These solutions disclose certain unexpected relationships between the Maclaurin sequence and the self-adjoint Riemann sequence $x = +1$ (cf. § 48 (c)).

We shall restrict our present considerations to the case when the only non-vanishing components of Λ and Ω are those along the a_3-axis of the ellipsoid:

$$\Lambda_1 = \Lambda_2 = \Omega_1 = \Omega_2 = 0 \quad \text{and} \quad \Lambda_3 = \Lambda, \quad \text{and} \quad \Omega_3 = \Omega \text{ (say)}. \qquad (235)$$

Under these circumstances, equation (62) of Chapter 4, written out explicitly for the three components, gives

$$\frac{d^2 a_1}{dt^2} - a_1(\Omega^2 + \Lambda^2) + 2a_2 \Lambda \Omega = -2A_1 a_1 + \frac{2p_c}{\rho a_1}, \tag{236}$$

$$\frac{d^2 a_2}{dt^2} - a_2(\Omega^2 + \Lambda^2) + 2a_1 \Lambda \Omega = -2A_2 a_2 + \frac{2p_c}{\rho a_2}, \tag{237}$$

and

$$\frac{d^2 a_3}{dt^2} = -2A_3 a_3 + \frac{2p_c}{\rho a_3}. \tag{238}$$

These equations allow the integrals (cf. equations (69), (71), and (75) of Chapter 4)

$$E = \frac{1}{2} \sum_{i=1}^{3} \left(\frac{da_i}{dt}\right)^2 + \tfrac{1}{2}(a_1^2 + a_2^2)(\Lambda^2 + \Omega^2) - 2a_1 a_2 \Lambda \Omega - 2I, \tag{239}$$

$$\frac{5L}{M} = (a_1^2 + a_2^2)\Omega - 2a_1 a_2 \Lambda, \tag{240}$$

and

$$\frac{5C}{M} = (a_1^2 + a_2^2)\Lambda - 2a_1 a_2 \Omega, \tag{241}$$

where E, L, and C denote the conserved energy, angular momentum, and circulation, respectively.

In the case we are presently considering, equations (236)–(238), together with the integrals (239)–(241), suffice to make the solution determinate. Indeed, we can eliminate Λ and Ω, altogether, from the equations. Thus, from equations (240) and (241), we find

$$(a_1 - a_2)^2(\Omega + \Lambda) = \frac{5}{M}(L + C) = 2K_1 \text{ (say)} \tag{242}$$

and

$$(a_1 + a_2)^2(\Omega - \Lambda) = \frac{5}{M}(L - C) = 2K_2 \text{ (say)}; \tag{243}$$

and making use of these relations, we can rewrite equations (236) and (237) in the forms

$$\frac{d^2 a_1}{dt^2} - \frac{2K_1^2}{(a_1 - a_2)^3} - \frac{2K_2^2}{(a_1 + a_2)^3} = -2A_1 a_1 + \frac{2p_c}{\rho a_1} \tag{244}$$

and

$$\frac{d^2 a_2}{dt^2} + \frac{2K_1^2}{(a_1 - a_2)^3} - \frac{2K_2^2}{(a_1 + a_2)^3} = -2A_2 a_2 + \frac{2p_c}{\rho a_2}. \tag{245}$$

And the energy integral (239), expressed in terms of K_1 and K_2, is

$$E = \frac{1}{2} \sum_{i=1}^{3} \left(\frac{da_i}{dt}\right)^2 + \frac{K_1^2}{(a_1 - a_2)^2} + \frac{K_2^2}{(a_1 + a_2)^2} - 2I. \tag{246}$$

From equations (242) and (246) it follows that, *for a bounded system with a finite energy, the constant K_1 must be zero if at any time $a_1(t) = a_2(t)$.* Therefore, for bounded systems which at some time pass through a state in which $a_1 = a_2$, we must put

$$K_1 = 0, \quad \Omega = -\Lambda, \quad L = -C, \quad \text{and} \quad K_2 = K \text{ (say)}. \qquad (247)$$

In other words, *the only systems, which can pass through (or, have passed through) a spheroidal state in which $a_1 = a_2$, are those that can be classed as self-adjoint in the sense of Dedekind's theorem.*

When $K_1 = 0$ and $K_2 = K$, the relevant equations are

$$\frac{d^2a_1}{dt^2} = \frac{2K^2}{(a_1+a_2)^3} - 2A_1 a_1 + \frac{2p_c}{\rho a_1}, \qquad (248)$$

$$\frac{d^2a_2}{dt^2} = \frac{2K^2}{(a_1+a_2)^3} - 2A_2 a_2 + \frac{2p_c}{\rho a_2}, \qquad (249)$$

and

$$\frac{d^2a_3}{dt^2} = -2A_3 a_3 + \frac{2p_c}{\rho a_3}, \qquad (250)$$

while the energy integral takes the form

$$E = \frac{1}{2}\sum_{i=1}^{3}\left(\frac{da_i}{dt}\right)^2 + \frac{K^2}{(a_1+a_2)^2} - 2I, \qquad (251)$$

where it may be recalled that

$$(a_1+a_2)^2\Omega = K. \qquad (252)$$

(a) Configurations with extremum or minimum energy

The form of the expression (251) for the energy E allows the following interpretation:

$$\frac{1}{2}\sum_{i=1}^{3}\left(\frac{da_i}{dt}\right)^2 = T = \text{kinetic energy} \qquad (253)$$

and

$$\frac{K^2}{(a_1+a_2)^2} - 2I = W = \text{potential energy}. \qquad (254)$$

From this interpretation of the energy, we may conclude on general grounds that equilibrium (or steady state) configurations are possible only when the potential energy W is an extremum for variations of a_1, a_2, and a_3 which preserve the constancy of the volume, i.e. of the product $a_1 a_2 a_3$. And, further, that these equilibrium configurations may be expected to be stable only if W is a true local minimum. It should, however, be pointed out that "stability" in this context means stability only for perturbations that are compatible with the various restrictive assumptions (such as the continued maintenance of an ellipsoidal figure and the absence of any component of Ω or Λ except in the direction of the x_3-axis) under which the basic equations (248)–(250) have been derived.

We shall now find the conditions when W will be an extremum, i.e. when

$$\left(\frac{\partial W}{\partial a_1}\right)_{a_1a_2a_3=\text{Const.}} = 0 \quad \text{and} \quad \left(\frac{\partial W}{\partial a_2}\right)_{a_1a_2a_3=\text{Const.}} = 0. \tag{255}$$

First, we observe that if $f(a_1, a_2, a_3)$ is explicitly a function of a_1, a_2, and a_3, then

$$\left(\frac{\partial f}{\partial a_1}\right)_{a_1a_2a_3=\text{Const.}} = \frac{\partial f}{\partial a_1} - \frac{a_3}{a_1}\frac{\partial f}{\partial a_3} \quad \text{and} \quad \left(\frac{\partial f}{\partial a_2}\right)_{a_1a_2a_3=\text{Const.}} = \frac{\partial f}{\partial a_2} - \frac{a_3}{a_2}\frac{\partial f}{\partial a_3}. \tag{256}$$

Evaluating the partial derivatives of W in accordance with (256) and making use of the relation (equation (23) of Chapter 3)

$$\frac{\partial I}{\partial a_i} = \frac{1}{a_i}(I - a_i^2 A_i) \quad (i = 1, 2, 3), \tag{257}$$

we find that

$$\left(\frac{\partial W}{\partial a_1}\right)_{a_1a_2a_3=\text{Const.}} = -\frac{2K^2}{(a_1+a_2)^3} + \frac{2}{a_1}(a_1^2 A_1 - a_3^2 A_3) \tag{258}$$

and

$$\left(\frac{\partial W}{\partial a_2}\right)_{a_1a_2a_3=\text{Const.}} = -\frac{2K^2}{(a_1+a_2)^3} + \frac{2}{a_2}(a_2^2 A_2 - a_3^2 A_3). \tag{259}$$

Therefore, at an extremum of W, we must have

$$\frac{K^2}{(a_1+a_2)^3} = \frac{1}{a_1}(a_1^2 A_1 - a_3^2 A_3) = \frac{1}{a_2}(a_2^2 A_2 - a_3^2 A_3). \tag{260}$$

There are *two* distinct ways in which the triangle of equalities (260) can be satisfied.

If we suppose that $a_1 = a_2$, the second equality in (260) is identically satisfied and the first requires

$$K^2 = 8a_1^2(a_1^2 A_1 - a_3^2 A_3) = 8a_1^2(a_1^2 - a_3^2)B_{13}, \tag{261}$$

or, in view of equation (252),

$$\Omega^2 = \tfrac{1}{2}(1 - a_3^2/a_1^2)B_{13} = \tfrac{1}{4}\Omega_{\text{Mc}}^2. \tag{262}$$

These configurations are, therefore, none other than the Maclaurin spheroids viewed from a frame of reference rotating with an angular velocity $\tfrac{1}{2}\Omega_{\text{Mc}}$.

On the other hand, *if we suppose that* $a_1 \neq a_2$, then the second equality in (260) requires

$$a_2 A_2 - a_1 A_1 = \frac{a_1 - a_2}{a_1 a_2}a_3^2 A_3; \tag{263}$$

and this locus, we observe, is the same as equation (72) which defines

the self-adjoint sequence $x = +1$. But it should also be verified that
the value of K given by the first equality in (260) is consistent with this
identification. With K given by equation (252), the equality gives

$$(a_1+a_2)\Omega^2 = a_1 A_1 - \frac{a_3^2}{a_1} A_3, \tag{264}$$

or, making use of equation (263), we have

$$\Omega^2 = \frac{1}{(a_1+a_2)}\left[a_1 A_1 + \frac{a_2}{a_1-a_2}(a_1 A_1 - a_2 A_2)\right] = \frac{a_1^2 A_1 - a_2^2 A_2}{a_1^2 - a_2^2} = B_{12}; \tag{265}$$

and this is one of the relations (74) that obtains on the self-adjoint
sequence $x = +1$. The remaining relations, that ensure the self-adjoint
character of the locus, follow from the present equality of Ω and $-\Lambda$.

We have thus proved that, consistent with the basic assumptions that
led to the equations (248)–(250), *equilibrium configurations, with an
extremum of the potential energy W, occur only along the Maclaurin sequence
and along the self-adjoint sequence $x = +1$.* A significant fact in this
connection is that the self-adjoint sequence $x = +1$ bifurcates from the
Maclaurin sequence at the point of onset of dynamical instability.

Turning next to the question whether the potential energy W, at its
extremum, is a true local minimum, we consider the change δW in W
caused by increments δa_1 and δa_2 in a_1 and a_2. The change, to the first
order in the increments, clearly vanishes at an extremum; to the second
order it is given by

$$\delta W = \frac{1}{2}\left(\frac{\partial^2 W}{\partial a_1^2}\delta a_1^2 + 2\frac{\partial^2 W}{\partial a_1 \partial a_2}\delta a_1 \delta a_2 + \frac{\partial^2 W}{\partial a_2^2}\delta a_2^2\right)_{a_1 a_2 a_3 = \text{Const.}} \tag{266}$$

The second derivatives in this expression for δW must first be evaluated
with the aid of equations (258) and (259) in accordance with the formulas
(256); and only after their evaluation in this fashion, should the equalities
(260), characterizing the extremum, be taken into account. We shall
restrict this examination to the Maclaurin line $a_1 = a_2$.

Along the Maclaurin line it is evident on symmetry grounds that

$$\left(\frac{\partial^2 W}{\partial a_1^2}\right)_{a_1 a_2 a_3 = \text{Const.};\ a_1 = a_2} = \left(\frac{\partial^2 W}{\partial a_2^2}\right)_{a_1 a_2 a_3 = \text{Const.};\ a_1 = a_2}. \tag{267}$$

Therefore, in this case we may rewrite equation (266) in the form

$$\delta W = \frac{1}{4}\left(\frac{\partial^2 W}{\partial a_1^2} - \frac{\partial^2 W}{\partial a_1 \partial a_2}\right)_{a_1 a_2 a_3 = \text{Const.};\ a_1 = a_2}(\delta a_1 - \delta a_2)^2 +$$

$$+ \frac{1}{4}\left(\frac{\partial^2 W}{\partial a_1^2} + \frac{\partial^2 W}{\partial a_1 \partial a_2}\right)_{a_1 a_2 a_3 = \text{Const.};\ a_1 = a_2}(\delta a_1 + \delta a_2)^2. \tag{268}$$

Now writing
$$\left(\frac{\partial W}{\partial a_1}\right)_{a_1 a_2 a_3 = \text{Const.}} = -\frac{2K^2}{(a_1 + a_2)^3} + \frac{2H}{a_1}, \tag{269}$$

where
$$H = a_1^2 A_1 - a_3^2 A_3, \tag{270}$$

we find
$$\left(\frac{\partial^2 W}{\partial a_1^2} - \frac{\partial^2 W}{\partial a_1 \partial a_2}\right)_{a_1 a_2 a_3 = \text{Const.}} = \frac{2}{a_1^2}\left[a_1\left(\frac{\partial H}{\partial a_1} - \frac{\partial H}{\partial a_2}\right) - H\right] \tag{271}$$

and
$$\left(\frac{\partial^2 W}{\partial a_2^2} + \frac{\partial^2 W}{\partial a_1 \partial a_2}\right)_{a_1 a_2 a_3 = \text{Const.}} = \frac{12K^2}{(a_1 + a_2)^4} + \frac{2}{a_1^2}\left[a_1\left(\frac{\partial H}{\partial a_1} + \frac{\partial H}{\partial a_2}\right) - \right.$$
$$\left. - H - a_3\left(1 + \frac{a_1}{a_2}\right)\frac{\partial H}{\partial a_3}\right]. \tag{272}$$

We readily verify that
$$\frac{\partial H}{\partial a_1} = \frac{1}{a_1}(3a_1^2 B_{11} - a_3^2 B_{13}), \qquad \frac{\partial H}{\partial a_3} = -\frac{1}{a_3}(3a_3^2 B_{33} - a_1^2 B_{13}),$$

and
$$\frac{\partial H}{\partial a_2} = \frac{1}{a_2}(a_1^2 B_{12} - a_3^2 B_{23}). \tag{273}$$

Inserting the relations (271)–(273) in equation (268) and noting that along the Maclaurin line ($a_1 = a_2$), we may, in accordance with equation (261), replace the term in K^2 on the right-hand side of equation (272) by $6H/a_1^2$, we find

$$\delta W = \frac{1}{2a_1^2}\{\tfrac{1}{2}a_1^2(4B_{11} - \Omega_{\text{Mc}}^2)(\delta a_1 - \delta a_2)^2 +$$
$$+ [4(a_1^2 B_{11} - a_3^2 B_{13}) + 6a_3^2 B_{33}](\delta a_1 + \delta a_2)^2\}. \tag{274}$$

Since
$$a_1^2 B_{11} - a_3^2 B_{13} = a_1 a_2 a_3 (a_1^2 - a_3^2)\int_0^\infty \frac{u^2\,du}{\Delta^3} \geqslant 0, \tag{275}$$

it is manifest that *the expression* (274) *for* δW *is positive-definite so long as*

$$\Omega_{\text{Mc}}^2 \leqslant 4B_{11}. \tag{276}$$

The potential energy W for these Maclaurin spheroids is, therefore, a true local minimum. But this is not the case for $\Omega_{\text{Mc}}^2 > 4B_{11}$: there exist increments δa_1 and δa_2 for which $\delta W < 0$. We may, therefore, conclude that Maclaurin spheroids with $\Omega_{\text{Mc}}^2 > 4B_{11}$ are unstable; and, further, that the spheroids with $\Omega_{\text{Mc}}^2 \leqslant 4B_{11}$ are stable with respect to the permitted perturbations. These conclusions are in agreement with what has been established in Chapter 5 (§ 33) by a direct determination of the characteristic frequencies of oscillation. The present manner of deducing the *instability* of the Maclaurin spheroids for $\Omega_{\text{Mc}}^2 > 4B_{11}$ is due to Riemann. The result is included in his 1860 paper.

A further consequence of the preceding analysis, which will be of subsequent interest, is the following.

Consider the equation

$$\frac{K_{\mathrm{Mc}}^2}{(a_1+a_2)^2} - 2I = W_{\mathrm{Mc}}, \tag{277}$$

where $\quad W_{\mathrm{Mc}} = -2(a_1^2 A_1 + 2a_3^2 A_3)_{\mathrm{Mc}} \quad$ and $\quad K_{\mathrm{Mc}} = 2(\Omega a_1^2)_{\mathrm{Mc}} \tag{278}$

denote the energy and the angular momentum of a chosen Maclaurin spheroid. Since W_{Mc} is a true local minimum for $\Omega_{\mathrm{Mc}}^2 \leqslant 4B_{11}$, the locus defined by equation (277), in these cases, reduces to the single point at Ω_{Mc}, in the neighborhood of the Maclaurin line. On the other hand, for $\Omega_{\mathrm{Mc}}^2 > 4B_{11}$, the locus (277) "opens out" into a lemniscate-shaped curve with its vertex at Ω_{Mc}: the reason is that, now, the Maclaurin spheroid is not located at the bottom of a potential valley; instead, it is perched

TABLE XIV

The directions of null change, δW

e	$\delta a_1/\delta a_2$		$\delta a_3/\delta a_2$	
0·9529	−1·0000	−1·00000	0	0
0·96	−1·2879	−0·77648	0·08060	−0·06259
0·97	−1·5260	−0·65529	0·12788	−0·08380
0·98	−1·8004	−0·55542	0·15928	−0·08847
0·99	−2·2362	−0·44719	0·17438	−0·07798
0·995	−2·6728	−0·37413	0·16708	−0·06251
0·999	−3·6967	−0·27051	0·12057	−0·03262
0·9999	−4·8742	−0·20516	0·05479	−0·01124

on a potential ridge which slopes downwards in some directions. And the two directions of null change, along which the two arms of the lemniscate fork out at Ω_{Mc}, are determined by the equation (cf. equation (274))

$$\left(\frac{\delta a_1 + \delta a_2}{\delta a_1 - \delta a_2}\right)^2 = \frac{\frac{1}{2}a_1^2(\Omega_{\mathrm{Mc}}^2 - 4B_{11})}{4(a_1^2 B_{11} - a_3^2 B_{13}) + 6a_3^2 B_{33}}. \tag{279}$$

Table XIV provides a brief listing of the directions of null change determined by equation (279). The table also includes the directions $\delta a_3/\delta a_2$ determined by the equation

$$\frac{\delta a_3}{\delta a_2} = -\frac{a_3}{a_1}\left(\frac{\delta a_1}{\delta a_2} + 1\right). \tag{280}$$

(b) Spheroidal oscillations

There is one specially simple case (considered by Dirichlet in his posthumous paper of 1860) for which the energy integral (251) suffices to determine the solution of equations (248)–(250). This is the case when, during the entire evolution, the object retains a spheroidal shape, so that

$$a_1(t) \equiv a_2(t). \tag{281}$$

TABLE XV

The range of $a_3/\bar{a}$ during a spheroidal oscillation of finite amplitude

$a_1(\text{Mc})/\bar{a}$	$a_3(\text{Mc})/\bar{a}$	q	$(a_3/\bar{a})_{\min}$	$(a_3/\bar{a})_{\max}$
		3×10^{-3}	0·884	1·13
1·00	1·0000	3×10^{-2}	0·673	1·48
		1×10^{-1}	0·470	2·07
		3×10^{-1}	0·226	3·97
		3×10^{-3}	0·496	0·605
1·35	0·5487	3×10^{-2}	0·391	0·737
		1×10^{-1}	0·281	0·920
		3×10^{-1}	0·136	1·28
		3×10^{-6}	0·389	0·392
1·60	0·3906	3×10^{-4}	0·379	0·403
		3×10^{-3}	0·354	0·429
		3×10^{-6}	0·159	0·161
2·50	0·1600	3×10^{-4}	0·155	0·165
		3×10^{-3}	0·144	0·177

From the constancy of the volume, which now requires

$$a_1^2 a_3 = \bar{a}^3 = \text{constant}, \tag{282}$$

it follows that

$$\frac{da_1}{dt} = -\frac{a_1}{2a_3} \frac{da_3}{dt}; \tag{283}$$

and the energy integral gives

$$\frac{1}{2}\left(1 + \frac{\bar{a}^3}{2a_3^3}\right)\left(\frac{da_3}{dt}\right)^2 + \frac{K^2}{4\bar{a}^3} a_3 - 2I = E. \tag{284}$$

Equation (284) can be integrated for various initially assigned values of K and E. In practice, it is convenient to assign to K the value K_{Mc}, appropriate to some chosen Maclaurin spheroid, and prescribe for E values that are in excess of W_{Mc} (defined in equations (278)) by different fractional amounts

$$q = \frac{E - W_{\text{Mc}}}{-W_{\text{Mc}}}. \tag{285}$$

It is clear that so long as $q < 1$, a_3 will exhibit an oscillatory behavior. In Table XV (provided by Rossner) we list the maximum and the minimum amplitudes attained in such oscillations for some assigned values of $a_1(\text{Mc})/\bar{a}$ and q.

(c) *Ellipsoidal oscillations*

Equations (248)–(250) have been integrated by Rossner for a number of representative cases. In the numerical procedure used by him, it was convenient to supplement the equations (248)–(250) by an equation for $2p_c/\rho$, rather than to eliminate it and use the constancy of the product $a_1 a_2 a_3$, instead. To obtain this supplementary equation, we first combine equations (248)–(250) to give

$$\sum_{i=1}^{3} \frac{1}{a_i} \frac{d^2 a_i}{dt^2} = \frac{2K^2}{a_1 a_2 (a_1 + a_2)^2} - 4 + \frac{2p_c}{\rho} \sum_{i=1}^{3} \frac{1}{a_i^2}; \tag{286}$$

then, by using the relation

$$0 = \frac{d}{dt}\left(\sum_{i=1}^{3} \frac{1}{a_i} \frac{da_i}{dt} \right) = \sum_{i=1}^{3} \frac{1}{a_i} \frac{d^2 a_i}{dt^2} - \sum_{i=1}^{3} \left(\frac{1}{a_i} \frac{da_i}{dt} \right)^2, \tag{287}$$

obtain
$$\frac{2p_c}{\rho} = \frac{1}{\displaystyle\sum_{i=1}^{3} a_i^{-2}} \left[\sum_{i=1}^{3} \left(\frac{1}{a_i} \frac{da_i}{dt} \right)^2 - \frac{2K^2}{a_1 a_2 (a_1 + a_2)^2} + 4 \right]. \tag{288}$$

Equations (248)–(250), together with equation (288), provide a complete set of equations. However, since, in deriving equation (288), the constancy of $a_1 a_2 a_3$ was used in the form (287), spurious effects, arising from the allowed linear increase of $\log(a_1 a_2 a_3)$ with t, may vitiate the integration of the equations. But one can guard against such effects by requiring that, at all times, $a_1 a_2 a_3$ has the same constant value.

The index symbols A_i, by their dependence on the a_i's, are also functions of time. But instead of evaluating them, at each instant, by making use of the formulas expressing them in terms of the a_i's, it is more convenient to determine them as solutions of the differential equations

$$\frac{dA_i}{dt} = -\left(3a_i A_{ii} \frac{da_i}{dt} + a_j A_{ij} \frac{da_j}{dt} + a_k A_{ik} \frac{da_k}{dt} \right), \tag{289}$$

which follow from the relations

$$\frac{\partial A_i}{\partial a_i} = \frac{1}{a_i}(A_i - 3a_i^2 A_{ii}) \quad \text{and} \quad \frac{\partial A_i}{\partial a_j} = \frac{1}{a_j}(A_i - a_j^2 A_{ij}) \tag{290}$$

($i \neq j \neq k$; and no summation over repeated indices in equations (289) and (290)).

In integrating the differential equations of the problem, we assign to K a value appropriate to some chosen Maclaurin spheroid and prescribe for E values that are in excess of W_{Mc} (defined in equations (278)) by different fractional amounts q (see equation (285)).

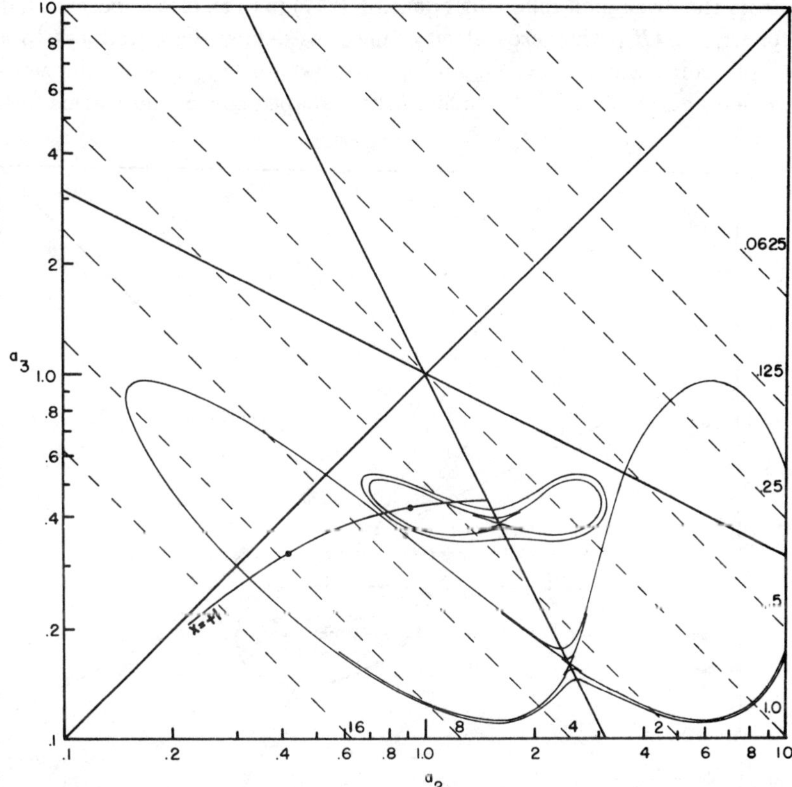

FIG. 18a. The zero-velocity curves (defined by equation (291)) for the two cases $a_1 = 1{\cdot}60\bar{a}$ and $2{\cdot}50\bar{a}$, for $q = 3 \times 10^{-3}$, 3×10^{-4}, and 3×10^{-6}. Both cases correspond to Maclaurin spheroids that are initially unstable. Notice that, for $q \to 0$, the curves tend towards a lemniscate shape.

The ordinate and the abscissa are the semi-axes a_2 and a_3 measured in the unit $(a_1 a_2 a_3)^{\frac{1}{3}}$. The solid lines of slopes 1, $-\frac{1}{2}$, and -2 are the loci $a_2 = a_3$, $a_1 = a_3$, and $a_1 = a_2$, respectively.

The curve identified by "$x = +1$" defines the self-adjoint Riemann sequence; and the two dots on this curve locate the particular Riemann ellipsoids that have the same energies as the Maclaurin spheroids at the vertices of the two lemniscates for $a_1 = 1{\cdot}60\bar{a}$ and $2{\cdot}50\bar{a}$.

Since $a_1 a_2 a_3 = \bar{a}^3$, the solution curves can be displayed as *trajectories* in the $(a_2/\bar{a}, a_3/\bar{a})$-plane. It is clear that in this plane the trajectory, for an assigned E and K_{Mc}, must be confined to the interior of the *zero-velocity curve* (cf. equation (277))

$$\frac{K_{\text{Mc}}^2}{(a_1+a_2)^2} - 2I = E. \tag{291}$$

From the discussion following equation (277) in § (a), it is apparent that for $\Omega^2_{Mc} \leqslant 4B_{11}$, the zero-velocity curve, as defined here, reduces to a single point when $E = W_{Mc}$ and $q = 0$; but for $\Omega^2_{Mc} > 4B_{11}$ the zero-velocity curve, for $q = 0$, is a lemniscate-shaped curve whose arms fork

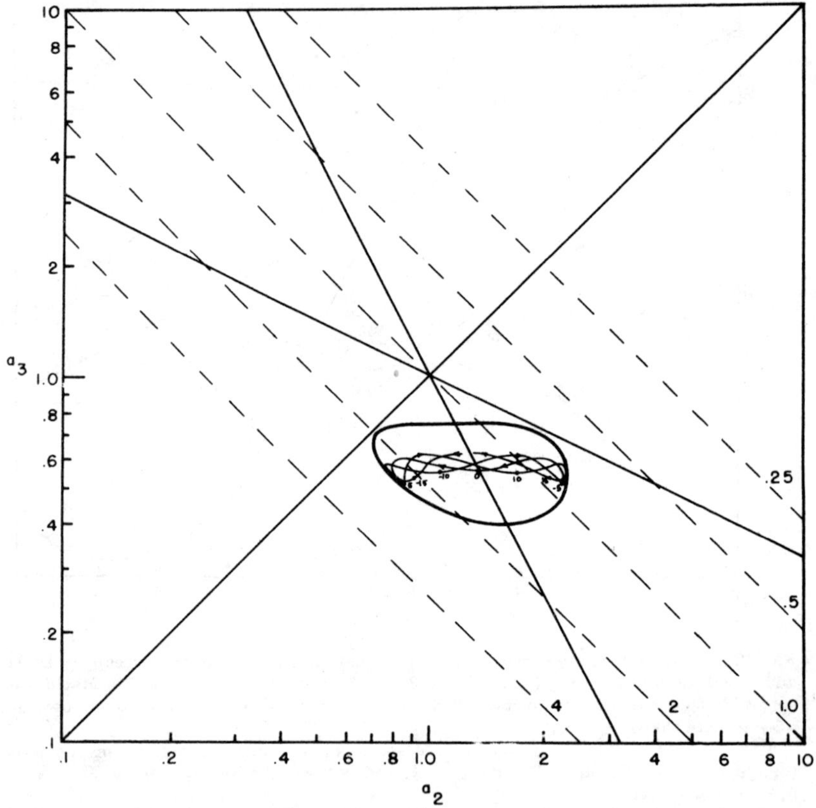

FIG. 18b. The trajectory for an initially stable Maclaurin spheroid, with $a_1 = 1\cdot35\bar{a}$, for $q = 3\times10^{-2}$. See Fig. 18c for the trajectory of the same spheroid for $q = 3\times10^{-1}$. (For an explanation of the various lines see the legend for Fig. 18a.)

out along directions determined by equations (279) and (280). Examples of zero-velocity curves (integrated by Rossner) are displayed in Fig. 18a; they are in accord with the theoretical deductions.

In Figs. 18b, 18c, and 18d, we display examples of trajectories integrated by Rossner.

Figs. 18b and 18c display the trajectories for an initially stable

Maclaurin spheroid with a major axis $a_1 = 1\cdot35\bar{a}$ and for values of $q = 3\times10^{-2}$ and 3×10^{-1}.

Fig. 18d displays the trajectories for two initially unstable Maclaurin spheroids with $a_1 = 1\cdot60\bar{a}$ and $a_1 = 2\cdot50\bar{a}$ and for a value of $q = 3\times10^{-3}$

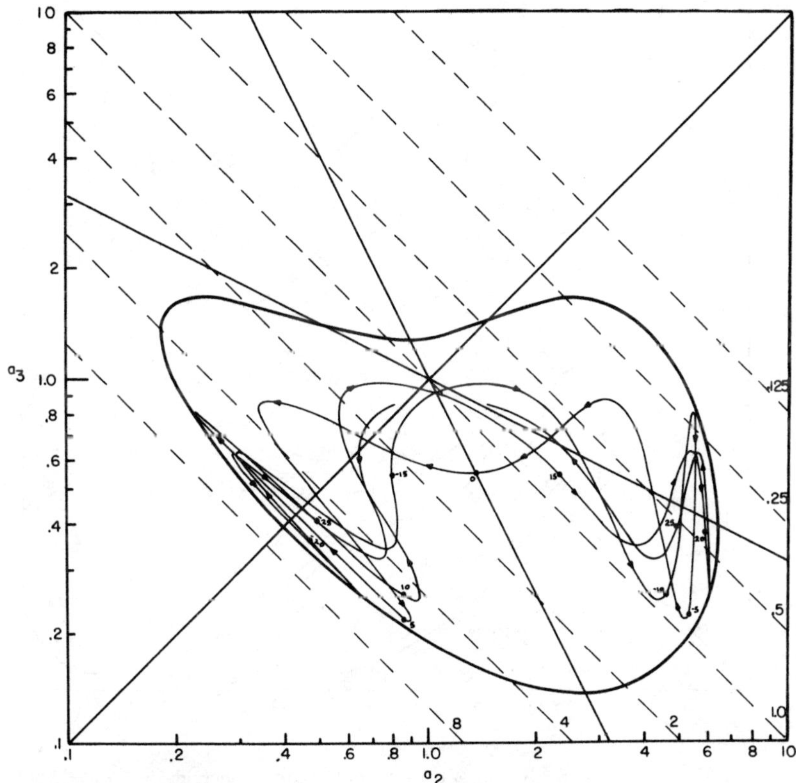

FIG. 18c. The trajectory for an initially stable Maclaurin spheroid, with $a_1 = 1\cdot35\bar{a}$, for $q = 3\times10^{-1}$. See Fig. 18b for the trajectory of the same spheroid for $q = 3\times10^{-3}$. (For an explanation of the various lines see the legend for Fig. 18a.)

in both cases. In the case $a_1 = 1\cdot60\bar{a}$, the "neck" of the zero-velocity curve is not sufficiently narrow to "trap" the trajectory. But in the second case, the trajectory does appear to be effectively trapped; indeed, it is seen to hover about the *equilibrium position* appropriate for a self-adjoint Riemann ellipsoid (on the sequence $x = +1$) having the same energy as the perturbed Maclaurin spheroid. This last fact suggests that a Maclaurin spheroid, perched on the potential ridge beyond $\Omega^2_{\mathrm{Mc}} = 4B_{11}$, has the tendency to slide down the slope into the valley that follows the course of the self-adjoint sequence $x = +1$.

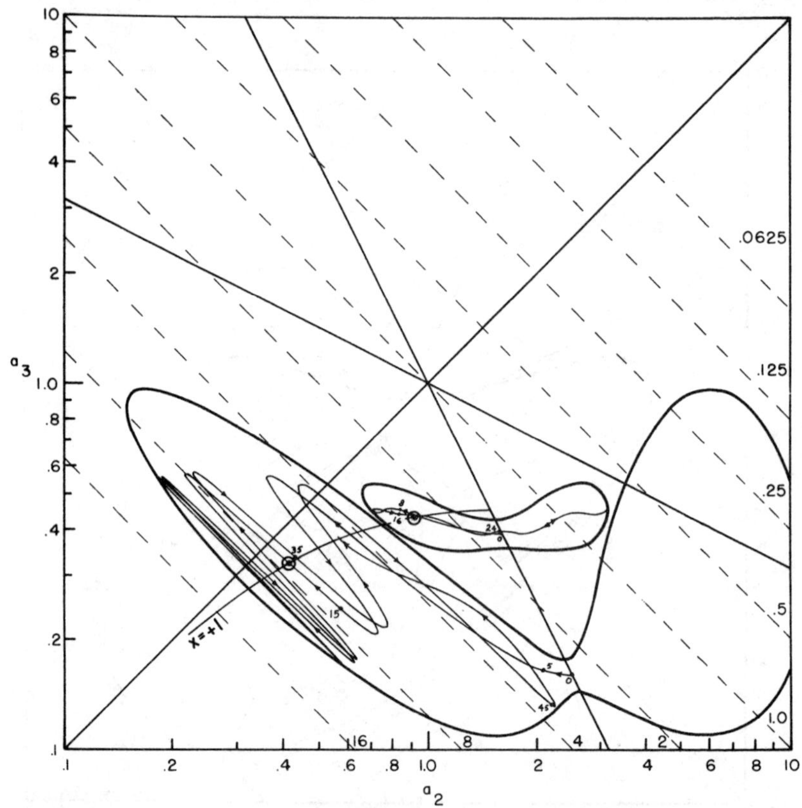

Fig. 18d. The trajectories for two initially unstable Maclaurin spheroids, with $a_1 = 1{\cdot}60\bar{a}$ and $2{\cdot}50\bar{a}$, for $q = 3 \times 10^{-3}$. The zero-velocity curves appropriate for these two cases are illustrated in Fig. 18a. (For an explanation of the various lines see the legend for Fig. 18a.) The two open circles on the self-adjoint sequence "$x = +1$" locate the particular equilibrium Riemann ellipsoids that have the same energies as the perturbed Maclaurin spheroids that describe the trajectories shown.

BIBLIOGRAPHICAL NOTES

Riemann's paper "Ein Beitrag zu den Untersuchungen über die Bewegung eines flüssigen gleichartigen Ellipsoides," communicated to *Der Königlichen Gesellschaft der Wissenschaften zu Göttingen* on December 8, 1860, is remarkable for the wealth of new results it contains and for the breadth of its comprehension of the entire range of problems. In the present writer's view this much neglected paper —it merits less than a sentence in Weber's biographical notice and none in Lewy's;

and there are no references to it in any of the writings of Poincaré, Darwin, or Jeans—deserves to be included among the other great papers of Riemann that are well known. Nevertheless, as we have remarked in the introductory section (§ 46) to this chapter, the paper contains some very surprising lapses and some definitely erroneous conclusions. In view of Riemann's unique place in science, a critical appraisal of this paper is perhaps justified.

To place the lapses and the errors in their proper perspective, it is necessary, first, to take measure of the paper's accomplishments. In their enumeration, which follows, the language and the terminology of this chapter will be used, even where they do not coincide with Riemann's.

(i) Dirichlet's problem is formulated in its entire generality and the basic set of nine equations (equivalent to the matrix-equation (57) of Chapter 4) is derived; and the integrals of energy, angular momentum, and circulation are isolated.

(ii) Dedekind's theorem is then deduced by what amounts to observing the equivalence of transposing the matrix-equation (57) to interchanging the roles of Λ^* and Ω^*. As Riemann says: "In this remark is contained the reciprocity theorem of Dedekind."

(iii) It is proved that under stationary conditions ellipsoidal figures are possible under precisely two conditions: *either* ζ and Ω are parallel, in which case they must lie along a principal axis of the ellipsoid *or* ζ and Ω are not parallel, in which case they must lie in a principal plane of the ellipsoid. (We have enunciated this result in § 47 as "Riemann's theorem.") Riemann further recognized that the equilibrium figures one obtains in the two cases are essentially different kinds of objects.

(iv) The analysis of the S-type ellipsoids (as we have called them) is complete to the extent that the basic equations determining the various parameters are derived and the fact recognized that their domain of occupancy is limited by the Maclaurin sequence and the two self-adjoint sequences. The irrotational sequence is briefly referred to in the context of his considerations of stability.

(v) The specification of the domain of occupancy of the ellipsoids of types I, II, and III—these are also Riemann's designations—is exceptionally complete. In particular, it is clearly and explicitly stated that the ellipsoids of type I adjoin the Maclaurin sequence; that the ellipsoids of type II are limited by the requirement $\Pi \geqslant 0$; and, finally, that one of the boundaries, limiting the domain of the ellipsoids of type III, is a sequence of S-type ellipsoids.

(vi) A substantial part of the paper is devoted to a discussion of the time-dependent equations (244) and (245) that are applicable to the case when Λ and Ω remain parallel to the a_3-axis. In particular, it is shown how in the special case $\Lambda = -\Omega$, the equations can be cast in the standard Lagrangian form. And the Lagrangian form of the equations enables a simple demonstration of the stability of the S-type ellipsoids to perturbations that are compatible with the restrictive assumptions underlying the equations. This result is equivalent to the stability of the S-type ellipsoids to the even second-harmonic modes of oscillation treated in § 49 (c).

(vii) And finally, by a discussion, equivalent to that given in § 53 (c), Riemann established (for the first time) the dynamical instability of the Maclaurin spheroids for $\Omega_{\mathrm{Mc}}^2 > 4B_{11}$. (But Riemann did not mention the relationship of these considerations to the fact that the self-adjoint sequence $x = +1$ bifurcates from the Maclaurin sequence at the same critical point $\Omega_{\mathrm{Mc}}^2 = 4B_{11}$.)

These, then, are the positive accomplishments of the paper. Certainly, few papers, if any, that have been written on this subject have comparable content or scope. But where Riemann went wrong was in his general considerations relative to the stability of his ellipsoids. Lebovitz (in Paper XXIX in the list on p. 246) has analyzed these parts of Riemann's paper and located the origin of his errors. The reader should refer to Lebovitz's paper for the analytical details. Here we shall restrict ourselves to bare statements of what Riemann's conclusions were and how they differ from the ones that have been arrived at in the text (and in Papers XXV and XXVIII) from a direct determination of the characteristic frequencies of oscillation.

It should first be stated that Riemann's discussion was restricted to perturbations that are linear in the coordinates, i.e. to the onset of instability via modes belonging to the second harmonics. Riemann clearly recognized this restriction; indeed, he refers to the importance of a "more general discussion." (The analysis in § 50 provides, for the S-type ellipsoids, this more general discussion.)

As we have already noted (in remark (vi) above), Riemann's conclusion with respect to the stability of the S-type ellipsoids for the even modes of oscillation (which do not perturb the initial directions of the vorticity and the angular velocity) is correct. But his general conclusion was that all S-type ellipsoids, in the domain included between the irrotational sequence $f = -2$ and the self-adjoint sequence $x = -1$, are unstable. This conclusion is incorrect: the S-type ellipsoids are unstable only in the smaller domain included between the locus (132) and the self-adjoint sequence $x = -1$.

Riemann also concluded that all the ellipsoids of types I, II, and III are unstable. Again, his conclusions with respect to the ellipsoids of types I and III are incorrect: among the ellipsoids of type I there are two distinct domains of stability—a domain adjoining the stable part of the Maclaurin sequence and another domain including disklike objects (see Fig. 17); and among the ellipsoids of type III, there is a fringe of stable configurations bordering on the locus, that these ellipsoids have in common with the S-type ellipsoids.

The surprising element in Riemann's errors is that his conclusions are in direct contradiction with some of his principal results; and it is difficult to understand why he failed to notice the contradictions.

Thus, consider his conclusions with respect to the instability of the ellipsoids of types III and S. Riemann had explicitly noted that these two types of ellipsoids have a sequence in common. The conclusion that, along this common sequence, the S-type ellipsoids *must* be characterized by a neutral mode of oscillation would appear inescapable; and, moreover, that an exchange of stability occurs along the same sequence would appear also very reasonable. How is it that Riemann failed to correlate the existence of a common sequence between the two types of ellipsoids with their stabilities?

Or, consider his conclusion with respect to the instability of *all* ellipsoids of type I. Riemann had explicitly noted that these ellipsoids adjoin the entire Maclaurin sequence; and he had also demonstrated that the Maclaurin spheroids are stable for $\Omega^2_{Mc} \leqslant 4B_{11}$. The conclusion that the known stability along that part of the Maclaurin line *must* extend into the domain of the type I ellipsoids would appear inescapable. How is it that Riemann overlooked this obvious consideration?

To conclude this appraisal of Riemann's great paper, the writer finds in its errors—so clearly at variance with some of its major findings—a tragic element. Were they connected in some way with Riemann's anxiety to contribute a suitable memorial to Dirichlet in "respectful gratitude?" We have already quoted (see p. 8) the eulogy to Dirichlet with which he began his paper. And the paper ends with this eulogy:

"The investigation of the conditions under which this may happen [i.e. the instability of the ellipsoids may arise] can be carried out by known methods since we shall be led only to linear differential equations. But the investigation of this question is beyond the scope of this paper which is devoted only to a further development of the beautiful ideas with which Dirichlet has crowned his scientific contributions."

We may briefly refer to the following three published accounts of the contributions of Dirichlet, Dedekind, and Riemann:

W. M. HICKS, "Recent progress in hydrodynamics," *Reports to the British Association* (1882), pp. 57–61.

A. BASSET, *A Treatise on Hydrodynamics*, **2** (Cambridge, England, Deighton, Bell and Company, 1888; reprint ed. New York, Dover Publications, 1961).

SIR HORACE LAMB, *Hydrodynamics* (Cambridge, England, Cambridge University Press, 1932), pp. 722–23.

Hicks's report is of interest as it presents a contemporaneous evaluation; and the enumeration of the results and the enunciation of the theorems of Dirichlet, Dedekind, and Riemann do occupy a central place in this report. Indeed, Hicks's report appears to be the only extant account which includes a statement of Dedekind's "remarkable reciprocal law." But the statement in the report, as to what Riemann's conclusions were concerning the stability of his ellipsoids, is erroneous.

In Chapter 15 of his *Hydrodynamics*, Basset attempts to give an adequate account of Dirichlet's and Riemann's investigations. On the formal side, the account is, indeed, adequate; and Dirichlet's investigation bearing on the solutions of equations (244) and (245) is particularly complete. But it must be admitted that in his account of Riemann's work he fails. Thus, while the basic equations and their integrals are derived in detail, there is no reference to Dedekind's theorem or to the fact that two states of motion are compatible with the same evolution of the external figure. Similarly, while Riemann's analysis, pertaining to the domains of occupancy of the ellipsoids of types I, II, and III, is set out in full, there is no reference to the relationships of these ellipsoids to the ellipsoids of type S or to the Maclaurin sequence. However, Basset's derivation of Riemann's criterion for the stability of the Maclaurin spheroid is clear and explicit: Riemann's original account is extremely terse.

Lamb's account appears to be a simple abridgement of Basset's (including its deficiencies).

Finally, reference may also be made to the following two papers by Greenhill:

A. G. GREENHILL, "On the rotation of a liquid ellipsoid about its mean axis," *Proc. Camb. Phil. Soc.*, **3** (1879), 233–46.

A. G. GREENHILL, "On the general motion of a liquid ellipsoid under gravitation of its own parts; continuation of a paper on the rotation of a liquid ellipsoid," *Proc. Camb. Phil. Soc.*, **4** (1880), 4–14.

The first paper contains a somewhat tedious derivation of the equations governing the S-type ellipsoids; these equations are then specialized appropriately to define the Jacobi, the Dedekind, and the irrotational sequences. In the second paper, Greenhill derives Riemann's basic equations with his choice of "moving frames" (see Chapter 4). But in neither paper is there any reference to Dedekind's theorem or to its implications for his problems.

As far as the present chapter is concerned, it is, except for § 53 (c), an abridged, but a more logically ordered, presentation of the author's two papers on the Riemann ellipsoids (Papers XXV and XXVIII). Reference to Lebovitz's account (Paper XXVII), in connection with his formulation of Riemann's equations, has already been made in the Bibliographical Notes for Chapter 4.

The results described in § 53 (c) are due to Rossner (Paper XXXVIII).

THE ROCHE ELLIPSOIDS

54. Introduction

So far, we have concerned ourselves with the ellipsoidal figures which a homogeneous mass can assume when, in addition to the action of its own gravity, centrifugal and Coriolis forces, derived from fluid motions, are operative. And we have seen in the preceding four chapters how the theory of these configurations, both as regards their equilibrium and as regards their stability, can be developed with complete mathematical exactitude. In this last chapter, we shall consider a further class of ellipsoidal figures that arise when, in addition to the forces already mentioned, the tidal effects of a neighboring mass are included in a certain approximation. The simplest problem in this more general context was first formulated by Roche in 1850 in a series of papers written, as Darwin has described, "in a style of singular modesty."

In spite of the approximate framework of the general theory, the study of these ellipsoids of Roche discloses certain novel relationships between dynamical and secular instability (cf. Chapter 5, § 37); and these relationships appear to be of some general interest.

55. Roche's problem

Roche's problem is concerned with the equilibrium and the stability of a homogeneous body ("the primary"—the body in which we are interested) rotating about another body ("the secondary"—the body which is the origin of the tidal effects on the primary) in such a manner that their relative dispositions remain the same.

Let the masses of the primary and the secondary be M and M', respectively; let the distance between their centers of mass be R; and let the constant angular velocity of rotation about their common center of mass be Ω.

Choose a coordinate system in which the origin is at the center of mass of the primary, the x_1-axis points to the center of mass of the secondary, and the x_3-axis is parallel to the direction of Ω. In this coordinate system, the equation of motion governing the fluid elements of M is

$$\rho\frac{du_i}{dt} = -\frac{\partial p}{\partial x_i} + \rho\frac{\partial}{\partial x_i}\left\{\mathfrak{B}+\mathfrak{B}'+\tfrac{1}{2}\Omega^2\left[\left(x_1-\frac{M'R}{M+M'}\right)^2+x_2^2\right]\right\}+2\rho\Omega\epsilon_{il3}\,u_l,$$

$$(1)$$

where $\mathfrak{B}'$ denotes the tidal potential due to M'.

In Roche's particular problem, the secondary is treated as a rigid sphere. Then, over the primary, the tide-generating potential $\mathfrak{B}'$ can be expanded in the form

$$\mathfrak{B}' = \frac{GM'}{R}\left(1+\frac{x_1}{R}+\frac{x_1^2-\tfrac{1}{2}x_2^2-\tfrac{1}{2}x_3^2}{R^2}+\cdots\right);\qquad(2)$$

and the approximation which underlies this theory is to retain, in this expansion for $\mathfrak{B}'$, only the terms which have been explicitly written down and ignore all the terms which are of higher order. On this assumption, the equation of motion becomes

$$\rho\frac{du_i}{dt} = -\frac{\partial p}{\partial x_i}+\rho\frac{\partial}{\partial x_i}\Bigg[\mathfrak{B}+\tfrac{1}{2}\Omega^2(x_1^2+x_2^2)+\mu(x_1^2-\tfrac{1}{2}x_2^2-\tfrac{1}{2}x_3^2)+$$
$$+\left(\frac{GM'}{R^2}-\frac{M'R}{M+M'}\Omega^2\right)x_1\Bigg]+2\rho\Omega\epsilon_{il3}\,u_l,\quad(3)$$

where we have introduced the abbreviation

$$\mu = \frac{GM'}{R^3}.\qquad(4)$$

So far, we have left Ω^2 unspecified. If we now let Ω^2 have the "Keplerian value"

$$\Omega^2 = \frac{G(M+M')}{R^3} = \mu\left(1+\frac{M}{M'}\right),\qquad(5)$$

the "unwanted" term in x_1, on the right-hand side of equation (3), vanishes and we are left with

$$\rho\frac{du_i}{dt} = -\frac{\partial p}{\partial x_i}+\rho\frac{\partial}{\partial x_i}[\mathfrak{B}+\tfrac{1}{2}\Omega^2(x_1^2+x_2^2)+\mu(x_1^2-\tfrac{1}{2}x_2^2-\tfrac{1}{2}x_3^2)]+2\rho\Omega\epsilon_{il3}\,u_l.$$

$$(6)$$

This is the basic equation of this theory; and Roche's problem is concerned with the equilibrium and the stability of homogeneous masses governed by equation (6).

(a) *The second-order virial equation appropriate to Roche's problem*

In treating Roche's problem, we shall continue to use the same methods based on the virial theorem and its extensions. However, the presence of the term in μ in equation (6) modifies the usual form of the virial equation to

$$\frac{d}{dt}\int_V \rho u_i x_j \, d\mathbf{x} = 2\mathfrak{T}_{ij}+\mathfrak{W}_{ij}+(\Omega^2-\mu)I_{ij}-\Omega^2\delta_{i3}\,I_{3j}+3\mu\delta_{i1}\,I_{1j}+$$
$$+2\Omega\epsilon_{il3}\int_V \rho u_l x_j \, d\mathbf{x} +\delta_{ij}\,\Pi. \qquad (7)$$

56. The Roche ellipsoids: the equilibrium figures

When no fluid motions are present in the frame of reference considered and hydrostatic equilibrium prevails, equation (7) becomes

$$\mathfrak{W}_{ij}+(\Omega^2-\mu)I_{ij}-\Omega^2\delta_{i3}\,I_{3j}+3\mu\delta_{i1}\,I_{1j} = -\delta_{ij}\,\Pi. \qquad (8)$$

The diagonal elements of this equation give

$$\mathfrak{W}_{11}+(\Omega^2+2\mu)I_{11} = \mathfrak{W}_{22}+(\Omega^2-\mu)I_{22} = \mathfrak{W}_{33}-\mu I_{33} = -\Pi, \qquad (9)$$

while the non-diagonal elements give

$$\mathfrak{W}_{12}+(\Omega^2+2\mu)I_{12} = \mathfrak{W}_{21}+(\Omega^2-\mu)I_{21} = 0,$$
$$\mathfrak{W}_{13}+(\Omega^2+2\mu)I_{13} = \mathfrak{W}_{31}-\mu I_{31} = 0,$$
$$\mathfrak{W}_{23}+(\Omega^2-\mu)I_{23} = \mathfrak{W}_{32}-\mu I_{32} = 0. \qquad (10)$$

In view of the symmetry of the tensors I_{ij} and $\mathfrak{W}_{ij}$, we may conclude from equations (10) that

$$\mathfrak{W}_{ij} = 0 \quad \text{and} \quad I_{ij} = 0 \quad \text{for } i \neq j. \qquad (11)$$

In other words, *in the chosen coordinate system the tensors $\mathfrak{W}_{ij}$ and I_{ij} are necessarily diagonal.*

Equations (9) and (11) are entirely general: they are in no way dependent on any constitutive relations that may exist.

Letting $\qquad p = M/M' \quad$ so that $\quad \Omega^2 = (1+p)\mu \qquad (12)$

and substituting for $\mathfrak{W}_{ij}$ and I_{ij} their known expressions for homogeneous ellipsoids, we obtain

$$(3+p)\mu a_1^2-2A_1 a_1^2 = p\mu a_2^2-2A_2 a_2^2 = -\mu a_3^2-2A_3 a_3^2. \qquad (13)$$

(In equation (13), Ω^2 and μ are measured in the unit $\pi G\rho$; and this convention will be adopted in the rest of this chapter.)

From equations (13) we obtain the pair of equations

$$[(3+p)a_1^2+a_3^2]\mu = 2(A_1 a_1^2-A_3 a_3^2) = 2(a_1^2-a_3^2)B_{13} \qquad (14)$$

and $\qquad (pa_2^2+a_3^2)\mu = 2(A_2 a_2^2-A_3 a_3^2) = 2(a_2^2-a_3^2)B_{23}. \qquad (15)$

Accordingly, the relation between a_2/a_1 and a_3/a_1 (for an assigned p) is determined by the equation

$$\frac{(3+p)a_1^2+a_3^2}{pa_2^2+a_3^2} = \frac{(a_1^2-a_3^2)B_{13}}{(a_2^2-a_3^2)B_{23}}, \tag{16}$$

and the values of μ and Ω^2 [$= (1+p)\mu$] that are to be associated with a particular solution of equation (16) follow from equation (14) or (15).

While we have derived the equations determining the equilibrium figure of a Roche ellipsoid by an application of the integral properties provided by the virial theorem, it can be readily shown that these equations are both necessary and sufficient: they are the same as would follow from a direct integration of the equation of hydrostatic equilibrium and the requirement that the resulting expression for the pressure vanishes on the boundary of the ellipsoid.

TABLE XVI

The properties of the Roche ellipsoids for $p = 0$ and $p = 1$

$p = 0$				$p = 1$			
$\cos^{-1}a_3/a_1$	a_2/a_1	a_3/a_1	Ω^2	$\cos^{-1}a_3/a_1$	a_2/a_1	a_3/a_1	Ω^2
24°	0·93188	0·91355	0·022624	12°	0·98660	0·97815	0·009293
36°	0·84112	0·80902	0·047871	24°	0·94376	0·91355	0·036152
48°	0·70687	0·66913	0·074799	36°	0·86345	0·80902	0·076342
57°	0·57787	0·54464	0·088267	48°	0·73454	0·66913	0·118726
60°	0·53013	0·50000	0·089946	54°	0·64956	0·58779	0·134284
61°	0·51373	0·48481	0·090068	59°	0·56892	0·51504	0·140854
62°	0·49714	0·46947	0·089977	60°	0·55186	0·50000	0·141250
63°	0·48040	0·45399	0·089689	61°	0·53451	0·48481	0·141298
66°	0·42898	0·40674	0·087201	66°	0·44429	0·40674	0·135785
71°	0·34052	0·32557	0·077474	69°	0·38813	0·35837	0·127424
72°	0·32254	0·30902	0·074648	71°	0·35022	0·32557	0·119625
75°	0·26827	0·25882	0·064426	72°	0·33119	0·30902	0·115054
79°	0·19569	0·19081	0·047111	73°	0·31213	0·29237	0·110044
				75°	0·27405	0·25882	0·098753
				78°	0·21726	0·20791	0·078934
				81°	0·16126	0·15643	0·056499

Solutions of equation (16) for the two important cases $p = 0$ and $p = 1$ are given in Table XVI. And Fig. 19 exhibits the variation of Ω^2 along the Roche sequences for $p = 0, 1, 4, 20$, and 100.

It will be observed that in the limit

$$\mu \to 0, \quad p \to \infty, \quad \text{and} \quad \mu p \to \Omega^2 \tag{17}$$

equations (13) reduce to the equations determining the Maclaurin and the Jacobian sequences. And it is apparent from Fig. 19 that *in this limit, the Roche sequence tends to the combined Maclaurin–Jacobi sequence* —the Maclaurin sequence prior to the point of bifurcation and the Jacobian sequence subsequent to that point.

(a) *The Jeans spheroids and the tidal problem*

A variant of Roche's problem and one which may be regarded as a special case of it was considered in some detail by Jeans as representing a "pure" tidal problem. This special case arises when we consider the

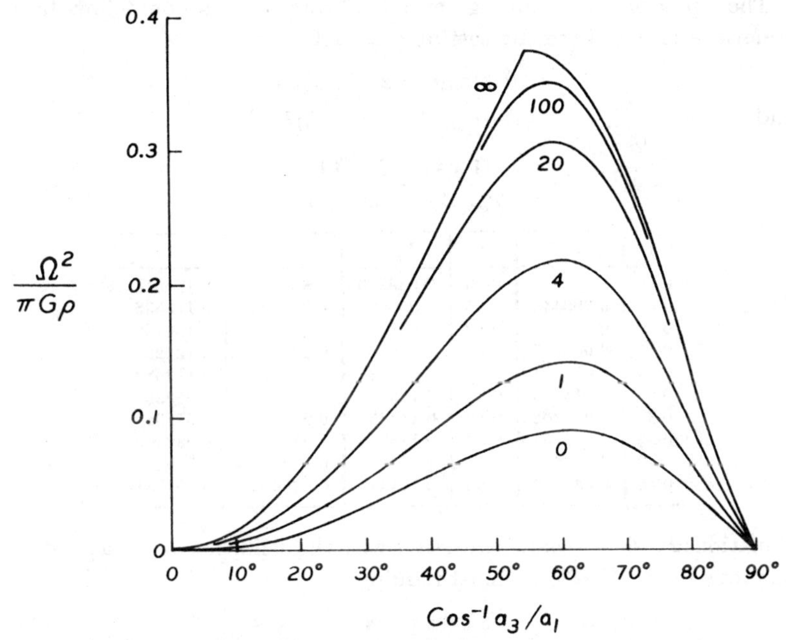

FIG. 19. The variation of Ω^2 along the Roche sequences. The curves are labeled by the values of p to which they belong; and the curve labeled by ∞ belongs to the combined Maclaurin–Jacobi sequence. The maxima of the curves define the Roche limit.

equation of motion (3) in an inertial, instead of in a rotating, frame. The equation then takes the form

$$\rho \frac{du_i}{dt} = -\frac{\partial p}{\partial x_i} + \rho \frac{\partial}{\partial x_i}[\mathfrak{B} + \mu(x_1^2 - \tfrac{1}{2}x_2^2 - \tfrac{1}{2}x_3^2)] + \rho \frac{GM'}{R^2}\delta_{i1}. \qquad (18)$$

We can now "neutralize" the constant force GM'/R^2 acting in the x_1-direction by referring the motion to a uniformly accelerated moving frame (in the sense described in Chapter 4, § 24). In such a moving frame, the center of gravity of M will always remain at the origin and equation (18) will become

$$\rho \frac{du_i}{dt} = -\frac{\partial p}{\partial x_i} + \rho \frac{\partial}{\partial x_i}[\mathfrak{B} + \mu(x_1^2 - \tfrac{1}{2}x_2^2 - \tfrac{1}{2}x_3^2)]. \qquad (19)$$

A comparison of equations (6) and (19) now shows that this tidal problem can be considered, formally, as a special case of Roche's problem for (cf. equation (12))

$$p = -1, \tag{20}$$

so that the meaning of p as M/M' should be abandoned in this case.

The equations determining the equilibrium figures now follow from equations (14) and (15) by setting $p = -1$. We have

$$(2a_1^2 + a_3^2)\mu = 2(a_1^2 - a_3^2)B_{13} \tag{21}$$

and

$$(a_3^2 - a_2^2)\mu = 2(a_2^2 - a_3^2)B_{23}. \tag{22}$$

TABLE XVII

The Jeans sequence

e	μ	e	μ	e	μ
0	0	0·50	0·046219	0·88	0·125514
0·05	0·000445	0·55	0·056209	0·8830265	0·125536
0·10	0·001781	0·60	0·067135	0·90	0·124760
0·15	0·004017	0·65	0·078864	0·92	0·121293
0·20	0·007164	0·70	0·091137	0·94	0·113683
0·25	0·011240	0·75	0·103451	0·96	0·099288
0·30	0·016262	0·80	0·114839	0·98	0·072040
0·35	0·022254	0·82	0·118754	0·995	0·030569
0·40	0·029235	0·84	0·122038	0·999	0·009236
0·45	0·037222	0·86	0·124419	0·9999	0·001381

Equation (22) shows that, compatible with the definition of μ as a positive quantity $(= GM'/R^3)$, we must require

$$a_2 = a_3. \tag{23}$$

The equilibrium figures are, therefore, *prolate spheroids*. And using the explicit expressions for the index symbols appropriate for this case (given in equations (38) of Chapter 3), we find from equation (21) that

$$\mu = \frac{1-e^2}{e^3}\log\frac{1+e}{1-e} - \frac{6}{e^2}\frac{1-e^2}{3-e^2}, \tag{24}$$

where

$$e = (1 - a_2^2/a_1^2)^{\frac{1}{2}} \tag{25}$$

denotes the eccentricity.

Table XVII gives the values of μ for various values of e; and the variation of μ with e is further illustrated in Fig. 20. We observe that, even as in the other Roche sequences, $\mu(e)$ attains a maximum value; it occurs at

$$e = 0·88303 \quad \text{where} \quad \mu = 0·12554. \tag{26}$$

(b) The arrangement of the solutions

The solutions for the different Roche sequences and the relationships which exist between these sequences and those of Maclaurin, Jacobi, and Jeans are most conveniently exhibited in a plane (first used for this purpose by Jeans) in which each equilibrium ellipsoid is represented by a point whose coordinates are

$$\bar{a}_1 = a_1/(a_1 a_2 a_3)^{\frac{1}{3}} \quad \text{and} \quad \bar{a}_2 = a_2/(a_1 a_2 a_3)^{\frac{1}{3}}. \tag{27}$$

By this normalization, the volume (and in view of the homogeneity, also the mass) of all the ellipsoids is the same.

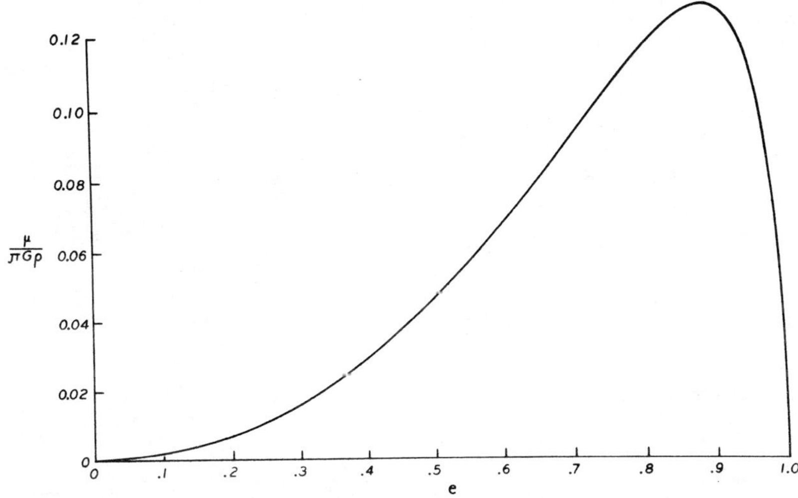

FIG. 20. The variation of $\mu/\pi G\rho$ along the Jeans sequence; μ attains its maximum at $e = 0\cdot 8830$ (where the Jeans spheroid becomes unstable).

In Fig. 21, the solutions for the various Roche sequences are exhibited in the $(\bar{a}_1, \bar{a}_2)$-plane. In this plane, the undistorted sphere is represented by the point $\bar{a}_1 = \bar{a}_2 = 1$; this is the point S in Fig. 21.

The Jeans spheroids, being prolate, are represented by the pseudo-hyperbolic locus

$$\bar{a}_1 \bar{a}_2^2 = 1; \tag{28}$$

this is the curve ST.

The Maclaurin spheroids, being oblate, are represented by the straight line

$$\bar{a}_1 = \bar{a}_2 \quad (\geqslant 1); \tag{29}$$

this is the straight line SM_3''. The Jacobian sequence bifurcates from the Maclaurin sequence at the point M_2 where

$$\bar{a}_1 = \bar{a}_2 = 1\cdot 19723; \tag{30}$$

at this point $\Omega^2 = 0\cdot 37423$.

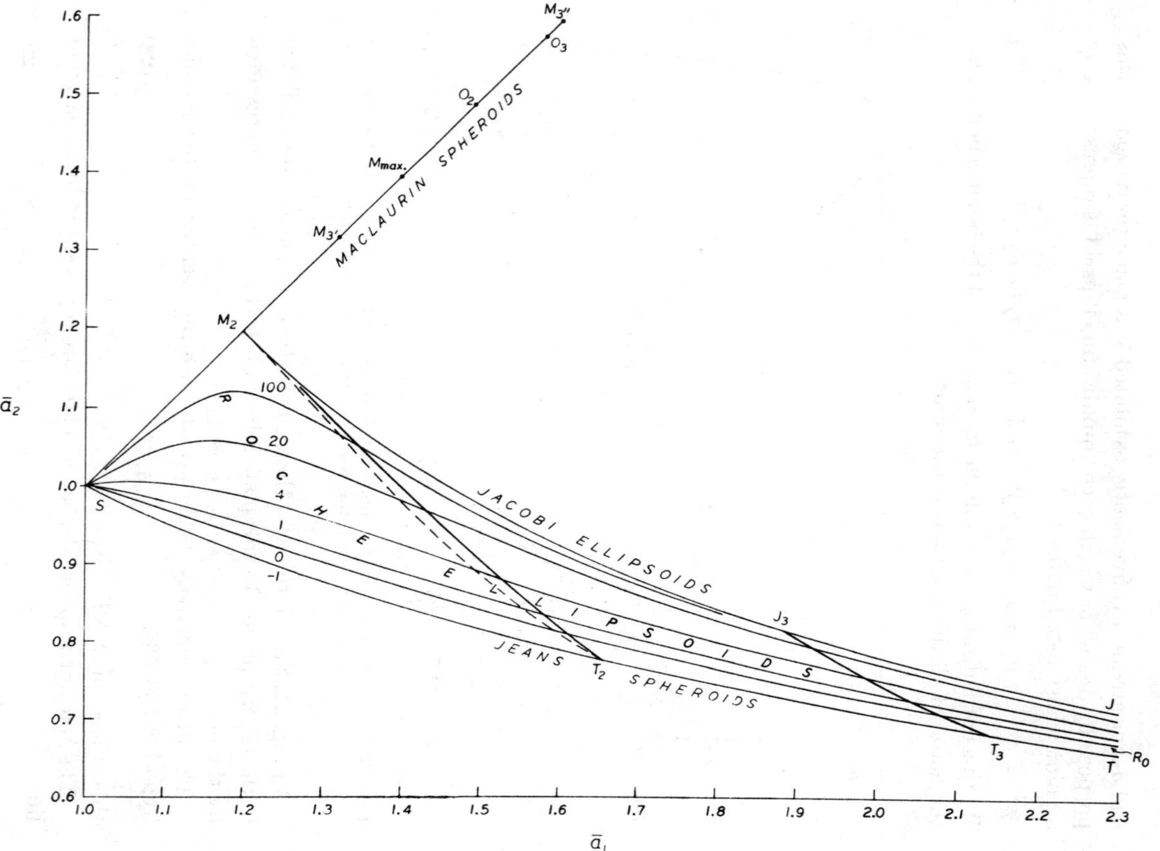

Fig. 21

(See opposite for explanation)

The Jacobian sequence is represented by the curve $M_2 J$. Since the Jacobian sequence eventually becomes prolate, the locus $M_2 J$ becomes asymptotic to ST as $\bar{a}_1 \to \infty$. Also, the Jacobi ellipsoids become unstable at the point J_3 where the pear-shaped sequence branches off.

The Roche sequences for the different p's are represented by a one-parameter family of curves bounded by the part SM_2 of the Maclaurin sequence, the Jacobian sequence $M_2 J$, and the Roche sequence SR_0 for $p = 0$. It follows from our earlier remarks, and it is now manifest from Fig. 21, that in the limit $p \to \infty$, the Roche sequence becomes the combined Maclaurin–Jacobi sequence represented by the broken curve $SM_2 J$.

(c)   The Roche limit

We have seen that Ω^2 and μ attain maxima, simultaneously, at some determinate point along a Roche sequence. Therefore, for distances R, less than what corresponds to the maximum value of μ attained, no equilibrium figures are possible. And this limit, defining the distance of closest approach for equilibrium to be possible, is called the *Roche limit*.

The Roche limit is most conveniently determined by a method similar to the one described in § 32 for locating the maximum of Ω^2 along the Maclaurin sequence.

Explanation to Fig. 21

Fig. 21. The relationships among the Maclaurin, the Jacobi, the Jeans, and the Roche sequences. The ordinates and the abscissae are the normalized values, $\bar{a}_1 = a_1/(a_1 a_2 a_3)^{\frac{1}{3}}$ and $\bar{a}_2 = a_2/(a_1 a_2 a_3)^{\frac{1}{3}}$, of two of the principal axes of the equilibrium ellipsoid (or spheroid). The undistorted sphere is represented by S, the Maclaurin sequence by SM_3'', the Jacobian sequence by $M_2 J$, the Jeans sequence by ST, and the Roche sequences (labeled by the values of $p = M/M'$ to which they belong) are confined to the domain bounded by the combined Maclaurin–Jacobi sequence, $SM_2 J$, and the Roche sequence, SR_0, for $p = 0$. The first point of bifurcation along the Maclaurin sequence occurs at M_2 ($e = 0{\cdot}8127$); at M_3' ($e = 0{\cdot}8993$) and M_3'' ($e = 0{\cdot}9694$) occur further neutral points belonging to the third harmonics. At the points O_2 ($e = 0{\cdot}9529$) and O_3 ($e = 0{\cdot}9670$) the Maclaurin spheroid becomes unstable by modes of overstable oscillation belonging to the second and the third harmonics; and at M_{max} ($e = 0{\cdot}9300$), Ω^2 attains its maximum values along SM_3''. The Jacobi ellipsoid becomes unstable by a mode of oscillation belonging to the third harmonics at J_3 ($\bar{a}_1 = 1{\cdot}8858$; $\bar{a}_2 = 0{\cdot}8150$) where the pear-shaped sequence branches off; also, along the entire sequence $M_2 J$ the Jacobi ellipsoids are characterized by a neutral mode of oscillation belonging to the second harmonics. At the points T_2 ($e = 0{\cdot}8830$) and T_3 ($e = 0{\cdot}9477$) the Jeans spheroids become unstable by modes of oscillation belonging to the second and the third harmonics. The Roche limit where Ω^2 and μ attain their maxima along the different Roche sequences is represented by the dashed curve joining M_2 and T_2; and the locus of points where instability sets in by a mode of oscillation belonging to the second harmonics is shown by the heavy curve joining M_2 and T_2. The locus of the neutral point (belonging to the third harmonics) is shown by the curve joining T_3 and J_3. Note that the Jacobi ellipsoids are to be considered unstable in the limit $p \to \infty$.

Since the structure of the Roche ellipsoids is uniquely determined by equations (9), it is clear that at the point where Ω^2 and μ attain their maxima, not only must these equations be satisfied, their first variations, for a displacement which preserves the ellipsoidal figure and the same constant density, must also be satisfied. In other words, at the maximum, in addition to equations (9), the equations

$$\delta\mathfrak{W}_{11}+(\Omega^2+2\mu)V_{11} = \delta\mathfrak{W}_{22}+(\Omega^2-\mu)V_{22} = \delta\mathfrak{W}_{33}-\mu V_{33} \tag{31}$$

must also be satisfied for a displacement for which

$$\frac{V_{11}}{a_1^2}+\frac{V_{22}}{a_2^2}+\frac{V_{33}}{a_3^2} = 0. \tag{32}$$

This last condition ensures that the density retains the same constant value.

Equations (31) are equivalent to the pair of equations

$$(\Omega^2+2\mu)V_{11}-(\Omega^2-\mu)V_{22} = \delta\mathfrak{W}_{22}-\delta\mathfrak{W}_{11} \tag{33}$$

and

$$(\Omega^2+2\mu)V_{11}+(\Omega^2-\mu)V_{22}+2\mu V_{33} = -(\delta\mathfrak{W}_{11}+\delta\mathfrak{W}_{22}-2\delta\mathfrak{W}_{33}). \tag{34}$$

In view of equations (149) and (150) of Chapter 3, equations (32)–(34) become linear and homogeneous in V_{11}, V_{22}, and V_{33}; and the condition that the equations allow a non-trivial solution is found to be

$$\begin{vmatrix} \Omega^2+2\mu-3B_{11}+B_{12} & -\Omega^2+\mu+3B_{22}-B_{12} & B_{23}-B_{13} \\ \Omega^2+2\mu-(3B_{11}+B_{12}-2B_{13}) & \Omega^2-\mu-(3B_{22}+B_{12}-2B_{23}) & 2\mu+6B_{33}-B_{13}-B_{23} \\ \dfrac{1}{a_1^2} & \dfrac{1}{a_2^2} & \dfrac{1}{a_3^2} \end{vmatrix} = 0. \tag{35}$$

And this equation determines the point where Ω^2 and μ attain their maxima. By some elementary transformations, equation (35) can be brought to the somewhat simpler form

$$\begin{vmatrix} \Omega^2+2\mu-3B_{11}+B_{13} & B_{23}-B_{12} & \mu+3B_{33}-B_{13} \\ B_{13}-B_{12} & \Omega^2-\mu-3B_{22}+B_{23} & \mu+3B_{33}-B_{23} \\ \dfrac{1}{a_1^2} & \dfrac{1}{a_2^2} & \dfrac{1}{a_3^2} \end{vmatrix}$$

$$= \Delta_{\text{Roche limit}} = 0. \tag{36}$$

The points along the different Roche sequences where Ω^2 and μ attain their maxima, determined with the aid of equation (36), are listed in Table XVIII.

The point where μ attains its maximum along the Jeans sequence can be obtained from equation (36) by setting $\Omega^2 = 0$ and remembering that the Jeans spheroid is prolate. We find

$$\mu = \frac{2}{4a_1^2 - a_2^2}[3a_1^2 B_{11} + 2a_2^2 B_{22} - (a_1^2 + a_2^2)B_{12}]. \tag{37}$$

TABLE XVIII

The Roche limit and the constants of the critical ellipsoid
(Ω^2 and μ are listed in the unit $\pi G\rho$)

p	$\Omega^2_{\max}$	$\mu_{\max}$	$\bar{a}_1$	$\bar{a}_2$	$\bar{a}_3$
-1	0	0·125536	1·6558	0·7771	0·7771
0	0·090093	0·090093	1·5943	0·8152	0·7694
1	0·141322	0·070661	1·5565	0·8418	0·7632
4	0·216861	0·043372	1·4944	0·8913	0·7508
20	0·306396	0·014590	1·3989	0·9821	0·7279
100	0·350562	0·003471	1·3224	1·0642	0·7106
∞	0·374230	0	1·1972	1·1972	0·6977

In Fig. 21, the locus of points where Ω^2 (and, or μ) attain their maxima is the dashed curve joining the points T_2 and M_2. This locus defines the Roche limit.

57. The stability of the Roche ellipsoids with respect to the second-harmonic oscillations

We now turn to the question of the stability of the Roche ellipsoids with respect to oscillations belonging to the second harmonics. As in the earlier chapters, we shall consider this question with the aid of the linearized version of the appropriate second-order virial equations.

From equation (7), it is apparent that for the Roche ellipsoids, the equation we have to consider is

$$\lambda^2 V_{i;j} - 2\lambda\Omega\epsilon_{il3}V_{l;j} = \delta\mathfrak{W}_{ij} + (\Omega^2 - \mu)V_{ij} - \Omega^2\delta_{i3}V_{3j} + 3\mu\delta_{i1}V_{1j} + \delta_{ij}\delta\Pi, \tag{38}$$

where λ is the parameter whose characteristic values are to be determined.

The nine equations that equation (38) represents fall into two non-combining groups distinguished by their parity with respect to the index 3. The equations, odd in the index 3, are

$$\lambda^2 V_{3;1} = \delta\mathfrak{W}_{31}-\mu V_{13} = -(2B_{13}+\mu)V_{13}, \tag{39}$$

$$\lambda^2 V_{3;2} = \delta\mathfrak{W}_{32}-\mu V_{23} = -(2B_{23}+\mu)V_{23}, \tag{40}$$

$$\lambda^2 V_{1;3}-2\lambda\Omega V_{2;3} = \delta\mathfrak{W}_{13}+(\Omega^2+2\mu)V_{13} = -(2B_{13}-\Omega^2-2\mu)V_{13}, \tag{41}$$

$$\lambda^2 V_{2;3}+2\lambda\Omega V_{1;3} = \delta\mathfrak{W}_{23}+(\Omega^2-\mu)V_{23} = -(2B_{23}-\Omega^2+\mu)V_{23}, \tag{42}$$

where we have substituted for the $\delta\mathfrak{W}_{ij}$'s in accordance with equation (148) of Chapter 3. And similarly, the equations, even in the index 3, are

$$\lambda^2 V_{3;3} = \delta\mathfrak{W}_{33}-\mu V_{33}+\delta\Pi, \tag{43}$$

$$\lambda^2 V_{1;1}-2\lambda\Omega V_{2;1} = \delta\mathfrak{W}_{11}+(\Omega^2+2\mu)V_{11}+\delta\Pi, \tag{44}$$

$$\lambda^2 V_{2;2}+2\lambda\Omega V_{1;2} = \delta\mathfrak{W}_{22}+(\Omega^2-\mu)V_{22}+\delta\Pi, \tag{45}$$

$$\lambda^2 V_{1;2}-2\lambda\Omega V_{2;2} = \delta\mathfrak{W}_{12}+(\Omega^2+2\mu)V_{12} = -(2B_{12}-\Omega^2-2\mu)V_{12}, \tag{46}$$

$$\lambda^2 V_{2;1}+2\lambda\Omega V_{1;1} = \delta\mathfrak{W}_{21}+(\Omega^2-\mu)V_{12} = -(2B_{12}-\Omega^2+\mu)V_{12}. \tag{47}$$

(a) *The characteristic equation governing the odd modes of oscillation*

Adding equations (39) and (41) and similarly equations (40) and (42), we obtain

$$(\lambda^2+4B_{13}-\Omega^2-\mu)V_{13}-2\lambda\Omega V_{23}+2\lambda\Omega V_{3;2} = 0 \tag{48}$$

and

$$(\lambda^2+4B_{23}-\Omega^2+2\mu)V_{23}+2\lambda\Omega V_{13}-2\lambda\Omega V_{3;1} = 0. \tag{49}$$

Eliminating $V_{3;1}$ and $V_{3;2}$ from the foregoing equations with the aid of equations (39) and (40), we have

$$\lambda(\lambda^2+4B_{13}-\Omega^2-\mu)V_{13}-2\Omega(\lambda^2+2B_{23}+\mu)V_{23} = 0 \tag{50}$$

and

$$\lambda(\lambda^2+4B_{23}-\Omega^2+2\mu)V_{23}+2\Omega(\lambda^2+2B_{13}+\mu)V_{13} = 0; \tag{51}$$

and these two equations lead to the characteristic equation

$$\lambda^2(\lambda^2+4B_{13}-\Omega^2-\mu)(\lambda^2+4B_{23}-\Omega^2+2\mu)+$$
$$+4\Omega^2(\lambda^2+2B_{23}+\mu)(\lambda^2+2B_{13}+\mu) = 0. \tag{52}$$

It will be observed that by setting $\mu = 0$ in equation (52), we recover the equation governing the odd modes of oscillation of the Jacobi ellipsoid (or the Maclaurin spheroid if the indices 1 and 2 are not distinguished).

For the case of the Jeans spheroid ($\Omega^2 = 0$ and $a_2 = a_3$), equation (52) becomes

$$\lambda^2(\lambda^2+4B_{12}-\mu)(\lambda^2+4B_{22}+2\mu) = 0. \tag{53}$$

The squares of the characteristic frequencies of oscillation are, therefore,

$$0, \quad 4B_{12}-\mu, \quad \text{and} \quad 4B_{22}+2\mu. \tag{54}$$

(b) The characteristic equation governing the even modes of oscillation

Turning next to the even equations (43)–(47), and combining them so as to eliminate $\delta\Pi$, we obtain the four equations

$$(\lambda^2+4B_{12}-2\Omega^2-\mu)V_{12}+\lambda\Omega(V_{11}-V_{22}) = 0, \tag{55}$$

$$\lambda^2(V_{1;2}-V_{2;1}) = \lambda\Omega(V_{11}+V_{22})+3\mu V_{12}, \tag{56}$$

$$\tfrac{1}{2}\lambda^2(V_{11}-V_{22})-2\lambda\Omega V_{12} = \delta\mathfrak{W}_{11}-\delta\mathfrak{W}_{22}+(\Omega^2+2\mu)V_{11}-(\Omega^2-\mu)V_{22}, \tag{57}$$

$$\tfrac{1}{2}\lambda^2(V_{11}+V_{22})+2\lambda\Omega(V_{1;2}-V_{2;1})-\lambda^2V_{33} = \delta\mathfrak{W}_{11}+\delta\mathfrak{W}_{22}-2\delta\mathfrak{W}_{33}+$$
$$+(\Omega^2+2\mu)V_{11}+(\Omega^2-\mu)V_{22}+2\mu V_{33}. \tag{58}$$

Rearranging equation (57) and eliminating $(V_{1;2}-V_{2;1})$ from equation (58) with the aid of equation (56) (and rearranging), we obtain the pair of equations

$$(\tfrac{1}{2}\lambda^2-\Omega^2-2\mu)V_{11}-(\tfrac{1}{2}\lambda^2-\Omega^2+\mu)V_{22}-2\lambda\Omega V_{12} = \delta\mathfrak{W}_{11}-\delta\mathfrak{W}_{22}, \tag{59}$$

$$(\tfrac{1}{2}\lambda^2+\Omega^2-2\mu)V_{11}+(\tfrac{1}{2}\lambda^2+\Omega^2+\mu)V_{22}-(\lambda^2+2\mu)V_{33}+\frac{6\Omega\mu}{\lambda}V_{12}$$
$$= \delta\mathfrak{W}_{11}+\delta\mathfrak{W}_{22}-2\delta\mathfrak{W}_{33}. \tag{60}$$

In view of equations (149) and (150) of Chapter 3, equations (32), (55), (59), and (60) become linear and homogeneous in V_{11}, V_{22}, V_{33}, and V_{12}; and they lead to the characteristic equation (61) (see next page).

After some elementary transformations, equation (61) can be brought to the somewhat simpler form (62), where in passing from equation (61) to (62) a factor λ^2 has been suppressed (see equation (88) in § 59 below).

The equation appropriate to the Jeans spheroids can be obtained by setting $\Omega^2 = 0$ in equation (62) and not distinguishing the indices 2 and 3. We find that the resulting equation factorizes to give

$$[\tfrac{1}{2}\sigma^2(2a_1^2+a_2^2)-6a_1^2B_{11}-4a_2^2B_{22}+2(a_1^2+a_2^2)B_{12}+(4a_1^2-a_2^2)\mu]\times$$
$$\times(\sigma^2-4B_{12}+\mu)(\sigma^2-4B_{22}-2\mu) = 0, \tag{63}$$

where we have written $\sigma^2 = -\lambda^2$. From equation (63) it follows that the two non-vanishing roots given in (54) are repeated, while a third root is given by

$$\sigma^2 = \frac{2}{2a_1^2+a_2^2}[6a_1^2B_{11}+4a_2^2B_{22}-2(a_1^2+a_2^2)B_{12}-(4a_1^2-a_2^2)\mu]. \tag{64}$$

Comparison with equation (37) shows that σ^2 given by equation (64) vanishes at the point where μ attains its maximum; and, further, that $\sigma^2 < 0$ beyond this point. We conclude that *the Jeans spheroid is unstable beyond the point where μ attains its maximum.*

$$\begin{vmatrix} \tfrac{1}{2}\lambda^2-\Omega^2-2\mu+3B_{11}-B_{12} & -\tfrac{1}{2}\lambda^2+\Omega^2-\mu-3B_{22}+B_{12} & B_{13}-B_{23} & -2\lambda\Omega \\[4pt] \tfrac{1}{2}\lambda^2+\Omega^2-2\mu+3B_{11}+B_{12}-2B_{13} & \tfrac{1}{2}\lambda^2+\Omega^2+\mu+3B_{22}+B_{12}-2B_{23} & -\lambda^2-2\mu-6B_{33}+B_{13}+B_{23} & 6\Omega\mu/\lambda \\[4pt] \lambda\Omega & -\lambda\Omega & 0 & \lambda^2+4B_{12}-2\Omega^2-\mu \\[4pt] \dfrac{1}{a_1^2} & \dfrac{1}{a_2^2} & \dfrac{1}{a_3^2} & 0 \end{vmatrix} = 0. \quad (61)$$

$$(\lambda^2+4B_{12}-2\Omega^2-\mu)\begin{vmatrix} \tfrac{1}{2}\lambda^2-\Omega^2-2\mu+3B_{11}-B_{12} & B_{23}-B_{13} & -3\mu+3(B_{11}-B_{22})+B_{13}-B_{23} \\[4pt] \Omega^2+B_{12}-B_{13} & \tfrac{1}{2}\lambda^2+\mu+3B_{33}-B_{23} & \Omega^2+3(B_{22}-B_{33})+B_{12}-B_{13} \\[4pt] \dfrac{1}{a_1^2} & -\dfrac{1}{a_3^2} & \dfrac{1}{a_1^2}+\dfrac{1}{a_2^2}+\dfrac{1}{a_3^2} \end{vmatrix} +$$

$$+\,\Omega^2\begin{vmatrix} 2\lambda^2 & B_{23}-B_{13} & -3\mu+3(B_{11}-B_{22})+B_{13}-B_{23} \\[4pt] -(\lambda^2+3\mu) & \tfrac{1}{2}\lambda^2+\mu+3B_{33}-B_{23} & \Omega^2+3(B_{22}-B_{33})+B_{12}-B_{13} \\[4pt] 0 & -\dfrac{1}{a_3^2} & \dfrac{1}{a_1^2}+\dfrac{1}{a_2^2}+\dfrac{1}{a_3^2} \end{vmatrix} = 0. \quad (62)$$

(c) *The point of onset of dynamical instability along a Roche sequence*

The characteristic frequencies of oscillation, along the Jeans sequence and the Roche sequences for $p = 0$ and 1, are listed in Table XIX. And the variations of σ^2, along the different Roche sequences for the mode by which the Roche ellipsoids become unstable, are exhibited in Fig. 22. We observe how the mode, which for $p = -1$ is unstable beyond

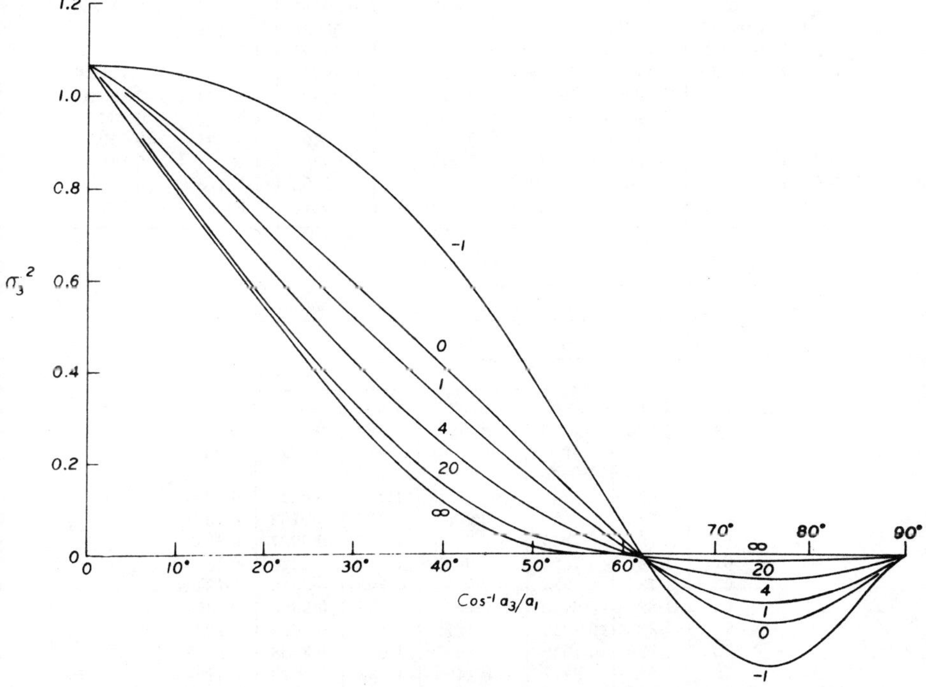

FIG. 22. The squares of the characteristic frequency σ_3^2 (measured in the unit $\pi G\rho$) belonging to the mode by which the Roche ellipsoids and the Jeans spheroids become unstable. The curves for the different Roche sequences are labeled by the values of p to which they belong; the curve labeled by ∞ belongs to the combined Maclaurin–Jacobi sequence and the curve labeled by -1 belongs to the Jeans sequence.

$\mu_{\max}$, becomes neutral along the entire Jacobian part of the combined Maclaurin–Jacobi sequence for $p \to \infty$.

In Table XX the points at which instability sets in along the different Roche sequences are listed; and in Fig. 21, the locus of these points of marginal stability is the full drawn curve joining M_2 and T_2.

It is manifest from Fig. 21 that *the Roche ellipsoids, for any finite $p \geqslant 0$,*

TABLE XIX

The squares of the characteristic frequencies belonging to the second harmonics (σ^2 is listed in the unit $\pi G\rho$)

Jeans sequence

e	σ_1^2	σ_2^2	σ_3^2	e	σ_1^2	σ_2^2	σ_3^2
0	1·06667	1·06667	1·06667	0·75	0·79312	1·52033	$+0\cdot42239$
0·05	1·06692	1·06805	1·06342	0·80	0·73210	1·59802	$+0\cdot28087$
0·10	1·06326	1·07332	1·06016	0·82	0·70341	1·63172	$+0\cdot21799$
0·15	1·05914	1·08165	1·05151	0·84	0·67162	1·66700	$+0\cdot15169$
0·20	1·05316	1·09343	1·03917	0·86	0·63609	1·70389	$+0\cdot08239$
0·25	1·04528	1·10881	1·02273	0·88	0·59593	1·74242	$+0\cdot01092$
0·30	1·03535	1·12793	1·00167	0·8830265	0·58938	1·74839	0
0·35	1·02323	1·15098	0·97525	0·90	0·54987	1·78259	$-0\cdot06117$
0·40	1·00860	1·17823	0·94275	0·92	0·49595	1·82434	$-0\cdot13103$
0·45	0·99121	1·20996	0·90304	0·94	0·43090	1·86754	$-0\cdot19319$
0·50	0·97060	1·24656	0·85493	0·96	0·34854	1·91189	$-0\cdot23617$
0·55	0·94627	1·28844	0·79684	0·98	0·23394	1·95671	$-0\cdot23051$
0·60	0·91751	1·33612	0·72689	0·995	0·09355	1·98961	$-0\cdot12693$
0·65	0·88337	1·39020	0·64289	0·999	0·02782	1·99798	$-0\cdot04361$
0·70	0·84255	1·45134	0·54229	0·9999	0·00415	1·99980	$-0\cdot00709$

Roche sequences

$\cos^{-1}\dfrac{a_3}{a_1}$	Even modes			Odd modes		
	σ_1^2	σ_2^2	σ_3^2	σ_4^2	σ_5^2	σ_6^2
$p = 0$						
0°	1·067	1·0667	$+1\cdot0667$	1·067	1·0667	0
24	1·323	1·1239	$+0\cdot7007$	1·272	0·9370	0·0241
36	1·429	1·1393	$+0\cdot4866$	1·412	0·8718	0·0551
48	1·569	1·0445	$+0\cdot2574$	1·577	0·7757	0·0954
57	1·697	0·8788	$+0\cdot0836$	1·708	0·6640	0·1234
60	1·740	0·8070	$+0\cdot0285$	1·752	0·6168	0·1301
61	1·754	0·7814	$+0\cdot0109$	1·767	0·5997	0·1318
62	1·767	0·7552	$-0\cdot0063$	1·781	0·5822	0·1333
63	1·783	0·7284	$-0\cdot0232$	1·793	0·5642	0·1344
66	1·824	0·6442	$-0\cdot0695$	1·833	0·5063	0·1356
71	1·886	0·4948	$-0\cdot1278$	1·893	0·3987	0·1284
72	1·898	0·4640	$-0\cdot1357$	1·903	0·3758	0·1253
75	1·928	0·3711	$-0\cdot1495$	1·933	0·3049	0·1125
79	1·962	0·2482	$-0\cdot1416$	1·963	0·2076	0·0865
$p = 1$						
0°	1·067	1·0667	$+1\cdot0667$	1·067	1·0667	0
12	1·280	1·0333	$+0\cdot8740$	1·181	0·9799	0·0094
24	1·392	1·1398	$+0\cdot6294$	1·313	0·9078	0·0380
36	1·494	1·1901	$+0\cdot4045$	1·464	0·8440	0·0854
48	1·602	1·1410	$+0\cdot1952$	1·625	0·7610	0·1440
54	1·673	1·0468	$+0\cdot1011$	1·705	0·6994	0·1713
59	1·738	0·9309	$+0\cdot0303$	1·770	0·6321	0·1884
60	1·752	0·9040	$+0\cdot0171$	1·782	0·6166	0·1909
61	1·765	0·8760	$+0\cdot0044$	1·795	0·6004	0·1930
66	1·830	0·7210	$-0\cdot0521$	1·853	0·5088	0·1965
72	1·901	0·5127	$-0\cdot0976$	1·914	0·3771	0·1803
78	1·955	0·3000	$-0\cdot1054$	1·961	0·2298	0·1354
81	1·975	0·1999	$-0\cdot0905$	1·978	0·1559	0·1017

do not become dynamically unstable at the Roche limit where Ω^2 and μ attain their maxima.

That the two points, the point where Ω^2 and μ attain their maxima and the point where dynamical instability sets in, are distinct follows

TABLE XX

The point at which instability sets in along the Roche sequences
(Ω^2 and μ are listed in the unit $\pi G\rho$)

p	Ω^2	μ	$\bar{a}_1$	$\bar{a}_2$	$\bar{a}_3$
-1	0	0·125536	1·6558	0·77712	0·77712
0	0·090067	0·090067	1·6107	0·81096	0·76557
1	0·141229	0·070614	1·5805	0·83502	0·75771
4	0·216581	0·043316	1·526	0·8808	0·7439
20	0·305937	0·014568	1·430	0·9683	0·7222
100	0·350303	0·003468	1·343	1·0523	0·7078
∞	0·374230	0	1·197	1·1972	0·6977

from a comparison of equation (36), which determines the former point, and the equation

$$(4B_{12}-2\Omega^2-\mu)\times$$

$$\begin{vmatrix} -\Omega^2-2\mu+3B_{11}-B_{12} & B_{23}-B_{13} & -3\mu+3(B_{11}-B_{22})+B_{13}-B_{23} \\ \Omega^2+B_{12}-B_{13} & \mu+3B_{33}-B_{23} & \Omega^2+3(B_{22}-B_{33})+B_{12}-B_{13} \\ \dfrac{1}{a_1^2} & -\dfrac{1}{a_3^2} & \dfrac{1}{a_1^2}+\dfrac{1}{a_2^2}+\dfrac{1}{a_3^2} \end{vmatrix}$$

$$-3\mu\Omega^2\begin{vmatrix} B_{13}-B_{23} & -3\mu+3(B_{11}-B_{22})+B_{13}-B_{23} \\ \dfrac{1}{a_3^2} & \dfrac{1}{a_1^2}+\dfrac{1}{a_2^2}+\dfrac{1}{a_3^2} \end{vmatrix} = \Delta_{\text{Dyn.Inst.}} = 0, \tag{65}$$

which determines the latter point. That the two equations are necessarily distinct can be seen more directly from a comparison of equations (33) and (34) and the limiting forms of equations (59) and (58), respectively, for $\lambda \to 0$. Thus, while equation (59), in the limit $\lambda = 0$, is the same as equation (33), this is not the case with equations (34) and (58). The difference in the latter case arises from the circumstance that, by virtue of equations (55) and (56), the additional term in equation (58), namely,

$$2\lambda\Omega(V_{1;2}-V_{2;1}) = 2\Omega^2(V_{11}+V_{22})+\frac{6\Omega\mu}{\lambda}V_{12}$$

$$= 2\Omega^2(V_{11}+V_{22})-6\Omega^2\mu\,\frac{V_{11}-V_{22}}{\lambda^2+4B_{12}-2\Omega^2-\mu}, \tag{66}$$

does not tend to zero as $\lambda \to 0$; it tends, instead, to a finite limit. In other words, the equations one obtains, by considering *ab initio* a quasi-static deformation and by considering a normal mode with the time-dependence $e^{\lambda t}$ and then passing to the limit $\lambda \to 0$, are not the same. The question, as to what "really" happens at the point of onset of the instability (by the criterion (65)), has been resolved by Lebovitz (see Paper XX in the list on p. 232) by a separate investigation of the modes that emerge as "neutral" from an analysis in which a dependence on time of the displacement is presupposed to be $e^{\lambda t}$.

We have already remarked that, in passing from equation (61) to equation (62), a factor λ^2 was suppressed. Accordingly, it would appear that $\lambda = 0$ is a characteristic root of multiplicity 2. On the other hand, by setting $\lambda = d/dt = 0$ in equations (32) and (55)–(58), we find that there is only one uniquely determinable stationary mode: it belongs to a proper displacement for which

$$V_{11} = V_{22} = V_{33} = V_{12} = 0 \quad \text{and} \quad V_{1;2} = -V_{2;1} = \text{constant.} \quad (67)$$

This mode represents a simple rotation of the ellipsoid as a rigid body (about the axis of rotation). Since this stationary mode is unique, we may infer that the second mode that is attributed $\lambda = 0$ in the normal-mode analysis must have a linear dependence on time. By direct substitution in equations (32) and (55)–(58) (in which λ is now taken to be the operator d/dt) we find that such a mode exists; and, further, that the displacement which belongs to it (given in equation (69b) below) deforms the ellipsoid into an infinitesimally adjacent ellipsoid, along the equilibrium sequence, with an angular velocity appropriate to it. This second mode becomes stationary at the Roche limit (and only at the Roche limit).

Returning to the point of onset of the instability by the criterion (65), we first observe that the normal-mode analysis leads one to suppose that here $\lambda = 0$ is a root of multiplicity 4. By our earlier considerations, only one of these can be strictly stationary; and one of them must have a linear dependence on time. The remaining two modes must accordingly have quadratic and cubic dependences on time. To find these modes, we make the substitution

$$V_{i;j} = V_{i;j}^{(0)} + V_{i;j}^{(1)} t + V_{i;j}^{(2)} t^2 + V_{i;j}^{(3)} t^3 \qquad (68)$$

in equations (32) and (55)–(58) (in which λ is again identified with d/dt). A straightforward calculation shows that the equations do, indeed, allow such solutions provided the condition (65), for the onset of instability, is satisfied; and, further, that the four linearly independent solutions, belonging to the four-fold root $\lambda = 0$, are

$$V_{11} = V_{22} = V_{33} = V_{12} = 0, \qquad V_{1;2} = -V_{2;1} = \text{constant}, \qquad (69a)$$

$$V_{11} = V_{11}^{(0)}, \qquad V_{22} = V_{22}^{(0)}, \qquad V_{33} = V_{33}^{(0)}, \qquad V_{12}^{(0)} = V_{1;2}^{(0)} + V_{2;1}^{(0)} = 0,$$

$$(V_{1;2} - V_{2;1}) = [V_{1;2}^{(0)} - V_{2;1}^{(0)}]t, \qquad (69b)$$

$$V_{11} = V_{11}^{(1)}t, \qquad V_{22} = V_{22}^{(1)}t, \qquad V_{33} = V_{33}^{(1)}t, \qquad V_{12} = V_{1;2}^{(0)} + V_{2;1}^{(0)},$$

$$(V_{1;2} - V_{2;1}) = [V_{1;2}^{(2)} - V_{2;1}^{(2)}]t^2, \qquad (69c)$$

and

$$V_{11} = V_{11}^{(0)} + V_{11}^{(2)}t^2, \qquad V_{22} = V_{22}^{(0)} + V_{22}^{(2)}t^2, \qquad V_{33} = V_{33}^{(0)} + V_{33}^{(2)}t^2,$$

$$V_{12} = [V_{1;2}^{(1)} + V_{2;1}^{(1)}]t, \qquad V_{1;2} - V_{2;1} = [V_{1;2}^{(3)} - V_{2;1}^{(3)}]t^3. \qquad (69d)$$

The constants appearing in the foregoing solutions are not arbitrary: they have determinate ratios. In solution (69c), for example, if $V_{1;2}^{(2)} - V_{2;1}^{(2)}$ is given, $V_{11}^{(1)}$, $V_{22}^{(1)}$, $V_{33}^{(1)}$, $V_{1;2}^{(0)}$, and $V_{2;1}^{(0)}$ are determined in terms of it. Solutions (69a) and (69b) obtain, not only at the point of onset of the instability, but along the entire Roche sequence.

It is now clear that, at the point of onset of the instability "an exchange of stabilities," in the usual sense, does *not* take place: prior to the onset of instability, the solutions have an oscillatory character; subsequent to the onset of instability, the solutions have an exponentially increasing or an exponentially decreasing dependence on time; and *at* the point of onset of the instability the solutions have the polynomial dependence on time given by equations (69c) and (69d).

58. The neutral point belonging to the third harmonics

As in the other cases we have studied, the neutral point belonging to the third harmonics can be located with the aid of the third-order virial equations.

First, we observe that the additional term in μ, due to the tidal potential, in equation (6) modifies the usual form of the third-order virial equation (equation (73) of Chapter 2) to

$$\frac{d}{dt} \int_V \rho u_i x_j x_k \, d\mathbf{x} = 2(\mathfrak{T}_{ij;k} + \mathfrak{T}_{ik;j}) + \mathfrak{W}_{ij;k} + \mathfrak{W}_{ik;j} +$$

$$+ (\Omega^2 - \mu)I_{ijk} - \Omega^2 \delta_{i3} I_{3jk} + 3\mu \delta_{i1} I_{1jk} +$$

$$+ 2\Omega \epsilon_{il3} \int_V \rho u_l x_j x_k \, d\mathbf{x} + \delta_{ij} \Pi_k + \delta_{ik} \Pi_j. \qquad (70)$$

Under stationary conditions and in the absence of internal motions, equation (70) gives

$$\mathfrak{W}_{ij;k}+\mathfrak{W}_{ik;j}+(\Omega^2-\mu)I_{ijk}-\Omega^2\delta_{i3}I_{3jk}+3\mu\delta_{i1}I_{1jk} = -\delta_{ij}\Pi_k-\delta_{ik}\Pi_j. \tag{71}$$

A necessary (and sufficient) condition for the occurrence of a neutral point belonging to the third harmonics is that a non-trivial Lagrangian displacement exists for which the first variation of equation (71) is satisfied. The equation we have to consider, then, is

$$\delta\mathfrak{W}_{ij;k}+\delta\mathfrak{W}_{ik;j}+(\Omega^2-\mu)V_{ijk}-\Omega^2\delta_{i3}V_{3jk}+3\mu\delta_{i1}V_{1jk} = -\delta_{ij}\delta\Pi_k-\delta_{ik}\delta\Pi_j, \tag{72}$$

and, as we know, the left-hand side of this equation is expressible as a linear combination of the V_{ijk}'s.

For our present purpose of isolating the neutral point, it will suffice to consider the equations (derived from equation (72)) that are odd in the index 1 and even in the indices 2 and 3. These equations are

$$2\delta\mathfrak{W}_{11;1}+(\Omega^2+2\mu)V_{111} = -2\delta\Pi_1, \tag{73}$$
$$\delta\mathfrak{W}_{22;1}+\delta\mathfrak{W}_{21;2}+(\Omega^2-\mu)V_{221} = -\delta\Pi_1, \tag{74}$$
$$\delta\mathfrak{W}_{33;1}+\delta\mathfrak{W}_{31;3}-\mu V_{331} = -\delta\Pi_1, \tag{75}$$
$$2\delta\mathfrak{W}_{12;2}+(\Omega^2+2\mu)V_{122} = 0, \tag{76}$$
$$2\delta\mathfrak{W}_{13;3}+(\Omega^2+2\mu)V_{133} = 0. \tag{77}$$

Eliminating $\delta\Pi_1$, appropriately, from equations (73)–(75), we obtain the pair of equations (cf. equations (22) and (23) of Chapter 6)

$$\delta S_{122}+(\Omega^2+2\mu)V_{111}-3\Omega^2 V_{122} = 0 \tag{78}$$

and
$$\delta S_{133}+(\Omega^2+2\mu)V_{111}-\Omega^2 V_{133} = 0, \tag{79}$$

where δS_{122} and δS_{133} have the same meanings as hitherto.

With the expressions for $\delta\mathfrak{W}_{12;2}$, $\delta\mathfrak{W}_{13;3}$, δS_{122}, and δS_{133} given by equations (153) and (157) of Chapter 3, equations (76)–(79) together with the solenoidal condition

$$\frac{V_{111}}{a_1^2}+\frac{V_{122}}{a_2^2}+\frac{V_{133}}{a_3^2} = 0 \tag{80}$$

provide a system of five linear homogeneous equations for V_{111}, V_{122}, and V_{133}. And the existence of a non-trivial neutral mode belonging to the third harmonics requires that, for some member of the Roche sequence considered, the 5×3-matrix of the system of equations (76)–(80) is, at most, of rank 2.

By evaluating (along a particular Roche sequence) the determinants of the six different sets of equations which we can form by considering equation (80) and any two of the four equations (76)–(79), we find that all six determinants (of order 3) vanish simultaneously at a determinate point. This result verifies that the rank of the system of equations (76)–(80) is, indeed, 2 for a particular member of the sequence. We conclude that a neutral point, of the kind sought, exists.

The remarks made in Chapter 6, following Table V (p. 106), in the context of analogous considerations relative to the Jacobian sequence, apply equally to the present. And questions of uniqueness that arise can be resolved similarly (if one so desired).

TABLE XXI

The neutral point along the Roche sequences belonging to the third harmonics
(Ω^2 and μ are listed in the unit $\pi G\rho$)

p	Ω^2	μ	$\bar{a}_1$	$\bar{a}_2$	$\bar{a}_3$
-1	0	0·10913	2·142	0·6833	0·6833
0	0·07754	0·07754	2·080	0·7091	0·6780
1	0·12073	0·06036	2·044	0·7258	0·6741
4	0·1825	0·03649	1·989	0·7535	0·6671
20	0·2496	0·01188	1·923	0·7912	0·6572
∞	0·2840	0	1·886	0·8150	0·6507

In Table XXI we list the neutral points along the different Roche sequences determined in the manner we have described. In Fig. 21, the locus of these points is the curve joining T_3 and J_3.

59. The effect of viscous dissipation on the stability of the Roche ellipsoid

In § 57, we have seen how the Roche ellipsoid does not become unstable at its distance of closest approach (to the secondary) compatible with equilibrium; it becomes unstable only at a subsequent point along the sequence where its distance is somewhat greater. But this result was obtained from an analysis based on the assumption that the fluid is inviscid and that no dissipative mechanism is operative. We shall now show that if this assumption is relaxed and the fluid is considered viscid, the Roche ellipsoid becomes secularly unstable, by a purely viscous mode, between the Roche limit and the point of onset of the dynamical instability.

In considering the secular instability of the Roche ellipsoid, we shall suppose (as in § 37) that we are dealing with stresses derived from ordinary viscosity defined in terms of a coefficient of kinematic viscosity

v. Then equation (38) will be replaced by (cf. equations (121) and (122) of Chapter 5)

$$\lambda^2 V_{i;j} - 2\lambda\Omega\epsilon_{il3} V_{l;j} = \delta\mathfrak{W}_{ij} + (\Omega^2 - \mu)V_{ij} - \Omega^2\delta_{i3} V_{3j} +$$
$$+ 3\mu\delta_{i1} V_{1j} + \delta_{ij}\delta\Pi - \delta\mathfrak{P}_{ij}, \qquad (81)$$

where $\delta\mathfrak{P}_{ij}$ is the first variation of the shear-energy tensor (defined in equations (111) and (115) of Chapter 5).

The considerations of § 37 (b) apply equally in the present context and the expression for $\delta\mathfrak{P}_{ij}$, in a low Reynolds-number approximation, is (by equation (124) of Chapter 5)

$$\delta\mathfrak{P}_{ij} = 5\lambda\nu\left(\frac{V_{i;j}}{a_j^2} + \frac{V_{j;i}}{a_i^2}\right) \qquad (82)$$

(no summation over repeated indices).

We shall consider the effect of viscous dissipation only on the even modes. (A separate investigation of the odd modes shows that they are all damped by viscosity.)

With the addition of the term $\delta\mathfrak{P}_{ij}$ in equation (81), equations (59), (58), (55), and (56), respectively, become

$$\left(\tfrac{1}{2}\lambda^2 - \Omega^2 - 2\mu + \frac{5\lambda\nu}{a_1^2}\right)V_{11} - \left(\tfrac{1}{2}\lambda^2 - \Omega^2 + \mu + \frac{5\lambda\nu}{a_2^2}\right)V_{22} - 2\lambda\Omega V_{12} = \delta\mathfrak{W}_{11} - \delta\mathfrak{W}_{22},$$
$$(83)$$

$$(\tfrac{1}{2}\lambda^2 - \Omega^2 - 2\mu)V_{11} + (\tfrac{1}{2}\lambda^2 - \Omega^2 + \mu)V_{22} - \left(\lambda^2 + 2\mu + \frac{15\lambda\nu}{a_3^2}\right)V_{33} + 2\lambda\Omega(V_{1;2} - V_{2;1})$$
$$= \delta\mathfrak{W}_{11} + \delta\mathfrak{W}_{22} - 2\delta\mathfrak{W}_{33}, \quad (84)$$

$$(\lambda^2 + 4B_{12} - 2\Omega^2 - \mu)V_{12} + \lambda\Omega(V_{11} - V_{22}) + 10\lambda\nu\left(\frac{V_{1;2}}{a_2^2} + \frac{V_{2;1}}{a_1^2}\right) = 0, \quad (85)$$

and
$$\lambda^2(V_{1;2} - V_{2;1}) - \lambda\Omega(V_{11} + V_{22}) - 3\mu V_{12} = 0. \qquad (86)$$

With the known expressions for $\delta\mathfrak{W}_{11} - \delta\mathfrak{W}_{22}$ and $\delta\mathfrak{W}_{11} + \delta\mathfrak{W}_{22} - 2\delta\mathfrak{W}_{33}$ equations (83)–(86), together with the solenoidal condition (32), become a set of five linear, homogeneous equations for V_{11}, V_{22}, V_{33}, $V_{1;2}$, and $V_{2;1}$; and we are led to the characteristic equation (87). By suitably combining the rows and the columns of the characteristic determinant (87), we can reduce it to the form (88).

From equation (88) it is manifest that, while in the absence of viscosity

$$\lambda^2 = 0 \qquad (89)$$

is a root, in the presence of viscosity only

$$\lambda = 0 \qquad (90)$$

is a root.

$$\left|
\begin{array}{ccccc}
\tfrac{1}{2}\lambda^2-\Omega^2-2\mu+3B_{11}-B_{12}+ & -\tfrac{1}{2}\lambda^2+\Omega^2-\mu-3B_{22}+B_{12}- & B_{13}-B_{23} & -2\lambda\Omega & -2\lambda\Omega \\
\;\;+5\lambda\nu/a_1^2 & \;\;-5\lambda\nu/a_2^2 & & & \\[4pt]
\tfrac{1}{2}\lambda^2-\Omega^2-2\mu+3B_{11}+B_{12}-2B_{13} & \tfrac{1}{2}\lambda^2-\Omega^2+\mu+3B_{22}+B_{12}-2B_{23} & -\lambda^2-2\mu-6B_{33}+B_{13}+B_{23}- & +2\lambda\Omega & -2\lambda\Omega \\
& & \;\;-15\lambda\nu/a_3^2 & & \\[4pt]
+\lambda\Omega & -\lambda\Omega & 0 & \lambda^2+4B_{12}-2\Omega^2-\mu & \lambda^2+4B_{12}-2\Omega^2-\mu \\
& & & \;\;+10\lambda\nu/a_2^2 & \;\;+10\lambda\nu/a_1^2 \\[4pt]
-\lambda\Omega & -\lambda\Omega & 0 & \lambda^2-3\mu & -\lambda^2-3\mu \\[4pt]
\dfrac{1}{a_1^2} & \dfrac{1}{a_2^2} & \dfrac{1}{a_3^2} & 0 & 0
\end{array}
\right| = 0. \quad (87)$$

$$\left|
\begin{array}{ccccc}
\tfrac{1}{2}\lambda^2-\Omega^2-2\mu+3B_{11}-B_{12}+ & -3\mu+3B_{11}-3B_{22}+B_{13}-B_{23}+ & B_{13}-B_{23} & -2\lambda\Omega & 0 \\
\;\;+5\lambda\nu/a_1^2 & \;\;+5\lambda\nu(a_1^{-2}-a_2^{-2}) & & & \\[4pt]
\Omega^2+B_{12}-B_{13} & \Omega^2+3B_{22}-3B_{33}+B_{12}-B_{13}+ & -\tfrac{1}{2}\lambda^2-\mu-3B_{33}+B_{23}- & \Omega(\lambda+3\mu/\lambda) & 0 \\
& \;\;+5\lambda\nu(a_2^{-2}-a_3^{-2}) & \;\;-5\lambda\nu/a_3^2 & & \\[4pt]
+\lambda\Omega & 0 & 0 & \lambda^2+4B_{12}-2\Omega^2-\mu+ & 5\lambda\nu(a_2^{-2}-a_1^{-2}) \\
& & & \;\;+5\lambda\nu(a_1^{-2}+a_2^{-2}) & \\[4pt]
-\lambda\Omega & -2\lambda\Omega & 0 & -3\mu & \lambda^2 \\[4pt]
\dfrac{1}{a_1^2} & \dfrac{1}{a_1^2}+\dfrac{1}{a_2^2}+\dfrac{1}{a_3^2} & \dfrac{1}{a_3^2} & 0 & 0
\end{array}
\right| = 0. \quad (88)$$

As we have seen in § 57 (c), of the two modes that in the absence of viscosity belong to $\lambda = 0$, only one is stationary, while the other has a linear dependence on time. The stationary mode, as one must expect, is unaffected by viscosity since it only leads to a simple rotation of the ellipsoid as a rigid body. Therefore, *the mode, that has a linear dependence on time in the absence of viscosity, acquires a finite characteristic value in the presence of viscosity.* We shall call this the *viscous mode*.

(a) The viscous mode

In considering equation (88), we must remember that it is valid only to the first order in ν; and that only those roots are admissible that are expressible in the form

$$\lambda = \lambda_0 + \text{constant } \nu + O(\nu^2), \tag{91}$$

where λ_0 is a characteristic root of the inviscid problem. Accordingly, in expanding the secular determinant (88) by the elements of the last column, we may suppress the terms in ν which occur in the minor of $5\lambda\nu(a_2^{-2} - a_1^{-2})$. We thus obtain (92), where we have excluded the root $\lambda = 0$.

It is now manifest from equation (92) that, consistently in the framework of the present low Reynolds-number approximation, we may obtain the characteristic value belonging to the viscous mode by letting λ and ν both tend to zero in the first determinant (which occurs as a multiplicand of λ) and letting λ tend to zero in the second determinant (which occurs as a multiplicand of ν). In this manner, we obtain (after some further elementary reductions of the determinants)

$$\lambda = -15\mu\left(\frac{1}{a_2^2} - \frac{1}{a_1^2}\right)\frac{\Delta_{\text{Roche Limit}}}{\Delta_{\text{Dyn. Inst.}}}\nu, \tag{93}$$

where $\Delta_{\text{Roche Limit}}$ and $\Delta_{\text{Dyn. Inst.}}$ are the same determinants that are defined in equations (36) and (65). As these determinants have been defined, they are positive before the respective limits are reached and negative after they have been surpassed. Therefore, it follows that

$\lambda < 0$ *before the Roche limit and after the point of onset of the dynamical instability* $\tag{94}$

and $\lambda > 0$ *only in the interval between the Roche limit and the point of onset of the dynamical instability.* $\tag{95}$

Thus, *the Roche ellipsoids become unstable by this purely viscous mode precisely in the interval (and, only in the interval) that separates the Roche limit and the point of onset of the dynamical instability.*

$$
\lambda
\begin{vmatrix}
\tfrac{1}{2}\lambda^2-\Omega^2-2\mu+3B_{11}-B_{12}+5\lambda\nu/a_1^2 & -3\mu+3B_{11}-3B_{22}+B_{13}-B_{23}+5\lambda\nu(a_1^{-2}-a_2^{-2}) & B_{13}-B_{23} & -2\lambda\Omega \\
\Omega^2+B_{12}-B_{13} & \Omega^2+3B_{22}-3B_{33}+B_{12}-B_{13}+5\lambda\nu(a_2^{-2}-a_3^{-2}) & -\tfrac{1}{2}\lambda^2-\mu-3B_{33}+B_{23}-5\lambda\nu/a_3^2 & \Omega(\lambda+3\mu/\lambda) \\
+\lambda\Omega & 0 & 0 & \lambda^2+4B_{12}-2\Omega^2-\mu+5\lambda\nu(a_1^{-2}+a_2^{-2}) \\
\dfrac{1}{a_1^2} & \dfrac{1}{a_1^2}+\dfrac{1}{a_2^2}+\dfrac{1}{a_3^2} & \dfrac{1}{a_3^2} & 0
\end{vmatrix}
$$

$$
-\,5\nu\left(\frac{1}{a_2^2}-\frac{1}{a_1^2}\right)
\begin{vmatrix}
\tfrac{1}{2}\lambda^2-\Omega^2-2\mu+3B_{11}-B_{12} & -3\mu+3B_{11}-3B_{22}+B_{13}-B_{23} & B_{13}-B_{23} & -2\lambda\Omega \\
\Omega^2+B_{12}-B_{13} & \Omega^2+3B_{22}-3B_{33}+B_{12}-B_{13} & -\tfrac{1}{2}\lambda^2-\mu-3B_{33}+B_{23} & \Omega(\lambda+3\mu/\lambda) \\
-\lambda\Omega & -2\lambda\Omega & 0 & -3\mu \\
\dfrac{1}{a_1^2} & \dfrac{1}{a_1^2}+\dfrac{1}{a_2^2}+\dfrac{1}{a_3^2} & \dfrac{1}{a_3^2} & 0
\end{vmatrix}=0. \quad (92)
$$

TABLE XXII

The characteristic values λ for the viscous mode for the Roche sequences $p = 0$ and $p = 1$

(λ/ν is listed in the unit a_1^{-2})

$p = 0$				$p = 1$			
a_2/a_1	a_3/a_1	$\Omega^2/\pi G\rho$	λ/ν	a_2/a_1	a_3/a_1	$\Omega^2/\pi G\rho$	λ/ν
0·500	0·472081	0·090039	−26·430	0·526	0·477388	0·141192	−39·907
0·502	0·473929	0·090056	−50·432	0·528	0·479131	0·141224	−247·23
0·503480†	0·475298	0·090067	±∞	0·528319†	0·479409	0·141229	±∞
0·504	0·475779	0·090070	+111·02	0·529	0·480003	0·141239	+105·98
0·506	0·477628	0·090081	+16·438	0·530	0·480875	0·141252	+38·998
0·508	0·479479	0·090088	+5·6499	0·532	0·482620	0·141276	+14·302
0·510	0·481330	0·090092	+1·5500	0·534	0·484366	0·141294	+7·0568
0·511345‡	0·482576	0·090093	0	0·536	0·486113	0·141308	+3·6324
0·512	0·483182	0·090093	−0·57191	0·538	0·487861	0·141317	+1·6593
0·514	0·485035	0·090090	−1·8452	0·540	0·489610	0·141321	+0·39176
0·516	0·486888	0·090084	−2·6773	0·540814‡	0·490322	0·141322	0
0·518	0·488742	0·090074	−3·2512	0·541	0·490485	0·141322	−0·082010
0·520	0·490596	0·090062	−3·6612	0·542	0·491360	0·141321	−0·48056

† Point of onset of dynamical instability.
‡ Roche limit.

In Table XXII the values of λ/ν derived from equation (93) are listed for the Roche sequences $p = 0$ and $p = 1$ in the short interval in which the mode, from being damped, becomes undamped, and then goes back to being damped once again.

(b) *The effect of viscous dissipation on the remaining three even modes*

The effect of viscous dissipation on the remaining three even modes can be determined by writing in equation (92)

$$\lambda = \lambda_0 + \delta\lambda, \tag{96}$$

where λ_0 is a particular characteristic value of the inviscid problem, and expanding, to the first order in $\delta\lambda$ and ν, the determinant which occurs as a multiplicand of λ (which is set equal to λ_0) while setting $\lambda = \lambda_0$ in the determinant which occurs as a multiplicand of ν. We shall clearly find that $\delta\lambda$ is proportional to ν; and the formula determining the constant

TABLE XXIII

The effect of viscous dissipation on the mode which becomes dynamically unstable

	$p = 0$			$p = 1$	
a_2/a_1	σ^2	$\delta\lambda/\nu$	a_2/a_1	σ^2	$\delta\lambda/\nu$
0·501	$-0\!\cdot\!002559$	$+8\!\cdot\!490$	0·526	$-0\!\cdot\!001628$	$+13\!\cdot\!39$
0·502	$-0\!\cdot\!0015290$	$+16\!\cdot\!95$	0·528	$-0\!\cdot\!0002246$	$+117\!\cdot\!11$
0·503	$-0\!\cdot\!0004960$	$+60\!\cdot\!75$	0·528319†	0	..
0·503480†	0	..	0·529	$+0\!\cdot\!0004802$	$-59\!\cdot\!48$
0·504	$+0\!\cdot\!0005391$	$-63\!\cdot\!71$	0·530	$+0\!\cdot\!0011871$	$-25\!\cdot\!97$
0·505	$+0\!\cdot\!0015765$	$-24\!\cdot\!47$			

† Point of onset of dynamical instability.

of proportionality can be readily written down. By using the formula so derived, it is found that the two even modes that are stable along the entire Roche sequence are always damped by viscosity. But the mode which becomes dynamically unstable is damped only prior to the onset of instability: it is undamped after that point is surpassed. This behavior is apparent from Table XXIII in which $\delta\lambda/\nu$ for this mode is listed.

60. The Roche–Riemann ellipsoids

There is clearly no difficulty in allowing, in the framework of Roche's problem, internal motions of uniform vorticity to be present and investigating the equilibrium and the stability of the resulting *Roche–Riemann* ellipsoids by the methods of Chapter 7. This problem has been investigated by Aizenman in some generality. The present account will, however, be restricted to the special case when the vorticity ζ of the internal motions is parallel to the direction of $\mathbf{\Omega}$. In this case, it is evident, on symmetry grounds, that the orientation of the principal axes of the ellipsoid will be the same as in Roche's problem considered in § 56. And the virial equations determining the equilibrium figures can be written down directly by combining the terms in equation (9), appropriate to the simple Roche ellipsoids, and equation (26) of Chapter 7, appropriate to the Riemann ellipsoids of type S. We have

$$a_2^2 Q_1^2 + a_1^2(\Omega^2 + 2Q_2\Omega) + 2a_1^2\mu - 2A_1 a_1^2$$
$$= a_1^2 Q_2^2 + a_2^2(\Omega^2 - 2Q_1\Omega) - a_2^2\mu - 2A_2 a_2^2 = -\mu a_3^2 - 2A_3 a_3^2, \quad (97)$$

where
$$Q_1 = -\frac{a_1^2}{a_1^2 + a_2^2}\zeta \quad \text{and} \quad Q_2 = +\frac{a_2^2}{a_1^2 + a_2^2}\zeta. \quad (98)$$

Inserting for Q_1 and Q_2 their values in equations (97), we obtain the pair of equations

$$\frac{a_1^2 a_2^2}{(a_1^2 + a_2^2)^2}\zeta^2 + (\Omega^2 + 2\mu) + \frac{2a_2^2}{a_1^2 + a_2^2}\zeta\Omega + \frac{a_3^2}{a_1^2}\mu = \frac{2}{a_1^2}(A_1 a_1^2 - A_3 a_3^2) \quad (99)$$

and

$$\frac{a_1^2 a_2^2}{(a_1^2 + a_2^2)^2}\zeta^2 + (\Omega^2 - \mu) + \frac{2a_1^2}{a_1^2 + a_2^2}\zeta\Omega + \frac{a_3^2}{a_2^2}\mu = \frac{2}{a_2^2}(A_2 a_2^2 - A_3 a_3^2). \quad (100)$$

Letting
$$x = \frac{a_1 a_2}{a_1^2 + a_2^2}\frac{\zeta}{\Omega} \quad (101)$$

as in equation (40) of Chapter 7, we can rewrite equations (99) and (100) in the forms

$$x^2 + 2\frac{a_2}{a_1}x + 1 + \frac{1}{1+p}\left(\frac{a_3^2}{a_1^2} + 2\right) = \frac{2}{\Omega^2}\left[\frac{1}{a_1^2}(A_{12} a_1^2 a_2^2 - A_3 a_3^2) + B_{12}\right] \quad (102)$$

and

$$x^2 + 2\frac{a_1}{a_2}x + 1 + \frac{1}{1+p}\left(\frac{a_3^2}{a_2^2} - 1\right) = \frac{2}{\Omega^2}\left[\frac{1}{a_2^2}(A_{12} a_1^2 a_2^2 - A_3 a_3^2) + B_{12}\right]. \quad (103)$$

By subtraction, we obtain from these two equations

$$\frac{2}{\Omega^2} = \frac{1}{a_3^2 A_3 - a_1^2 a_2^2 A_{12}}\left[\frac{1}{1+p}\left(\frac{3a_1^2 a_2^2}{a_1^2 - a_2^2} - a_3^2\right) - 2a_1 a_2 x\right]. \quad (104)$$

And finally, eliminating Ω^2 from equation (102) (or (103)) with the aid of equation (104) and simplifying, we are left with (cf. equation (43) of Chapter 7)

$$x^2 + \frac{2a_1 a_2 B_{12}}{a_3^2 A_3 - a_1^2 a_2^2 A_{12}}x + \frac{p}{1+p} +$$
$$+ \frac{1}{(1+p)(a_3^2 A_3 - a_1^2 a_2^2 A_{12})}\left[a_3^2 B_{12} - \frac{3a_1^2(a_2^2 - a_3^2)}{a_1^2 - a_2^2}B_{23}\right] = 0. \quad (105)$$

For assigned values of p and x, equation (105) determines the ratios of the axes of the ellipsoid that are compatible with equilibrium; and the values of ζ, Ω^2, and μ, that are to be associated with a particular solution of equation (105), follow from equations (12), (101), and (104).

For a solution of equations (104) and (105) to be admissible, it is necessary, *first*, that the two roots of x given by equation (105) are real and, *second*, that the right-hand side of equation (104) is positive. In the analogous case of the Riemann ellipsoids of type S, we had a condition equivalent to the first—the condition, in fact, defined the two bounding self-adjoint sequences $x = +1$ and $x = -1$; but we had no condition equivalent to the second since the reality of x ensured the reality of Ω (as is evident from equations (41) of Chapter 7). Under the present circumstances, however, it can happen that while one of the two roots for x leads to a real Ω, the other does not. In other words, while an equilibrium figure, determined by equation (105), will in general be compatiblo with two physically distinct situations, there is no simple theorem, analogous to Dedekind's, in this case.

(a) The solution for the case $p = 0$

For the case $p = 0$, equations (104) and (105) become

$$\frac{2}{\Omega^2} = \frac{1}{a_3^2 A_3 - a_1^2 a_2^2 A_{12}}\left(\frac{3a_1^2 a_2^2}{a_1^2 - a_2^2} - a_3^2 - 2a_1 a_2 x\right) \tag{106}$$

and

$$x^2 + \frac{1}{a_3^2 A_3 - a_1^2 a_2^2 A_{12}}\left[2a_1 a_2 B_{12} x + a_3^2 B_{12} - \frac{3a_1^2(a_2^2 - a_3^2)}{a_1^2 - a_2^2} B_{23}\right] = 0. \tag{107}$$

The equilibrium figures that follow from equations (106) and (107) have been determined by Aizenman. His results on the domain of occupancy of these ellipsoids in the $(a_2/a_1, a_3/a_1)$-plane are exhibited in Fig. 23. It will be observed that the Dedekind ellipsoids appear among them. It is natural that they do since they represent permissible equilibrium figures at "infinity" where only internal motions are allowed. It will also be observed that the domains of the ellipsoids derived from the two roots of equation (107)—distinguished by S_+ and S_-—are different.

The stability of these Roche–Riemann ellipsoids of type S can be investigated along the lines of the treatment of the Riemann ellipsoids in § 49. The results of such a stability analysis are presented in Figs. 24a and 24b. It will be observed that along certain neutral loci, other more general types of configurations bifurcate in a manner analogous to the bifurcation of the Riemann ellipsoids of type III from the ellipsoids of type S. For a description of the properties of these more general classes of the Roche–Riemann ellipsoids reference should be made to Aizenman's paper.

61. Darwin's problem

Darwin's problem differs from Roche's problem in only one respect: it seeks to allow for the mutual tidal interactions between the primary

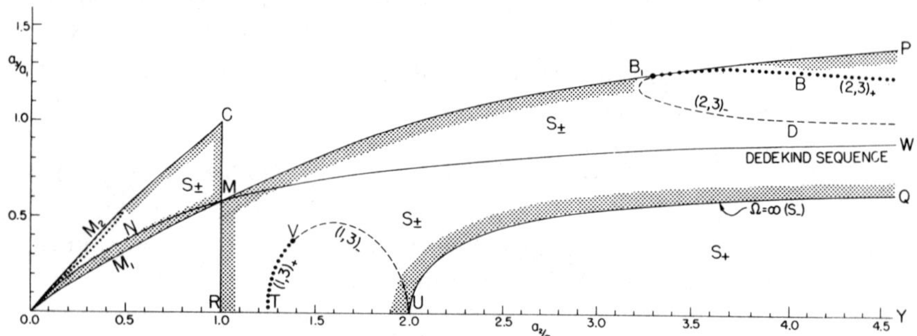

Fig. 23. The domain of occupancy of the Roche–Riemann ellipsoids in the $(a_2/a_1,\ a_3/a_1)$-plane. The two configurations that result from the two roots of equation (107) are designated ellipsoids of types S_+ and S_-.

For $a_2 < a_1$, the ellipsoids of both types occupy the domain $OM_1 MCM_2$. For $a_2 > a_1$, the ellipsoids of type S_+ occupy the domain $YRMP$; but the ellipsoids of type S_- are restricted to the domain $QURMP$: along QU the angular velocity of the S_- ellipsoids tends to infinity.

The Maclaurin sequence (represented by the line CR) is not included in the domain of occupancy. The Dedekind sequence (represented by the curve $ONMW$) bifurcates from the Maclaurin line. The Dedekind ellipsoids are limiting members of these Roche–Riemann ellipsoids at infinity.

The curve $BB_1 D$ is a locus of marginal stability for the S_+ ellipsoids by an odd mode of second-harmonic oscillation. Along this curve a new class of Roche–Riemann ellipsoids, in which the directions of vorticity (ζ) and angular velocity (Ω) do not coincide but lie in the (a_2, a_3)-plane of the ellipsoid, bifurcates. Similarly, the curve UVT is a curve along which a class of ellipsoids, in which ζ and Ω lie in the (a_1, a_3)-plane of the ellipsoid, bifurcates from the S_- ellipsoids.

and the secondary. Thus, while the starting points of the two problems are the same, namely equation (1), in Darwin's problem an attempt is made to take into account the tidal distortion of the secondary (by the primary) in evaluating *its* tidal potential $\mathfrak{B}'$ over the primary.

In evaluating $\mathfrak{B}'$, we shall find it convenient to consider it in a coordinate system (X_1, X_2, X_3) whose origin is at the center of mass of M' and whose orientation with respect to M' is the same as that of the coordinate

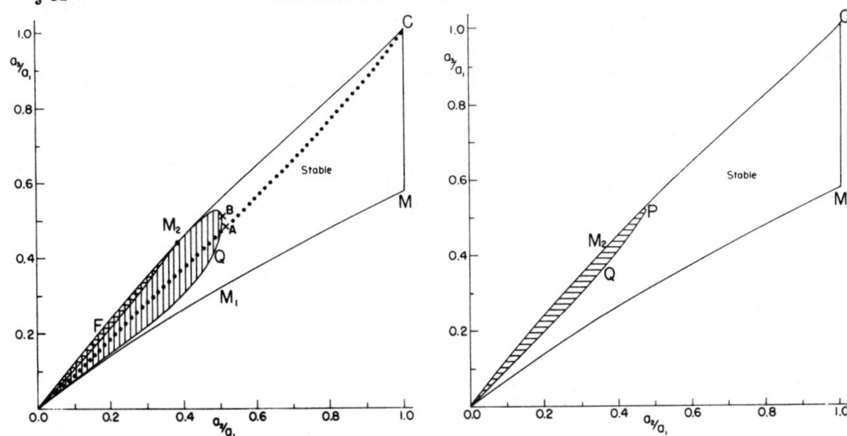

FIG. 24a. The loci of marginal stability of the ellipsoids of types S_+ (the figure to the left) and S_- (the figure to the right) in the $(a_2/a_1, a_3/a_1)$-plane, for $a_2 < a_1$. In the regions hatched by vertical lines, instability sets in via an even mode of oscillation, while in the regions hatched by horizontal lines, instability sets in via an odd mode of oscillation.

The dotted curve OAC represents the Roche sequence for $p = 0$. The Roche limit occurs at A and dynamical instability sets in at the point where the curve OAC intersects the vertically hatched area. The ellipsoid closest to the secondary, in this domain, occurs at B.

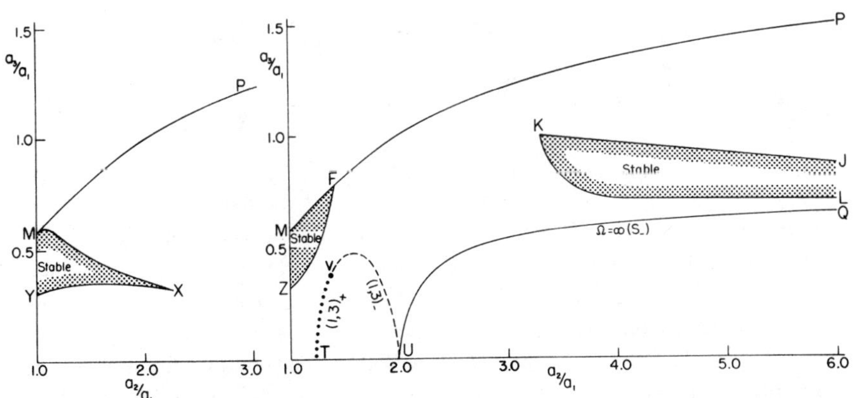

FIG. 24b. The loci of marginal stability of the ellipsoids of types S_+ (the figure to the left) and S_- (the figure to the right) in the $(a_2/a_1, a_3/a_1)$-plane for $a_2 > a_1$. Except in the regions indicated, the ellipsoids are unstable. (For the meaning of the curve UVT, see the legend for Fig. 23.)

system (x_1, x_2, x_3) with respect to M. The transformation between the two systems is

$$x_1 = -(X_1 - R), \qquad x_2 = -X_2, \quad \text{and} \quad x_3 = X_3. \qquad (108)$$

We now expand $\mathfrak{B}'(X_1, X_2, X_3)$ in a Taylor expansion about $(R, 0, 0)$;

and in writing this expansion, we shall suppose that the secondary has triplanar symmetry. Then

$$\mathfrak{B}'(X_1, X_2, X_3) = \mathfrak{B}'(R, 0, 0) + (X_1 - R)\left(\frac{\partial \mathfrak{B}'}{\partial X_1}\right)_{R,0,0} +$$

$$+ \tfrac{1}{2}(X_1 - R)^2\left(\frac{\partial^2 \mathfrak{B}'}{\partial X_1^2}\right)_{R,0,0} + \tfrac{1}{2}X_2^2\left(\frac{\partial^2 \mathfrak{B}'}{\partial X_2^2}\right)_{R,0,0} + \tfrac{1}{2}X_3^2\left(\frac{\partial^2 \mathfrak{B}'}{\partial X_3^2}\right)_{R,0,0} + ...; \quad (109)$$

and we shall not consider the terms beyond those that we have written.

Inserting the expansion (109) in equation (1), we can write, in view of equations (108),

$$\rho\frac{du_i}{dt} = -\frac{\partial p}{\partial x_i} + \rho\frac{\partial}{\partial x_i}\left\{\mathfrak{B} + \frac{1}{2}\sum_{j=1}^{3} x_j^2\left(\frac{\partial^2 \mathfrak{B}'}{\partial X_j^2}\right)_{R,0,0} + \tfrac{1}{2}\Omega^2(x_1^2 + x_2^2) - \right.$$

$$\left. - x_1\left[\left(\frac{\partial \mathfrak{B}'}{\partial X_1}\right)_{R,0,0} + \frac{M'R}{M+M'}\Omega^2\right]\right\} + 2\rho\Omega\epsilon_{il3}u_l. \quad (110)$$

If we now let Ω^2 (hitherto unspecified) have the value

$$\Omega^2 = -\frac{M+M'}{M'R}\left(\frac{\partial \mathfrak{B}'}{\partial X_1}\right)_{R,0,0}, \quad (111)$$

equation (110) reduces to the form

$$\rho\frac{du_i}{dt} = -\frac{\partial p}{\partial x_i} + \rho\frac{\partial}{\partial x_i}\left(\mathfrak{B} + \tfrac{1}{2}\sum_{k=1}^{3} \beta_{kk} x_k^2\right) + 2\rho\Omega\epsilon_{il3}u_l, \quad (112)$$

where
$$\beta_{11} = \Omega^2 + \left(\frac{\partial^2 \mathfrak{B}'}{\partial X_1^2}\right)_{R,0,0}, \qquad \beta_{22} = \Omega^2 + \left(\frac{\partial^2 \mathfrak{B}'}{\partial X_2^2}\right)_{R,0,0}, \quad (113)$$

and
$$\beta_{33} = \left(\frac{\partial^2 \mathfrak{B}'}{\partial X_3^2}\right)_{R,0,0} = -\left(\frac{\partial^2 \mathfrak{B}'}{\partial X_1^2} + \frac{\partial^2 \mathfrak{B}'}{\partial X_2^2}\right)_{R,0,0}. \quad (114)$$

It should be noted that, in deriving the equation of motion (112) governing the fluid elements of M, we have made the assumption that the mass M' is static; otherwise, Ω^2 defined by equation (111) would not be a constant (see § 63, below).

It is manifest that equation (112) will allow stationary solutions, with no internal motions, leading to ellipsoidal figures of equilibrium. On the other hand, since the treatment has to be symmetric with respect to both components (in the stationary state), it is clear that stationary solutions derived from equation (112) can be considered valid only if the choice of the frame of reference, which enabled the reduction of equation (1) to equation (112) (via the definition (111)), is independent of the particular component that we may be considering. This is clearly not the case: the expression (111) for Ω^2 is not symmetric in M and M'. Indeed, it is apparent that the determination of Ω^2 by equation (111) is strictly admissible only in two cases: the case when $M = M'$ and the two components are in addition congruent and the "singular" case $M/M' \to 0$ when the distortion of the secondary by the primary (of infinitesimal mass!) can be ignored. In all other cases, the values of Ω^2, which will eliminate the "unwanted" terms in x_1 and X_1, in equation (110) for M and in the analogous equation for M', are different. In other words, only in the two special cases mentioned has Darwin's problem, as formulated, a consistent solution.

(a) The second-order virial equation appropriate to Darwin's problem

On the assumption that β_{11}, β_{22}, and β_{33} are constants, it is manifest that the second-order virial equation that will follow from equation (112) is

$$\frac{d}{dt} \int_V \rho u_i x_j \, d\mathbf{x} = 2\mathfrak{T}_{ij} + \mathfrak{W}_{ij} + (\boldsymbol{\beta}\mathbf{I})_{ij} + \delta_{ij}\,\Pi + 2\Omega\epsilon_{il3} \int_V \rho u_l x_j \, d\mathbf{x},$$

$$(115)$$

where

$$\boldsymbol{\beta} = \begin{vmatrix} \beta_{11} & 0 & 0 \\ 0 & \beta_{22} & 0 \\ 0 & 0 & \beta_{33} \end{vmatrix}.$$

$$(116)$$

Under conditions of hydrostatic equilibrium, equation (115) gives

$$\mathfrak{W}_{ij} + (\boldsymbol{\beta}\mathbf{I})_{ij} = -\Pi\delta_{ij}.$$

$$(117)$$

The non-diagonal components of equation (117) will be trivially satisfied; and the diagonal components give

$$\mathfrak{W}_{11} + \beta_{11}\,I_{11} = \mathfrak{W}_{22} + \beta_{22}\,I_{22} = \mathfrak{W}_{33} + \beta_{33}\,I_{33} = -\Pi.$$

$$(118)$$

Equations (118) will suffice to determine the equilibrium figures.

62. The Darwin ellipsoids

We shall show how the equilibrium figures, for the two cases mentioned in § 61, can be constructed.

(a) The case $M/M' = 0$

In this case, the primary is of infinitesimal mass compared to the secondary; and we may consider the secondary as unaffected by the presence of the primary. The departure from sphericity of the secondary is, then, determined solely by the centrifugal potential. The secondary is accordingly a Maclaurin spheroid or a Jacobi ellipsoid. However, it will appear that solutions for Darwin's problem exist only for members of the Maclaurin sequence with eccentricities less than $0\cdot40504$. We need not, therefore, consider the Jacobian form for the secondary.

Consider then a Maclaurin spheroid with an eccentricity e. Its angular velocity of rotation is given by (equation (6) of Chapter 5)

$$\frac{\Omega^2}{\pi G \rho_{\text{Mc}}} = 2\frac{(1-e^2)^{\frac{1}{2}}}{e^3}(3-2e^2)\sin^{-1}e - \frac{6}{e^2}(1-e^2),\tag{119}$$

where ρ_{Mc} denotes the density of the spheroid. It will be noted that we have restored the factor $\pi G \rho_{\text{Mc}}$ (usually suppressed) in equation (119), since we are now dealing with two homogeneous objects (the secondary and the primary) whose densities need not be the same.

In order now that the "satellite" we are considering may rotate synchronously with the Maclaurin spheroid, it is necessary that it describe a circular orbit about it with the same angular velocity Ω. The orbit must accordingly be described at such a distance R that the dynamical condition (111) leads to the same value of Ω as equation (119). The equation which expresses this equality can be obtained as follows.

The gravitational potential, in the equatorial plane of a homogeneous spheroid, at a distance ϖ from the center is given by

$$\frac{\mathfrak{B}'}{\pi G \rho_{\text{Mc}}} = -\frac{(1-e^2)^{\frac{1}{2}}}{e}a_{\text{Mc}}^2\left[\left(\frac{\varpi^2}{e^2 a_{\text{Mc}}^2}-2\right)\sin^{-1}\left(\frac{ea_{\text{Mc}}}{\varpi}\right) - \frac{\varpi}{ea_{\text{Mc}}}\left(1-\frac{e^2 a_{\text{Mc}}^2}{\varpi^2}\right)^{\frac{1}{2}}\right],$$
$$\tag{120}$$

where a_{Mc} denotes the semi-major axis of the spheroid. From equations (111) and (120) we now find

$$\frac{\Omega^2}{\pi G \rho_{\text{Mc}}} = -\frac{1}{\pi G \rho_{\text{Mc}} R}\left(\frac{\partial \mathfrak{B}'}{\partial X_1}\right)_{R,0,0}$$

$$= \frac{2(1-e^2)^{\frac{1}{2}}}{e^3}\left[\sin^{-1}\left(\frac{ea_{\text{Mc}}}{R}\right) - \frac{ea_{\text{Mc}}}{R}\left(1-\frac{e^2 a_{\text{Mc}}^2}{R^2}\right)^{\frac{1}{2}}\right].\tag{121}$$

Since Ω^2 given by equations (119) and (121) must agree, we must have

$$\sin^{-1}\left(\frac{ea_{\text{Mc}}}{R}\right) - \frac{ea_{\text{Mc}}}{R}\left(1-\frac{e^2 a_{\text{Mc}}^2}{R^2}\right)^{\frac{1}{2}} = (3-2e^2)\sin^{-1}e - 3e(1-e^2)^{\frac{1}{2}}.\tag{122}$$

Equation (122) determines R/a_{Mc} along the Maclaurin sequence. It should be noted in this connection that if the mass of the spheroid is specified (as it is in the present context), a_{Mc} will vary along the sequence. What is constant, under these circumstances, is $a_{\text{Mc}}^3(1-e^2)^{\frac{1}{2}}$. Accordingly, a convenient unit in which to measure R is

$$\bar{a}_{\text{Mc}} = a_{\text{Mc}}(1-e^2)^{1/6}.\tag{123}$$

Returning to equation (120) and evaluating its second derivatives with respect to X_1 and X_2, we find that with Ω^2 given by equation (121), the coefficients β_{jj}, defined in equations (113) and (114) become

$$\beta_{11} = q\Omega^2, \quad \beta_{22} = 0, \quad \text{and} \quad \beta_{33} = -(q-2)\Omega^2, \tag{124}$$

where

$$q\frac{\Omega^2}{\pi G \rho_{\mathrm{Mc}}} = 4\left(\frac{a_{\mathrm{Mc}}}{R}\right)^3 \left(\frac{1-e^2}{1-e^2 a_{\mathrm{Mc}}^2/R^2}\right)^{\frac{1}{2}}. \tag{125}$$

With the β's given by equations (124), the virial equations (118) give

$$q\frac{\Omega^2}{\pi G \rho} a_1^2 - 2A_1 a_1^2 = -2A_2 a_2^2 = -(q-2)\frac{\Omega^2 a_3^2}{\pi G \rho} - 2A_3 a_3^2, \tag{126}$$

or, alternatively

$$\frac{q}{q-2}\frac{a_1^2}{a_3^2} = \frac{A_1 a_1^2 - A_2 a_2^2}{A_2 a_2^2 - A_3 a_3^2} \tag{127}$$

and

$$\frac{\Omega^2}{\pi G \rho} = 2\frac{A_1 a_1^2 - A_2 a_2^2}{q a_1^2}. \tag{128}$$

Equations (122), (125), (127), and (128) determine the orbit and the figure of an infinitesimal satellite rotating synchronously with a Maclaurin spheroid. The following algorism for solving these equations clarifies their relationships.

We start with a Maclaurin spheroid of some assigned eccentricity e. Its angular velocity of rotation is given by equation (119). The distance at which the satellite must circulate in order that it may rotate synchronously with the spheroid follows from equation (122); and equation (125) determines the value of q which appears in the subsequent formulas. Equation (127) then determines a relation between $(a_2/a_1, a_3/a_1)$ which must be satisfied if the satellite is to be in equilibrium. And the solution becomes determinate by the requirement that the value of Ω^2, which follows from equation (128), is in agreement with the value, appropriate to a Maclaurin spheroid of eccentricity e, with which we started.

The results summarized in Table XXIV, for the case $\rho = \rho_{\mathrm{Mc}}$, were obtained by the procedure described in the preceding paragraph. In Figs. 25a and 25b the results for these Darwin ellipsoids are compared with those for the corresponding Roche ellipsoids given in Table XVI. It will be observed that the results for the two sets of ellipsoids are very nearly the same. This near identity of the results for the two cases is due to the fact that the values of q for the Maclaurin spheroids, for the range of eccentricities of interest, differ from 3 (the value appropriate for a rigid sphere) by an amount that hardly exceeds $\frac{1}{2}$ per cent.

We observe that Darwin's problem, in this case, allows no solution for Maclaurin spheroids with eccentricities exceeding a certain maximum value $e_{\max}$ ($= 0 \cdot 405034$). The maximum angular velocity and the distance of closest approach occur for this same spheroid. And it can be

TABLE XXIV

The Darwin sequence for $M/M' = 0$

$$(\bar{a}_{\mathrm{Mc}} = a_{\mathrm{Mc}}(1-e^2)^{1/6}, \ \bar{a}_i = a_i/(a_1 a_2 a_3)^{1/3})$$

e	$\Omega^2/\pi G\rho$	$R/\bar{a}_{\mathrm{Mc}}$	q	$\bar{a}_1$	$\bar{a}_2$	$\bar{a}_3$
0·390	0·08287	2·5309	3·0153	1·3726	0·8790	0·8288
0·400	0·08727	2·4883	3·0168	1·4566	0·8537	0·8042
				1·7663	0·7732	0·7322
0·402	0·08816	2·4799	3·0171	1·4851	0·8455	0·7964
				1·7244	0·7830	0·7406
0·404	0·08906	2·4717	3·0174	1·5282	0·8333	0·7852
				1·6680	0·7967	0·7525
0·405034†	0·089529	2·4675	3·0175	1·5943	0·8155	0·7691
0·405	0·08951	2·4676	3·0175	1·5818	0·8188	0·7721
				1·6073	0·8121	0·7661
0·40498‡	0·089504	2·4677	3·0175	1·6110	0·8111	0·7653

† Roche limit.
‡ Point of onset of dynamical instability.

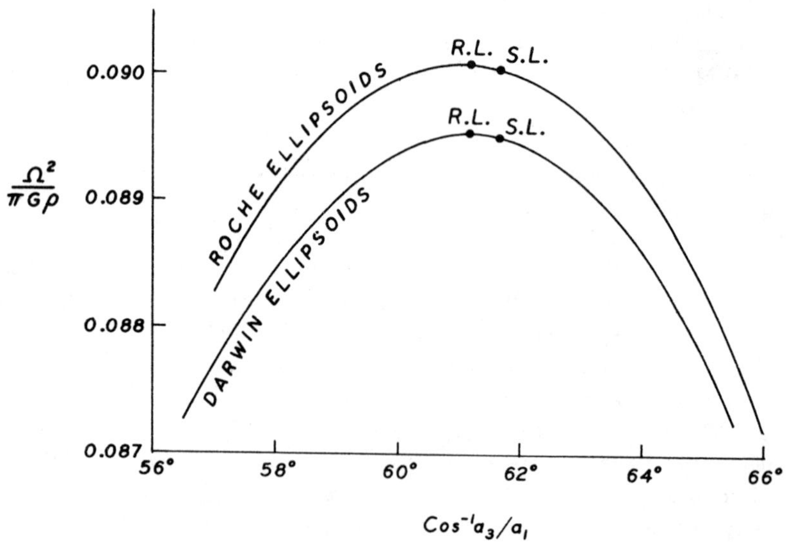

Fig. 25a. The variation of Ω^2 along the Darwin and the Roche sequences for the case $M/M' = 0$. The points $R.L.$ and $S.L.$ denote the Roche limit and the stability limit, respectively. Along both sequences the Roche limit occurs at the point where Ω^2 attains its maximum.

verified that the *Roche limit* (defined as the place along the sequence where a member can be deformed into an adjacent one by the application of an infinitesimal displacement without affecting equilibrium and preserving the ellipsoidal shape and the same constant density) occurs at the same place.

The stability of these Darwin ellipsoids can be determined with the aid of the linearized version of the virial equation (115). (No ambiguity, in treating the β's in this equation as constants, arises in this case; it

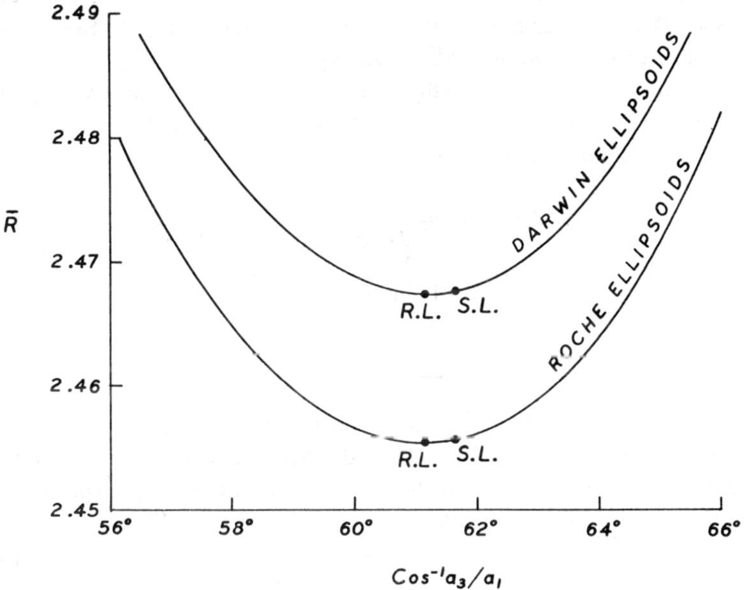

FIG. 25b. The variation of the distance between the centers of mass of the two components in the Roche and in the Darwin problems for the case $M/M' = 0$. Along both sequences the distance of separation is measured in the units of the mean radius of the central body. (In the Darwin problem the central body is a Maclaurin spheroid.) The points *R.L.* and *S.L.* have the same meanings as in Fig. 25a. Along both sequences the Roche limit occurs at the distance of closest approach.

does arise in the context of the problem considered in § (b) below.) The analysis is so similar to that described in § 57 for the case of the Roche ellipsoids that it need not be repeated. It is found, as one might have expected, that dynamical instability sets in at a point subsequent to the Roche limit (see the entries in the last line of Table XXIV). And again, it is clear that, as in the case of the Roche ellipsoids, secular instability will intervene between the two limits.

(b) The case of congruent components

We now consider the case when the masses and the densities of the two components are the same. The components of the system will then be congruent and the reduction to a stationary solution in a common rotating frame can be accomplished without any ambiguity. The problem then is to determine the geometry of the ellipsoidal figures of equilibrium (in the framework of the approximation described in § 61) when the centers of mass of the two components are at a distance R apart and they are rotating about each other with a constant angular velocity Ω consistent with their figures and their separation. The solution of this problem can be accomplished as follows.

Consider a homogeneous ellipsoid of mass M and semi-axes a_j. By equations (70) and (72) of Chapter 3, the gravitational potential $\mathfrak{B}$ at an external point $\mathbf{x}$ is given by

$$\mathfrak{B}(\mathbf{x}) = a_1 a_2 a_3 \int_\lambda^\infty \left(1 - \sum_{j=1}^3 \frac{x_j^2}{a_j^2 + u}\right) \frac{du}{\Delta(u)}, \tag{129}$$

where λ is the positive root of the equation

$$\sum_{j=1}^3 \frac{x_j^2}{a_j^2 + \lambda} = 1. \tag{130}$$

(In equation (129) a factor $\pi G \rho$ has been suppressed.)

For the application of the method described in § 61, we need the first and the second derivatives of $\mathfrak{B}$ with respect to the spatial coordinates at $(R, 0, 0)$. For such a point

$$\lambda = R^2 - a_1^2 \tag{131}$$

and

$$\Delta(\lambda) = R[(R^2 + a_2^2 - a_1^2)(R^2 + a_3^2 - a_1^2)]^{\frac{1}{2}}. \tag{132}$$

Also, in evaluating the derivatives of $\mathfrak{B}$ given by equation (129), we must allow for the fact that λ, as defined by equation (130), is, implicitly, a function of the coordinates and that

$$\frac{\partial \lambda}{\partial x_i} = \frac{2x_i}{a_i^2 + \lambda}\left[\sum_{j=1}^3 \frac{x_j^2}{(a_j^2 + \lambda)^2}\right]^{-1}. \tag{133}$$

By straightforward calculations, we now find

$$\left(\frac{\partial \mathfrak{B}}{\partial x_1}\right)_{R,0,0} = -2\alpha_1 R, \qquad \left(\frac{\partial \mathfrak{B}}{\partial x_2}\right)_{R,0,0} = \left(\frac{\partial \mathfrak{B}}{\partial x_3}\right)_{R,0,0} = 0, \tag{134}$$

$$\left(\frac{\partial^2 \mathfrak{B}}{\partial x_1^2}\right)_{R,0,0} = -2\alpha_1 + \frac{4a_1 a_2 a_3}{\Delta(\lambda)},$$

$$\left(\frac{\partial^2 \mathfrak{B}}{\partial x_2^2}\right)_{R,0,0} = -2\alpha_2, \quad \text{and} \quad \left(\frac{\partial^2 \mathfrak{B}}{\partial x_3^2}\right)_{R,0,0} = -2\alpha_3, \tag{135}$$

where
$$\alpha_j = a_1 a_2 a_3 \int_{R^2-a_1^2}^{\infty} \frac{du}{(a_j^2+u)\Delta(u)}. \tag{136}$$

The coefficients α_j defined in equation (136) are similar to the index symbols A_j defined in Chapter 3. They satisfy the relation

$$\alpha_1+\alpha_2+\alpha_3 = \frac{2a_1 a_2 a_3}{\Delta(\lambda)}; \tag{137}$$

and they are expressible in terms of the elliptic integrals, $E(\theta,\phi)$ and $F(\theta,\phi)$, of the two kinds with the arguments

$$\theta = \sin^{-1}\left(\frac{a_1^2-a_2^2}{a_1^2-a_3^2}\right)^{\frac{1}{2}} \quad \text{and} \quad \phi_R = \sin^{-1}\left(\frac{a_1^2-a_3^2}{R^2}\right)^{\frac{1}{2}}. \tag{138}$$

Thus,

$$\alpha_1 = \frac{2a_1 a_2 a_3}{(a_1^2-a_3^2)^{\frac{3}{2}}} \frac{1}{\sin^2\theta}[F(\theta,\phi_R)-E(\theta,\phi_R)], \tag{139}$$

$$\alpha_2 = \frac{2a_1 a_2 a_3}{(a_1^2-a_3^2)^{\frac{3}{2}}} \frac{1}{\sin^2\theta \cos^2\theta}\left[E(\theta,\phi_R)-F(\theta,\phi_R)\cos^2\theta - \frac{\sin^2\theta \sin\phi_R \cos\phi_R}{(1-\sin^2\theta \sin^2\phi_R)^{\frac{1}{2}}}\right], \tag{140}$$

$$\alpha_0 = \frac{2a_1 a_2 a_3}{(a_1^2-a_3^2)^{\frac{3}{2}}} \frac{1}{\cos^2\theta}[(1-\sin^2\theta \sin^2\phi_R)^{\frac{1}{2}} \tan\phi_R - E(\theta,\phi_R)]. \tag{141}$$

With the various derivatives of $\mathfrak{B}$ given by equations (134) and (135), equations (111), (113), and (114) now give

$$\Omega^2 = 4\alpha_1, \quad \beta_{11} = 2(2\alpha_1+\alpha_2+\alpha_3), \quad \beta_{22} = 2(2\alpha_1-\alpha_2),$$

and
$$\beta_{33} = -2\alpha_3. \tag{142}$$

With the coefficients β_{jj} given by equations (142), the virial equations (118) give

$$\alpha_1 = \frac{A_1 a_1^2 - A_3 a_3^2}{(2+\alpha_2/\alpha_1)a_1^2+(a_1^2+a_3^2)\alpha_3/\alpha_1} = \frac{A_2 a_2^2 - A_3 a_3^2}{(2-\alpha_2/\alpha_1)a_2^2+a_3^2 \alpha_3/\alpha_1}, \tag{143}$$

where it may be recalled that the index symbols A_j are also expressible in terms of the same elliptic integrals with, however, the arguments $\phi = \cos^{-1}a_3/a_1$ (which is different from ϕ_R) and θ (which is the same as in the equations defining the α_j's).

A convenient procedure for solving equations (143) and determining the equilibrium figures is the following.

We first assign an angle ϕ_R. Then for a chosen ϕ ($= \cos^{-1}a_3/a_1$) we determine, by a method of trial and error, the angle θ (which occurs in the expressions for the A's and the α's) that will satisfy the second of the two equalities in (143). For an arbitrarily chosen ϕ, the value of θ, which the first equality in (143) requires, will not agree with the value which the assigned ϕ_R and the determined θ will require according to equation (139). The problem then is to "adjust" ϕ such that for the assigned ϕ_R, the values of α_1 required by equations (139) and (143) agree.

The results summarized in Table XXV were obtained by the procedure described in the preceding paragraph.

TABLE XXV

The Darwin ellipsoids for the case of congruent components

$$\phi_R = \sin^{-1}\sqrt{[(a_1^2-a_3^2)/R^2]}, \quad \phi = \cos^{-1}a_3/a_1,$$

$$\theta = \sin^{-1}\sqrt{[(a_1^2-a_2^2)/(a_1^2-a_3^2)]}, \quad \overline{R} = R/(a_1a_2a_3)^{\frac{1}{3}}, \quad \bar{a}_i = a_i/(a_1a_2a_3)^{\frac{1}{3}}$$

ϕ_R	ϕ	θ	$\Omega^2/\pi G\rho$	$\overline{R}$	$\bar{a}_1$	$\bar{a}_2$	$\bar{a}_3$
5°	21°·898	53°·798	0·03023	4·4572	1·0416	0·9933	0·9665
6°	24°·408	54°·559	0·03721	4·1613	1·0526	0·9911	0·9585
7°	26°·751	55°·355	0·04424	3·9308	1·0643	0·9886	0·9504
8°	28°·957	56°·180	0·05125	3·7457	1·0767	0·9858	0·9421
9°	31°·050	57°·032	0·05819	3·5936	1·0899	0·9826	0·9338
10°	33°·047	57°·910	0·06502	3·4667	1·1039	0·9790	0·9253
11°	34°·961	58°·811	0·07169	3·3595	1·1187	0·9750	0·9168
12°	36°·802	59°·732	0·07817	3·2682	1·1343	0·9707	0·9082
13°	38°·578	60°·673	0·08441	3·1901	1·1508	0·9659	0·8996
14°	40°·296	61°·630	0·09037	3·1231	1·1683	0·9606	0·8910
15°	41°·962	62°·601	0·09603	3·0657	1·1867	0·9550	0·8824
16°	43°·580	63°·585	0·10134	3·0166	1·2062	0·9488	0·8738
17°	45°·154	64°·578	0·10628	2·9749	1·2268	0·9422	0·8651
18°	46°·688	65°·578	0·11082	2·9398	1·2485	0·9352	0·8564
19°	48°·184	66°·583	0·11494	2·9108	1·2715	0·9277	0·8478
20°	49°·647	67°·590	0·11860	2·8874	1·2959	0·9197	0·8391
21°	51°·077	68°·596	0·12179	2·8692	1·3216	0·9112	0·8304
22°	52°·476	69°·599	0·12449	2·8559	1·3489	0·9023	0·8216
23°	53°·848	70°·596	0·12668	2·8473	1·3778	0·8929	0·8128
24°	55°·193	71°·585	0·12836	2·8433	1·4085	0·8831	0·8040
24°·4035†	55°·728	71°·981	0·12888	2·8429	1·4214	0·8790	0·8004
24°·4065‡	55°·732	71°·984	0·12889	2·8429	1·4215	0·8790	0·8004

† The "Roche limit"; the point of closest approach.
‡ The point where the ellipsoids are in contact.

The most important feature of this Darwin sequence is that at a certain point along the sequence $R = 2a_1$; at this point the two components are in contact. Beyond this point $R < 2a_1$ and the two components overlap. These solutions are, therefore, not physically meaningful.

The calculations indicate that at the distance of closest approach the two components are almost, but not quite, in contact; they further show that, of the two congruent figures of equilibrium one formally obtains as solutions of the relevant equations for each separation, those with the greater elongations overlap; these second solutions are, therefore, not physically meaningful.

63. The stability of the congruent Darwin ellipsoids

As we have already remarked in § 62 (*a*), the discussion of the stability of the Darwin ellipsoids in the case $M/M' = 0$ presents no difficulties: for, in this case, any perturbations of the satellite, in view of its infinitesimal mass, will not affect the central Maclaurin spheroid. But in the case of the congruent components, a general treatment of the system presents special difficulties. Thus, while it is a simple matter to isolate the points of onset of the instability (dynamical or secular) via a natural mode of oscillation of either component by itself, with the other remaining fixed, the treatment of the general modes of coupled oscillations is not as simple. However, there exists a special class of coupled oscillations which is amenable to treatment by a generalization of the methods that we have adopted in other cases; and it will appear that one of the eight modes of oscillations belonging to this class excites instabilities beyond a certain point along the Darwin sequence.

(a) *A description of the nature of the oscillations considered*

We shall consider the perturbation problem in a frame rotating with the orbital angular velocity, Ω, of the equilibrium ellipsoids (cf. equation (142)). And we shall consider the motions of the fluid elements in the two ellipsoids, in Cartesian frames $[x_1^{(i)}, x_2^{(i)}, x_3^{(i)}; i = 1,2]$ located at the centers of the respective equilibrium figures, the $x_1 =$ axis of one pointing to the equilibrium center of the other. Also, we shall distinguish the quantities pertaining to the two ellipsoids by superscripts (1) and (2), referring, respectively, to the "primary" and the "secondary."

We shall suppose that the two ellipsoids are subjected to deformations described by the Lagrangian displacements

$$\xi_1^{(i)} = L_{1;0}^{(i)} + L_{1;1}^{(i)} x_1^{(i)} + L_{1;2}^{(i)} x_2^{(i)},$$

$$\xi_2^{(i)} = L_{2;0}^{(i)} + L_{2;1}^{(i)} x_1^{(i)} + L_{2;2}^{(i)} x_2^{(i)},$$

$$\xi_3^{(i)} = L_{3;3}^{(i)} x_3^{(i)} \qquad (i = 1,2), \qquad (144)$$

where a time-dependent factor $e^{\lambda t}$ in the coefficients $L_{j;k}^{(i)}$ has been suppressed.

By the application of the displacement (144), the boundary of the ellipsoid, specified by

$$\frac{(x_1^{(i)})^2}{a_1^2} + \frac{(x_2^{(i)})^2}{a_2^2} + \frac{(x_3^{(i)})^2}{a_3^2} = 1, \qquad (145)$$

becomes

$$\frac{(x_1^{(i)})^2}{a_1^2} + \frac{(x_2^{(i)})^2}{a_2^2} + \frac{(x_3^{(i)})^2}{a_3^2} - 2\left(\frac{\xi_1^{(i)} x_1^{(i)}}{a_1^2} + \frac{\xi_2^{(i)} x_2^{(i)}}{a_2^2} + \frac{\xi_3^{(i)} x_3^{(i)}}{a_3^2}\right) = 1; \qquad (146)$$

or, equivalently to the first order in the displacement,

$$\frac{(x_1^{(i)} - L_{1;0}^{(i)})^2}{a_1^2 (1 + 2 L_{1;1}^{(i)})} + \frac{(x_2^{(i)} - L_{2;0}^{(i)})^2}{a_2^2 (1 + 2 L_{2;2}^{(i)})} + \frac{x_3^2}{a_3^2 (1 + 2 L_{3;3}^{(i)})}$$

$$- 2 x_1 x_2 \left(\frac{L_{1;2}^{(i)}}{a_1^2} + \frac{L_{2;1}^{(i)}}{a_2^2} \right) = 1. \tag{147}$$

Equation (147) represents an ellipsoid with its center at $(L_{1;0}^{(i)}, L_{2;0}^{(i)}, 0)$ and semi-axes of lengths

$$a_1(1 + L_{1;1}^{(i)}), \ a_2(1 + L_{2;2}^{(i)}), \ a_3(1 + L_{3;3}^{(i)}); \tag{148}$$

and the major axis of the ellipsoid is inclined to the $x_1^{(i)}$-direction by the angle

$$\delta\theta^{(i)} = \left(\frac{L_{1;2}^{(i)}}{a_1^2} + \frac{L_{2;1}^{(i)}}{a_2^2} \right) \frac{a_1^2 a_2^2}{a_1^2 - a_2^2} \qquad (i = 1,2). \tag{149}$$

It will be convenient to have the foregoing effects of the displacements (144) on the geometry of the ellipsoid expressed in terms of the virials $V_j^{(i)}$ and $V_{jk}^{(i)}$. Thus, by equation (68) of Chapter 5, the lengths of the semi-axes of the ellipsoid are altered by the amounts

$$\frac{\delta a_j^{(i)}}{a_j} = \frac{5}{2M} \frac{V_{jj}^{(i)}}{a_j^2} \qquad (j = 1,2,3; i = 1,2), \tag{150}$$

while the ellipsoid is rotated about the $x_3^{(i)}$-axis by an angle,

$$\delta\theta^{(i)} = \frac{5}{M} \frac{V_{12}^{(i)}}{a_1^2 - a_2^2} \qquad (i = 1,2). \tag{151}$$

At the same time, the centroids of the equilibrium ellipsoids are displaced to the points

$$\frac{1}{M} \left(\int_V \rho\xi_1^{(i)} dx, \ \int_V \rho\xi_2^{(i)} dx, \ 0 \right) = \frac{1}{M} \left(V_1^{(i)}, V_2^{(i)}, 0 \right)$$

$$= \left(L_{1;0}^{(i)}, L_{2;0}^{(i)}, 0 \right). \tag{152}$$

The relative geometry of the two ellipsoids that results from the displacements considered is illustrated in Fig. 26. We observe that the centers of equilibrium ellipsoids, originally at S and P, have been displaced to S' and P'; and, further, that the inclination, $\delta\Theta^{(2)}$, of the major axis of the disoriented ellipsoid to the line joining S' and P (the equilibrium position of the primary) is given by

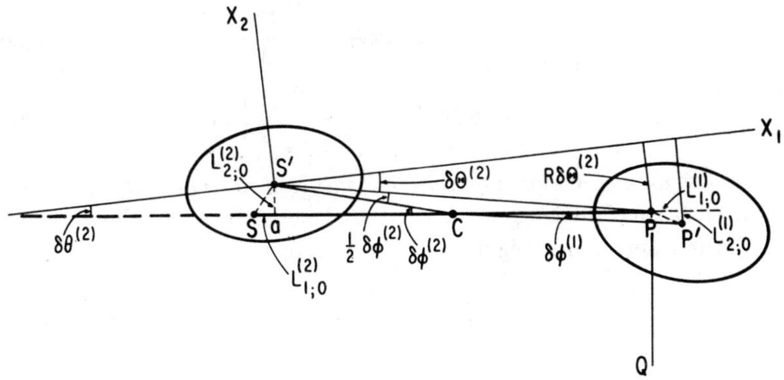

FIG. 26. The geometry of the system. The system is considered, both when it is in equilibrium and when it is perturbed, in a frame of reference rotating with the uniform constant angular velocity appropriate for equilibrium. The section illustrated is through the orbital plane. In equilibrium, the primary and the secondary, at relative rest, are at P and at S, respectively; their common center of mass is at C, and the distance between them is R. Lagrangian displacements of the kind considered will displace the centers of the primary and the secondary to positions such as P' and S', while rotating them about their vertical axes by amounts $\delta\theta^{(1)}$ and $\delta\theta^{(2)}$. Thus, in terms of the displacement considered (equation (144)), $Sa = L_{1;0}^{(2)}$ and $S'a = L_{2;0}^{(2)}$; and the secondary has been rotated by the angle $\delta\theta^{(2)}$. It is also clear from the illustration that $\delta R^{(1)} = -L_{1;0}^{(1)}$, $\delta R^{(2)} = -L_{1;0}^{(2)}$, $\delta\phi^{(1)} = 2L_{2;0}^{(1)}/R$, $\delta\phi^{(2)} = 2L_{2;0}^{(2)}/R$, and $\delta\Theta^{(2)} = \delta\theta^{(2)} + \tfrac{1}{2}\delta\phi^{(2)}$.

$$\measuredangle\, S'X_1, S'P = \delta\Theta^{(2)} = \delta\theta^{(2)} + \tfrac{1}{2}\,\delta\phi^{(2)}, \tag{153}$$

where
$$\delta\phi^{(i)} = 2\,\frac{L_{2;0}^{(i)}}{R} = 2\,\frac{V_2^{(i)}}{RM} \qquad (i = 1,2). \tag{154}$$

Accordingly, in view of equation (151),

$$\delta\Theta^{(2)} = \frac{1}{M}\left(5\frac{V_{12}^{(2)}}{a_1^2 - a_2^2} + \frac{V_2^{(2)}}{R}\right). \tag{155}$$

We shall also note here for later reference that

$$\measuredangle\, S'X_1, S'P' = \delta\Theta = \left(\delta\theta^{(2)} + \tfrac{1}{2}\,\delta\phi^{(1)} + \tfrac{1}{2}\,\delta\phi^{(2)}\right), \tag{156}$$

or
$$\delta\Theta = \frac{1}{M}\left(5\frac{V_{12}^{(2)}}{a_1^2 - a_2^2} + \frac{V_2^{(1)}}{R} + \frac{V_2^{(2)}}{R}\right). \tag{157}$$

and further, that

$$\delta R^{(i)} = -L_{1;0}^{(i)} = -\frac{V_1^{(i)}}{M} \qquad (i = 1,2), \tag{158}$$

$$SP - S'P = \delta R^{(2)} = -\frac{V_1^{(2)}}{M}, \tag{159}$$

and
$$SP - S'P' = \delta R^{(1)} + \delta R^{(2)} = -\frac{1}{M}\left(V_1^{(1)} + V_1^{(2)}\right). \tag{160}$$

(b) The equation of motion appropriate to the fluid elements in the primary

The equation of motion governing the fluid elements in the primary (in the coordinate frame whose origin is at P) is given by

$$\rho\frac{du_j^{(1)}}{dt} = -\frac{\partial p^{(1)}}{\partial x_j^{(1)}} + \rho\frac{\partial}{\partial x_j^{(1)}}\Bigg\{ \mathfrak{B}^{(1)}\left(x_1^{(1)}, x_2^{(1)}, x_3^{(1)}\right) + \mathfrak{B}^{(2)}\left(x_1^{(1)}, x_2^{(1)}, x_2^{(1)}\right)$$

$$- \tfrac{1}{2} R\Omega^2 x_1^{(1)} + \tfrac{1}{2}\Omega^2\left[\left(x_1^{(1)}\right)^2 + \left(x_2^{(1)}\right)^2\right]\Bigg\} + 2\rho\Omega\epsilon_{jl3}u_l^{(1)}, \tag{161}$$

where Ω denotes the constant Keplerian angular velocity of rotation of the equilibrium ellipsoids (given in equation (142)) and $\mathfrak{B}^{(2)}$ is the tidal potential of the displaced reoriented ellipsoid, of altered semi-axes at S', on the ellipsoid centered at P'. Consistent with the standard procedure of expressing the tidal potential over an object, we expand $\mathfrak{B}^{(2)}(x_1^{(1)}, x_2^{(1)}, x_3^{(1)})$ as a Taylor series about the instantaneous center of mass of the primary (at P' in Fig. 26) and retain only terms that are linear and quadratic in the displacement. Since the origin of the coordinate frame, in which we have chosen to consider the perturbation problem, is at the center $(L_{1;0}^{(1)}, L_{2;0}^{(1)}, 0)$ of the equilibrium ellipsoid (at P), the required expansion of the tidal potential is given by

$$\mathfrak{B}^{(2)}\left(x_1^{(1)}, x_2^{(1)}, x_3^{(1)}\right) = \mathfrak{B}^{(2)}(P') + \left(\frac{\partial \mathfrak{B}^{(2)}}{\partial x_i^{(1)}}\right)_{P'}\left(x_i^{(1)} - L_{i;0}^{(1)}\right)$$

$$+ \tfrac{1}{2}\left(\frac{\partial^2 \mathfrak{B}^{(2)}}{\partial x_i^{(1)}\,\partial x_j^{(1)}}\right)_{P'}\left(x_i^{(1)} - L_{i;0}^{(1)}\right)\left(x_j^{(1)} - L_{j;0}^{(1)}\right), \tag{162}$$

or, rearranging the terms, we can write to the first order in the displacement,

$$\mathfrak{B}^{(2)}\left(x_1^{(1)}, x_2^{(1)}, x_3^{(1)}\right) = \mathfrak{B}^{(2)}(P') - \left(\frac{\partial \mathfrak{B}^{(2)}}{\partial x_i^{(1)}}\right)_{P'} L_{i;0}^{(1)}$$

$$+ \left[\left(\frac{\partial \mathfrak{B}^{(2)}}{\partial x_i^{(1)}}\right)_{P'} - \left(\frac{\partial^2 \mathfrak{B}^{(2)}}{\partial x_i^{(1)}\,\partial x_j^{(1)}}\right)_{P'} L_{j;0}^{(1)}\right] x_i^{(1)} + \tfrac{1}{2}\left(\frac{\partial^2 \mathfrak{B}^{(2)}}{\partial x_i^{(1)}\,\partial x_j^{(1)}}\right)_{P'} x_i^{(1)} x_j^{(1)} + O(|\boldsymbol{\xi}|^2)$$

$$= \mathfrak{B}^{(2)}(P) + \left(\frac{\partial \mathfrak{B}^{(2)}}{\partial x_i^{(1)}}\right)_P x_i^{(1)} + \tfrac{1}{2}\left(\frac{\partial^2 \mathfrak{B}^{(2)}}{\partial x_i^{(1)}\,\partial x_j^{(1)}}\right)_{P'} x_i^{(1)} x_j^{(1)} + O(|\boldsymbol{\xi}|^2). \tag{163}$$

We observe that the first derivatives of the tidal potential exerted by the displaced secondary must be evaluated at P while the second derivatives must be evaluated at P'.

(c) *The potential of an ellipsoid at an external point on a line slightly displaced with respect to the major axis*

Since we contemplate a slight misalignment of the two ellipsoids (as in Fig. 26), we shall consider first the derivatives of the external potential, $\mathfrak{B}$, of an ellipsoid at a point $(R, -R\delta\Theta, 0)$ slightly off the continuation of the major axis. (The coordinate frame in which we shall consider this preliminary problem is the same that was adopted in Chapter 3 and in which the standard formulae of the theory have been written.)

From equations (129) and (130) it follows that to the first order in $R\delta\Theta$, the solution of equation (130) is given by

$$\lambda = R_1^2 - a_1^2 + O(R\delta\Theta)^2, \tag{164}$$

i.e., the value of λ to the required order is the same as for the point $(R,0,0)$ on the x_1-axis (cf. equation (131)). It can be readily deduced from this fact that the same is also true of $\partial\mathfrak{B}/\partial x_1$ and $\partial^2\mathfrak{B}/\partial x_1^2$; i.e., the expressions for these derivatives, to the first order in $R\delta\Theta$, are the same as those given in equations (134)–(136) for the point $(R,0,0)$. However, $\partial\mathfrak{B}/\partial x_2$ and $\partial^2\mathfrak{B}/\partial x_1\,\partial x_2$ differ from zero—as they are at $(R,0,0)$—by quantities of the first order in $R\delta\Theta$; and these we must retain. We have

$$\frac{\partial\mathfrak{B}}{\partial x_2} = -2a_1a_2a_3\,x_2 \int_\lambda^\infty \frac{du}{(a_2^2 + u)\,\Delta(u)} \tag{165}$$

and

$$\frac{\partial^2\mathfrak{B}}{\partial x_1\,\partial x_2} = \frac{2a_1a_2a_3\,x_2}{(a_2^2 + \lambda)\Delta(\lambda)}\,\frac{\partial\lambda}{\partial x_1}. \tag{166}$$

By making use of equation (133), equation (166) becomes

$$\frac{\partial^2\mathfrak{B}}{\partial x_1\,\partial x_2} = \frac{4\,a_1a_2a_3\,x_2x_1}{(a_1^2 + \lambda)(a_2^2 + \lambda)\,\Delta(\lambda)}\left[\sum_{j=1}^{3}\frac{x_j^2}{(a_j^2 + \lambda)^2}\right]^{-1}. \tag{167}$$

Hence to the order required,

$$\left(\frac{\partial\mathfrak{B}}{\partial x_2}\right)_{(R,-R\delta\Theta,0)} = 2\,\alpha_2\,R\delta\Theta + O(R\delta\Theta)^2 \tag{168}$$

and

$$\left(\frac{\partial^2\mathfrak{B}}{\partial x_1\,\partial x_2}\right)_{(R,-R\delta\Theta,0)} = -\beta_{12}\,R\delta\Theta + O(R\delta\Theta)^2, \tag{169}$$

where

$$\beta_{12} = \frac{4a_1a_2a_3}{(R^2 + a_2^2 - a_1^2)^{3/2}\,(R^2 + a_3^2 - a_1^2)^{1/2}}. \tag{170}$$

The remaining derivatives, to the required order, are given by the same equations (134–136).

(d) The required expansion for the tidal potential

The expressions for the first and the second derivatives of the tidal potential given in § *(c)* must be modified appropriately to apply to the situation illustrated in Fig. 26, remembering that while the first derivatives are to be evaluated at P the second derivatives must be evaluated at P'. The required modifications derive from two sources: *first*, the principal axes a_j and the distance R differ from their equilibrium values by amounts specified in equations (150), (159), and (160); and *second*, the orientation of the axes to which equations (168) and (169) refer (namely $S'X_1$ and $S'X_2$) is different from that to which equation (163) refers.

Allowance for the *first* feature requires that we replace the expressions for the first derivatives (given in equation (134) for $\partial\mathfrak{B}/\partial x_1$, and in equation (168) for $\partial\mathfrak{B}/\partial x_2$) by

$$\left(\frac{\partial\mathfrak{B}^{(2)}}{\partial X_1}\right)_P = -2\,\alpha_1\,R - 2\,[(\delta\alpha_1^{(2)})_P\,R + \alpha_1\,\delta R^{(2)}] \tag{171}$$

and

$$\left(\frac{\partial\mathfrak{B}^{(2)}}{\partial X_2}\right)_P = 2\,\alpha_2\,R\delta\Theta^{(2)}. \tag{172}$$

(Equation (172) follows from noting that the inclination between the directions $S'P$ and $S'X_1$ is given by $\delta\Theta^{(2)}$ given in equations (153) and (155).) Similarly, the expressions for the second derivatives given in equations (135) and (169) must be replaced by (cf. equation (158) and (159))

$$\left(\frac{\partial^2\mathfrak{B}^{(2)}}{\partial X_1\,\partial X_2}\right)_{P'} = -\beta_{12}\,R(\delta\Theta^{(2)} + \tfrac{1}{2}\,\delta\phi^{(1)}) = -\beta_{12}\,R\delta\Theta; \tag{173}$$

$$\left(\frac{\partial^2\mathfrak{B}^{(2)}}{\partial X_1^2}\right)_{P'} = 2\,(\alpha_2 + \alpha_3) + 2\,(\delta\alpha_2^{(2)} + \delta\alpha_3^{(2)})_{P'}; \tag{174}$$

$$\left(\frac{\partial^2\mathfrak{B}^{(2)}}{\partial X_2^2}\right)_{P'} = -2\,\alpha_2 - 2\,(\delta\alpha_2^{(2)})_{P'}; \quad \left(\frac{\partial^2\mathfrak{B}^{(2)}}{\partial X_3^2}\right) = -2\,\alpha_3 - 2\,(\delta\alpha_3^{(2)})_{P'}. \tag{175}$$

In equations (171) and (173)–(175), $\delta\alpha_j^{(2)}$ denotes the change in α_j consequent to the changes in the semi-axes (a_j) of the ellipsoid and in the separation (R). Accordingly,

$$\delta\alpha_j^{(2)} = \sum_{k=1}^{3}\frac{\partial\alpha_j}{\partial a_k}\,\delta a_k^{(2)} + \frac{\partial\alpha_j}{\partial R}\,\delta R, \tag{176}$$

where for $\delta\alpha_j^{(2)}$ evaluated at P and P',

$$(\delta R)_P = \delta R^{(2)} = -L_{1;0}^{(2)} = -\frac{1}{M}\,V_1^{(2)} \tag{177}$$

and
$$(\delta R)_{P'} = \delta R^{(1)} + \delta R^{(2)} = -\left(L^{(1)}_{1;0} + L^{(2)}_{1;0}\right) = -\frac{1}{M}\left(V^{(1)}_1 + V^{(2)}_1\right); \qquad (178)$$

and $\delta\, a^{(2)}_j$ is expressed in terms of the virials $V^{(2)}_{jj}$ in equation (150). From the definition of α_j in equation (136) it now follows directly that

$$(\delta\alpha^{(2)}_1)_P = \frac{1}{2M}\,[5\,(Q_{11}V^{(2)}_{11} - \alpha_{12}V^{(2)}_{22} - \alpha_{13}V^{(2)}_{33}) + 2q_1 R V^{(2)}_1] \qquad (179)$$

for insertion in equation (171), and

$$(\delta\alpha^{(2)}_1)_{P'} = \frac{1}{2M}\,[5\,(Q_{11}V^{(2)}_{11} - \alpha_{12}V^{(2)}_{22} - \alpha_{13}V^{(2)}_{33}) + 2q_1 R(V^{(1)}_1 + V^{(2)}_1)],$$

$$(\delta\alpha^{(2)}_2)_{P'} = \frac{1}{2M}\,[5\,(Q_{21}V^{(2)}_{11} - 3\alpha_{22}V^{(2)}_{22} - \alpha_{23}V^{(2)}_{33}) + 2q_2 R(V^{(1)}_1 + V^{(2)}_1)],$$

$$(\delta\alpha^{(2)}_3)_{P'} = \frac{1}{2M}\,[5\,(Q_{31}V^{(2)}_{11} - \alpha_{32}V^{(2)}_{22} - 3\alpha_{33}V^{(2)}_{33}) + 2q_3 R(V^{(1)}_1 + V^{(2)}_1)],$$

$$(180)$$

for insertion in equations (174) and (175), where

$$Q_{11} = q_1 - 3\alpha_{11}, \; Q_{21} - q_2 - \alpha_{21}, \; Q_{31} = q_3 \quad \alpha_{31}; \qquad (181)$$

and
$$q_j = \frac{2a_1 a_2 a_3}{(a^2_j + \lambda)\,\Delta(\lambda)} = \frac{\alpha_1 + \alpha_2 + \alpha_3}{R^2 + a^2_j - a^2_1}. \qquad (182)$$

Also, in the foregoing equations α_{jk} denotes the "two-index" symbol,

$$\alpha_{ij} = a_1 a_2 a_3 \int\limits_{R^2-a^2_1}^{\infty} \frac{du}{(a^2_i + u)\,(a^2_j + u)\,\Delta(u)}. \qquad (183)$$

Turning to the *second* feature arising from the different orientations of the axes $(S'X_1, S'X_2)$ and $(PS$ and $PQ)$, we must consider the transformation

$$\begin{aligned}
X_1 &= -(x^{(1)}_1 - R)\cos\delta\theta^{(2)} - x^{(1)}_2 \sin\delta\theta^{(2)} \\
&= -(x^{(1)}_1 - R) - x^{(1)}_2\,\delta\theta^{(2)} + O(\delta\theta^{(2)})^2, \\
X_2 &= +(x^{(1)}_1 - R)\sin\delta\theta^{(2)} - x^{(1)}_2 \cos\delta\theta^{(2)} \\
&= (x^{(1)}_1 - R)\,\delta\theta^{(2)} - x^{(1)}_2 + O(\delta\theta^{(2)})^2.
\end{aligned} \qquad (184)$$

In view of this transformation, the required Taylor expansion for $\mathfrak{B}_2$ can be written as

$$\mathfrak{B}^{(2)}(x^{(1)}_1, x^{(1)}_2, x^{(1)}_3) = \mathfrak{B}^{(2)}(P) - \left(\frac{\partial\mathfrak{B}^{(2)}}{\partial X_1}\right)_P [x^{(1)}_1 + x^{(1)}_2\,\delta\theta^{(2)}]$$

$$- \left(\frac{\partial \mathfrak{B}^{(2)}}{\partial X_2}\right)_P x_2^{(1)} + \tfrac{1}{2} \left(\frac{\partial^2 \mathfrak{B}^{(2)}}{\partial X_1^2}\right)_{P'} [(x_1^{(1)})^2 + 2\, x_1^{(1)} x_2^{(1)}\, \delta\theta^{(2)}]$$

$$+ \tfrac{1}{2} \left(\frac{\partial^2 \mathfrak{B}^{(2)}}{\partial X_2^2}\right)_{P'} [(x_2^{(1)})^2 - 2\, x_1^{(1)} x_2^{(1)}\, \delta\theta^{(2)}]$$

$$+ \tfrac{1}{2} \left(\frac{\partial^2 \mathfrak{B}^{(2)}}{\partial X_3^2}\right)_{P'} (x_3^{(1)})^2 + \left(\frac{\partial^2 \mathfrak{B}^{(2)}}{\partial X_1 \partial X_2}\right)_{P'} x_1^{(1)} x_2^{(1)}. \tag{185}$$

Inserting for the derivatives of $\mathfrak{B}^{(2)}$ from equations (171)–(175), we can rewrite the foregoing expansion for $\mathfrak{B}^{(2)}$ in the form,

$$\mathfrak{B}^{(2)}(x_1^{(1)}, x_2^{(1)}, x_3^{(1)}) - \tfrac{1}{2} R\Omega^2 x_1^{(1)} + \tfrac{1}{2}\Omega^2\left[(x_1^{(1)})^2 + (x_2^{(1)})^2\right] = \mathfrak{B}^{(2)}(P) + \delta\beta_1^{(2)} x_1^{(1)}$$

$$+ \delta\beta_2^{(2)} x_2^{(1)} + \tfrac{1}{2} \sum_{k=1}^{3} [\beta_{kk} + \delta\beta_{kk}^{(2)}] (x_k^{(1)})^2 + \delta\beta_{12}^{(2)} x_1^{(1)} x_2^{(1)}, \tag{186}$$

where (cf. equations (142))

$$\beta_{11} = 2\,(2\alpha_1 + \alpha_2 + \alpha_3); \quad \beta_{22} = 2\,(2\alpha_2 - \alpha_2); \quad \beta_{33} = -2\alpha_3; \tag{187}$$

and (cf. equations (155), (157), (179), and (186))

$$\delta\beta_1^{(2)} = 2\,[(\delta\alpha_1^{(2)})_P R + \alpha_1 \delta R^{(2)}]$$

$$= \frac{R}{M}\,[5\,(Q_{11}V_{11}^{(2)} - \alpha_{12}V_{22}^{(2)} - \alpha_{13}V_{33}^{(2)}) + 2V_1^{(2)}(q_1 R^2 - \alpha_1)]; \tag{188}$$

$$\delta\beta_2^{(2)} = 2R(\alpha_1 \delta\theta^{(2)} - \alpha_2 \delta\Theta^{(2)})$$

$$= -2\,\frac{R}{M}\left(5\alpha_{12}V_{12}^{(2)} + \alpha_2 \frac{V_2^{(2)}}{R}\right); \tag{189}$$

$$\delta\beta_{12}^{(2)} = 2\,(2\alpha_2 + \alpha_3)\,\delta\theta^{(2)} - \beta_{12} R\delta\Theta$$

$$= \frac{5}{M(a_1^2 - a_2^2)}\,[2\,(2\alpha_2 + \alpha_3) - R\beta_{12}]\,V_{12}^{(2)} - \frac{\beta_{12}}{M}\,(V_2^{(1)} + V_2^{(2)}); \tag{190}$$

$$\delta\beta_{11}^{(2)} = 2\,(\delta\alpha_2^{(2)} + \delta\alpha_3^{(2)})_{P'}$$

$$= \frac{1}{M}\left\{5\,[(Q_{21} + Q_{31})V_{11}^{(2)} - (3\alpha_{22} + \alpha_{32})V_{22}^{(2)} - (3\alpha_{33} + \alpha_{23})V_{33}^{(2)}]\right.$$

$$\left. + 2\,R\,(q_2 + q_3)\,(V_1^{(1)} + V_1^{(2)})\right\}; \tag{191}$$

$$\delta\beta_{22}^{(2)} = -2\,(\delta\alpha_2^{(2)})_{P'}$$

$$= -\frac{1}{M}\,[5\,(Q_{21}V_{11}^{(2)} - 3\alpha_{22}V_{22}^{(2)} - \alpha_{23}V_{33}^{(2)}) + 2q_2 R(V_1^{(1)} + V_1^{(2)})]; \tag{192}$$

$$\delta\beta_{33}^{(2)} = -2\,(\delta\alpha_3^{(2)})_{P'}$$

$$= -\frac{1}{M}\,[5\,(Q_{31}V_{11}^{(2)} - \alpha_{32}V_{22}^{(2)} - 3\alpha_{33}V_{33}^{(2)}) + 2q_3 R(V_1^{(1)} + V_1^{(2)})].$$

$$(193)$$

With the expression for the tidal potential given in equation (186), the equation of motion (161) governing the fluid elements of the component "1" becomes

$$\rho\,\frac{du_j^{(1)}}{dt} = -\frac{\partial p^{(1)}}{\partial x_j^{(1)}} + \rho\,\frac{\partial}{\partial x_j^{(1)}}\left\{\mathfrak{V}^{(1)}(x_1^{(1)},x_2^{(1)},x_3^{(1)}) + \delta\beta_1^{(2)}x_1^{(1)} + \delta\beta_2^{(2)}x_2^{(1)}\right.$$

$$+ \tfrac{1}{2}\sum_{k=1}^{3}[\beta_{kk} + \delta\beta_{kk}^{(2)}]\,(x_k^{(1)})^2 + \delta\beta_{12}^{(2)}x_1^{(1)}x_2^{(1)}\Bigg\}$$

$$+ 2\,\rho\Omega\,\epsilon_{jl3}u_l^{(1)}.$$

$$(194)$$

The equation of motion governing the fluid elements of component "2" can be obtained by simply interchanging the superscripts (1) and (2).

When the conditions are stationary, all the $\delta\beta$'s in equation (194) (and in the equations in which the superscripts (1) and (2) are interchanged) vanish; and we shall be led to the equilibrium figures described in § 62(h).

(e) The virial equations governing the coupled oscillations

Remembering that, for the perturbations considered, the direction of the x_3-axes remains invariant, we can readily derive from equation (194) the virial equations for the coupled oscillations we are considering (with the time dependence $e^{\lambda t}$) in the usual manner. Thus, the second-order virial equations which follow from equation (194) are:

$$\tfrac{1}{2}\lambda^2 V_{11}^{(1)} - 2\lambda\Omega V_{2;1}^{(1)} = \delta\mathfrak{W}_{11}^{(1)} + \beta_{11}V_{11}^{(1)} + \delta\beta_{11}^{(2)}I_{11} + \delta\Pi^{(1)}, \qquad (195)$$

$$\tfrac{1}{2}\lambda^2 V_{22}^{(1)} + 2\lambda\Omega V_{1;2}^{(1)} = \delta\mathfrak{W}_{22}^{(1)} + \beta_{22}V_{22}^{(1)} + \delta\beta_{22}^{(2)}I_{22} + \delta\Pi^{(1)}, \qquad (196)$$

$$\tfrac{1}{2}\lambda^2 V_{33}^{(1)} = \delta\mathfrak{W}_{33}^{(1)} + \beta_{33}V_{33}^{(1)} + \delta\beta_{33}^{(2)}I_{33} + \delta\Pi^{(1)}, \qquad (197)$$

$$\lambda^2 V_{1;2}^{(1)} - \lambda\Omega V_{22}^{(1)} = -(2B_{12} - \beta_{11})V_{12}^{(1)} + \delta\beta_{12}^{(2)}I_{22}, \qquad (198)$$

$$\lambda^2 V_{2;1}^{(1)} + \lambda\Omega V_{11}^{(1)} = -(2B_{12} - \beta_{22})V_{12}^{(1)} + \delta\beta_{12}^{(2)}I_{11}, \qquad (199)$$

where we have substituted for $\delta\mathfrak{W}_{12}$ its known expression.

After eliminating $\delta\Pi^{(1)}$ appropriately from the equations (195)–(197) and combining equations (198) and (199) suitably, we obtain

$$\tfrac{1}{2}\lambda^2(V_{11}^{(1)} + V_{22}^{(1)}) + 2\lambda\Omega(V_{1;2}^{(1)} - V_{2;1}^{(1)}) - \lambda^2 V_{33}^{(1)}$$

$$= \delta\mathfrak{W}_{11}^{(1)} + \delta\mathfrak{W}_{22}^{(1)} - 2\,\delta\mathfrak{W}_{33}^{(1)} + \beta_{11}V_{11}^{(1)} + \beta_{22}V_{22}^{(1)} - 2\beta_{33}V_{33}^{(1)}$$

$$+ \delta\beta_{11}^{(2)}I_{11} + \delta\beta_{22}^{(2)}I_{22} - 2\,\delta\beta_{33}^{(2)}I_{33}, \qquad (200)$$

$$\tfrac{1}{2}\lambda^2(V_{11}^{(1)} - V_{22}^{(1)}) - 2\lambda\Omega V_{12}^{(1)} = \delta\mathfrak{W}_{11}^{(1)} - \delta\mathfrak{W}_{22}^{(1)} + \beta_{11}V_{11}^{(1)} - \beta_{22}V_{22}^{(1)}$$
$$+ \ \delta\beta_{11}^{(2)}I_{11} - \delta\beta_{22}^{(2)}I_{22}, \tag{201}$$

$$(\lambda^2 + 4B_{12} - \beta_{11} - \beta_{22})V_{12}^{(1)} + \lambda\Omega(V_{11}^{(1)} - V_{22}^{(1)}) = \delta\beta_{12}^{(2)}(I_{11} + I_{22}), \tag{202}$$

$$\lambda^2(V_{1;2}^{(1)} - V_{2;1}^{(1)}) = \lambda\Omega(V_{11}^{(1)} + V_{22}^{(1)}) + (\beta_{11} - \beta_{22})V_{12}^{(1)} - \delta\beta_{12}^{(2)}(I_{11} - I_{22}). \tag{203}$$

Next eliminating $(V_{1;2}^{(1)} - V_{2;1}^{(1)})$ between equations (200) and (203) and substituting for $\delta\mathfrak{W}_{11}^{(1)} + \delta\mathfrak{W}_{22}^{(1)} - 2\delta\mathfrak{W}_{33}^{(1)}$ and $\delta\mathfrak{W}_{11}^{(1)} - \delta\mathfrak{W}_{22}^{(1)}$ their known expressions (Chapter 3, equations (149) and (150)), we obtain the following equations:

$$(\tfrac{1}{2}\lambda^2 + 3B_{11} + B_{12} - 2B_{13} + 2\Omega^2 - \beta_{11})V_{11}^{(1)}$$
$$+ \ (\tfrac{1}{2}\lambda^2 + 3B_{22} + B_{12} - 2B_{23} + 2\Omega^2 - \beta_{22})V_{22}^{(1)}$$
$$+ \ (-\lambda^2 - 6B_{33} + B_{13} + B_{23} + 2\beta_{33})V_{33}^{(1)}$$
$$+ \ 2\frac{\Omega}{\lambda}(\beta_{11} - \beta_{22})V_{12}^{(1)} - 2\frac{\Omega}{\lambda}(I_{11} - I_{22})\,\delta\beta_{12}^{(2)}$$
$$- \ \delta\beta_{11}^{(2)}I_{11} - \delta\beta_{22}^{(2)}I_{22} + 2\delta\beta_{33}^{(2)}I_{33} = 0, \tag{204}$$

$$(\tfrac{1}{2}\lambda^2 + 3B_{11} - B_{12} - \beta_{11})V_{11}^{(1)}$$
$$+ \ (-\tfrac{1}{2}\lambda^2 - 3B_{22} + B_{12} + \beta_{22})V_{22}^{(1)} - 2\lambda\Omega V_{12}^{(1)}$$
$$+ \ (B_{13} - B_{23})V_{33}^{(1)} - \delta\beta_{11}^{(2)}I_{11} + \delta\beta_{22}^{(2)}I_{22} = 0, \tag{205}$$

$$\lambda\Omega(V_{11}^{(1)} - V_{22}^{(1)}) + (\lambda^2 + 4B_{12} - \beta_{11} - \beta_{22})V_{12}^{(1)}$$
$$- \ \delta\beta_{12}^{(2)}(I_{11} + I_{22}) = 0. \tag{206}$$

Similarly, we find that the first-order virial equations which follow from equation (194) are

$$\lambda^2 V_1^{(1)} - 2\lambda\Omega V_2^{(1)} = M\delta\beta_1^{(2)} + \beta_{11}V_1^{(1)}, \tag{207}$$
$$\lambda^2 V_2^{(1)} + 2\lambda\Omega V_1^{(1)} = M\delta\beta_2^{(2)} + \beta_{22}V_2^{(1)}. \tag{208}$$

These equations must be supplemented by solenoidal condition

$$\sum_{k=1}^{3} \frac{V_{kk}^{(1)}}{a_k^2} = 0. \tag{209}$$

Equations (204)–(209), together with the equations obtained from them by interchanging the superscripts (1) and (2), provide a complete set of 12 homogeneous equations for the 12 virials $V_1^{(i)}$, $V_2^{(i)}$, $V_{11}^{(i)}$, $V_{22}^{(i)}$, $V_{33}^{(i)}$, and $V_{12}^{(i)}$ ($i = 1,2$); and the determinant of this set of equations will provide the characteristic equation for determining all the characteristic frequencies belonging to the coupled modes of oscillation considered.

It is clear from the structure of the equations (204)–(209) and those obtained from them by interchanging the superscripts (1) and (2) that the resulting modes of oscillation fall into two classes: the *congruent modes*, for which the V_j's and the V_{jk}'s belonging to the two components are equal; and the *anticongruent modes*, for which the V_j's and the V_{jk}'s belonging to the two components are equal in magnitude but opposite in signs.

(f) Numerical results

The characteristic frequencies which follow from equations (204)–(209) for the congruent and the anticongruent modes have been determined by Lebovitz; and his results are listed in Table XXVI. From these results it follows that the congruent Darwin ellipsoids become unstable when $R < 2.895$ by a congruent mode of oscillations. In contrast, the anticongruent modes do not exhibit any instability.

The treatment of the Darwin ellipsoids in this and in the preceding section has been severely restricted. In particular, we have not considered whether points along the Darwin sequence analogous to the Roche limit occur where one

TABLE XXVI

The squares of the characteristic frequencies (in the unit $\pi G\rho$) of the coupled modes of oscillation of the congruent Darwin ellipsoids

ϕ_R	Ω^2	R	Congruent modes				Anticongruent modes			
			σ_1^2	σ_2^2	σ_3^2	σ_4^2	σ_1^2	σ_2^2	σ_3^2	σ_4^2
5°	0·0302297	4·45720	1·37039	1·12596	0·669402	0·0296658	1·37048	1·13043	0·670347	0·0302297
6°	0·0372119	4·16125	1·39591	1·13797	0·621903	0·0361490	1·39584	1·14423	0·623191	0·0372119
7°	0·0442377	3·93080	1·41794	1·14896	0·577759	0·0424046	1·41760	1·15735	0·579384	0·0442377
8°	0·0512480	3·74568	1·43731	1·15873	0·536532	0·0482769	1·43659	1·16954	0·538428	0·0512480
9°	0·0581917	3·59361	1·45465	1·16696	0·497945	0·0535906	1·45345	1·18044	0·499955	0·0581917
10°	0·0650219	3·46666	1·47052	1·17335	0·461848	0·0581384	1·46875	1·18968	0·463698	0·0650219
11°	0·0716947	3·35946	1·48540	1·17765	0·428204	0·0616699	1·48301	1·19693	0·429452	0·0716947
12°	0·0781697	3·26820	1·49969	1·17960	0·397068	0·0638817	1·49670	1·20183	0·397055	0·0781697
13°	0·0844074	3·19011	1·51375	1·17901	0·368590	0·0644068	1·51022	1·20406	0·366371	0·0844074
14°	0·0903715	3·12314	1·52783	1·17574	0·342994	0·0628268	1·52389	1·20336	0·337290	0·0903715
15°	0·0960272	3·06570	1·54212	1·16972	0·320545	0·0587002	1·53795	1·19957	0·309719	0·0960272
16°	0·101341	3·01660	1·55675	1·16094	0·301475	0·0516338	1·55256	1·19259	0·283579	0·101341
17°	0·106284	2·97489	1·57177	1·14946	0·285883	0·0413814	1·56778	1·18241	0·258799	0·106284
18°	0·110825	2·93983	1·58718	1·13538	0·273644	0·0279293	1·58362	1·16907	0·235321	0·110825
19°	0·114939	2·91081	1·60298	1·11886	0·264400	0·0115120	1·60000	1·15269	0·213091	0·114939
ϕ_R	0·117268	2·89563	1·61296	1·10744	0·259936	0	..	..	..	..
20°	0·118602	2·88738	1·61910	1·10006	0·257620	−0·00745501	1·61686	1·13340	0·192065	0·118602
21°	0·121792	2·86917	1·63548	1·07917	0·252715	−0·0284822	1·63410	1·11135	0·172205	0·121792
22°	0·124492	2·85588	1·65206	1·05640	0·249124	−0·0510942	1·65161	1·08673	0·153477	0·124492
23°	0·126684	2·84731	1·66877	1·03196	0·246360	−0·0748749	1·66929	1·05971	0·135853	0·126684
24°	0·128358	2·84329	1·68552	1·00606	0·244025	−0·0994791	1·68703	1·03046	0·119313	0·128358

$\phi_R = 19\cdot6211°$ determines the point of instability along the Darwin sequence

may expect the onset of secular instability, as in the case of the Roche sequence (cf. § 59). The consideration of these matters requires the resolution of issues that are beyond the scope of this chapter.

BIBLIOGRAPHICAL NOTES

The references to the historic papers of Roche, Darwin, and Jeans have been given in the Bibliographical Notes for Chapter 1. For a perceptive account of this earlier work see:

E. A. MILNE, *Sir James Jeans, a Biography* (Cambridge, England, Cambridge University Press, 1952), pp. 110–12.

The treatment of the equilibrium and the stability of the Roche ellipsoids, in §§ 56, 57, and 58, is in large measure based on Paper XIX (which deals with all matters considered in these sections) and Paper XX (by Lebovitz, which gives the correct interpretation of the modes which are attributed vanishing characteristic frequencies in the normal-mode analysis).

The Jeans spheroids are considered in the text as special cases of the Roche ellipsoids. A treatment of these spheroids, independently of Roche's problem, will be found in Paper XV (see also Paper XVI).

Robe has extended Cartan's method (originally developed in the context of the stability of the Jacobi ellipsoid) to the treatment of the oscillations of the Jeans spheroids and the Roche ellipsoids:

H. ROBE, "Note sur les oscillations du sphéroïde de Jeans," *Bull. Acad. Roy. Belgique,* **49** (1963), 1148–55.
H. ROBE, "Note sur les oscillations de l'ellipsoïde de Roche," *Bull. Acad. Roy. Belgique,* **50** (1964), 1252–67.

Also, it was Robe who first discovered that the Roche ellipsoids become secularly unstable at the Roche limit:

H. ROBE (unpublished).

But the treatment in the text (§ 59), leading to the explicit evaluation of the characteristic e-folding times in a low Reynolds-number approximation, is taken from Paper XXXVI.

The Roche–Riemann ellipsoids (§ 60) have been considered in some detail by Aizenman (Paper XXXIX); but the account in the text is restricted to only the simplest class of them.

The treatment of Darwin's problem in §§ 61 and 62 is derived from Paper XXI. But the general formulation of the problem in these sections (and also in Paper XXI) is based on an account by Jeans:

J. H. JEANS, *Problems of Cosmogony and Stellar Dynamics* (Cambridge, England, Cambridge University Press, 1919), pp. 55–64, 85–86, 118–28, 134–36; and figs., pp. 50 and 86.

While the author has not made any attempt to relate the methods and results of § 62 to Darwin's, it is proper that an explicit reference is made here to his paper:

G. H. Darwin, "On the figure and stability of a liquid satellite," *Phil. Trans. R. Soc.* (*London*), **206** (1906), 161–248 ; see also *Scientific Papers*, **3** (Cambridge, England, 1910), 436.

The treatment of the stability of the congruent Darwin ellipsoids in § 63 has been revised and rewritten in its entirety. The analysis in the original edition, based on Paper XL, is vitiated by some oversights pointed out by

M. Tassoul, "On the stability of congruent Darwin ellipsoids," *Astrophys. J.*, **202** (1975), 803–8.

A more general treatment of the problem, allowing for both congruent and anti-congruent modes of oscillation (as distinguished in the text), was given in

S. Chandrasekhar, "On coupled second-harmonic oscillations of the congruent Darwin ellipsoids," *Astrophys. J.*, **202** (1975), 809–14.

Again, it was pointed out to the author by Professor Norman Lebovitz that the expansion of the tidal potential about the instantaneous center of mass of the primary is to be preferred to the expansion about the undisplaced center adopted in the foregoing paper.

The revised version owes a great deal to the advice and help I have received from Professor Lebovitz. I am also grateful to him for the calculations of the characteristic frequencies listed in Table XXVI.

The principal results of this section differ in detail and in substance from what has been written on this subject by earlier writers, particularly Darwin and Jeans (see papers already cited). For further information on related matters not included in the text, see Paper XXI.

EPILOGUE

Ex cept for Chapter 2, the present book has been concerned exclusively with homogeneous liquid masses. A critic may be disposed to side with Eddington† and say, "It is not enough to deal with theoretical liquid masses. The astronomer wants to know how these results are maintained when we take into account the non-homogeneous or gaseous condition of actual stars." And the critic is not likely to be swayed by Riemann's justification that the subject "has an interest for the mathematician even apart from its relevance to the forms of heavenly bodies which initially instigated these investigations." But the author may be permitted to state his reason for devoting a substantial part of nine years to the problems treated in this book.

The subject had attracted the attention of a long succession of distinguished mathematicians and astronomers—albeit of an earlier more relaxed age. But the subject, nevertheless, had been left in an incomplete state with many gaps and omissions and some plain errors and misconceptions. It seemed a pity that it should be allowed to remain in that destitute state. Whether the effort expended in its rehabilitation was worth the time, the author cannot presume to judge.

There is, however, another aspect to the matter. As stated in the introductory section to Chapter 2, the method of the virial is not restricted to homogeneous masses. Quite generally, the method consists in replacing the equation of motion governing a problem by its moments with respect to the coordinates; and these moment equations may serve the purpose of deducing certain relationships that constrain the solutions in some manifest way, or in developing a systematic method of approximation for the solution. The method has, indeed, found such applications in the theory of radial and non-radial oscillations of a gaseous mass.

And the method is not restricted to gravitational problems either. In

† The quotation is from Eddington's "Presidential Address" to the Royal Astronomical Society "on the award of the Gold Medal to Dr. James Hopwood Jeans for his contributions to the Theories of Cosmogony" (*M.N.R.A.S.*, **82** (1922), 279). In the same address, referring to the neglect of radiation pressure, Eddington says, "Personally, I find it difficult to resist the opinion that in dealing with the disruption of diffuse gaseous masses, compelled as we are by mathematical exigencies to leave out radiation-pressure, our playbill is like that which 'announced the tragedy of Hamlet, the character of the Prince of Denmark being left out'." And these remarks, in an address extolling Jeans's investigations on the equilibrium and stability of liquid masses!

the framework of hydromagnetics, the second-order virial equation takes the form

$$\frac{d}{dt} \int_V \rho u_i x_j \, d\mathbf{x} = 2\mathfrak{T}_{ij} + \mathfrak{W}_{ij} - 2\mathfrak{M}_{ij} + \delta_{ij}(\Pi + \mathfrak{M}), \qquad (1)$$

where

$$\mathfrak{M}_{ij} = \frac{1}{8\pi} \int H_i H_j \, d\mathbf{x} \qquad (2)$$

denotes the magnetic-energy tensor whose trace $\mathfrak{M}$ ($= \mathfrak{M}_{ii}$) is the energy in the magnetic field $\mathbf{H}$. (In equation (2) the integral may have to be extended over the whole of space.) Similarly, for a liquid drop held together by surface tension, the appropriate form of the equation is

$$\frac{d}{dt} \int_V \rho u_i x_j \, d\mathbf{x} = 2\mathfrak{T}_{ij} - 2\mathfrak{S}_{ij} + \Pi \delta_{ij}, \qquad (3)$$

where

$$\mathfrak{S}_{ij} = \tfrac{1}{2} T \int_S (\delta_{ij} - n_i n_j) \, dS, \qquad (4)$$

T is the surface tension, and $\mathbf{n}$ is the unit outward normal to the boundary S. In equation (3), $\mathfrak{S}_{ij}$ is the surface-energy tensor whose trace $\mathfrak{S}$ ($= \mathfrak{S}_{ii}$) is the surface energy of the drop. If the liquid drop should be charged, then, on the left-hand side of equation (3) we must include an additional term in the electrostatic energy (defined in exact analogy with $\mathfrak{W}_{ij}$). These more general virial equations have also found applications.

At one stage the author considered including in this book an account of the applications of the virial method in the larger contexts mentioned. But that would have enlarged the size of the book substantially; and it would also have entailed a loss of exactness, coherence, and completeness with respect to the problems to which it is addressed.

A selected list of references, that may serve as an introduction to these other topics, must, therefore, suffice.

SELECTED REFERENCES

EXAMPLES of applications of the virial method to problems in contexts other than homogeneous masses will be found in the following references.

COMPRESSIBLE MASSES:

P. LEDOUX, "On the radial pulsation of gaseous stars," *Astrophys. J.*, **102** (1945), 143–53.

P. LEDOUX, "Stellar stability," *Stellar Structure*, eds. L. H. Aller and D. B. McLaughlin (Chicago, University of Chicago Press, 1965), pp. 548–50, 558–61.

S. CHANDRASEKHAR and NORMAN R. LEBOVITZ, "On the oscillations and the stability of rotating gaseous masses," *Astrophys. J.*, **135** (1962), 248–60.

S. CHANDRASEKHAR and NORMAN R. LEBOVITZ, "On the oscillations and the stability of rotating gaseous masses. II. The homogeneous, compressible model," *Astrophys. J.*, **136** (1962), 1069–81.

S. CHANDRASEKHAR and NORMAN R. LEBOVITZ, "Non-radial oscillations and the convective instability of gaseous masses," *Astrophys. J.*, **138** (1963), 185–99.

MONIQUE TASSOUL and JEAN LOUIS TASSOUL, "Adiabatic pulsations and convective instability of gaseous masses. I," *Astrophys. J.*, **150** (1967), 213–21.

MONIQUE TASSOUL and JEAN LOUIS TASSOUL, "Adiabatic pulsations and convective instability of gaseous masses. II," *Astrophys. J.*, **150** (1967), 1031–40.

JEAN LOUIS TASSOUL and J. P. OSTRIKER, "On the oscillations and stability of rotating stellar modes. I," *Astrophys. J.*, **154** (1968), 613–26.

HYDROMAGNETICS:

S. CHANDRASEKHAR and E. FERMI, "Problems of gravitational stability in the presence of a magnetic field," *Astrophys. J.*, **118** (1953), 116–41.

S. CHANDRASEKHAR, *Hydrodynamic and Hydromagnetic Stability* (Clarendon Press, Oxford, 1961), pp. 577–87.

L. WOLTJER, "Emission nuclei in galaxies," *Astrophys. J.*, **130** (1959), 38–44.

L. WOLTJER, "On the magnetospheres of self-gravitating bodies," *Astrophys. J.*, **148** (1967), 291–93.

E. N. PARKER, "The dynamical state of the interstellar gas and field," *Astrophys. J.*, **145** (1966), 811–33.

E. N. PARKER, "The gross dynamics of a hydromagnetic gas cloud," *Astrophys. J. Suppl.*, **3** (1957), 51–76.

DONAT G. WENTZEL, "Hydromagnetic oscillations of a self-gravitating fluid," *Astrophys. J.*, **135** (1962), 593–98.

JEAN LOUIS TASSOUL, "Adiabatic pulsations and convective instability of gaseous masses, III," *M.N.R.A.S.*, **138** (1968), 123–36.

LIQUID DROPS:

S. CHANDRASEKHAR, "The stability of a rotating liquid drop," *Proc. Roy. Soc. (London)*, A, **286** (1965), 1–26.

CARL E. ROSENKILDE, "Surface-energy tensors," *J. Math. Phys.*, **8** (1967), 84–88.

CARL E. ROSENKILDE, "Surface-energy tensors for ellipsoids," *J. Math. Phys.*, **8** (1967), 88–97.

CARL E. ROSENKILDE, "Stability of axisymmetric figures of equilibrium of a rotating charged liquid drop," *J. Math. Phys.*, **8** (1967), 98–118.

GENERAL RELATIVITY :

S. CHANDRASEKHAR, "The post-Newtonian effects of general relativity on the equilibrium of uniformly rotating bodies. I. The Maclaurin spheroids and the virial theorem," *Astrophys. J.*, **142** (1965), 1513–18.

S. CHANDRASEKHAR, "The post-Newtonian effects of general relativity on the equilibrium of uniformly rotating bodies. II. The deformed figures of the Maclaurin spheroids," *Astrophys. J.*, **147** (1967), 334–52.

S. CHANDRASEKHAR, "The post-Newtonian effects of general relativity on the equilibrium of uniformly rotating bodies. III. The deformed figures of the Jacobi ellipsoids," *Astrophys. J.*, **148** (1967), 621–44.

APPENDIX: LIST OF PAPERS

THE presentation of the subject in this book is based on the following papers.

   I. S. CHANDRASEKHAR, "The virial theorem in hydromagnetics," *J. Math. Anal. Applic.*, **1** (1960), 240–52.

  II. NORMAN R. LEBOVITZ, "The virial tensor and its application to self-gravitating fluids," *Astrophys. J.*, **134** (1961), 500–36.

 III. S. CHANDRASEKHAR, "A theorem on rotating polytropes," *Astrophys. J.*, **134** (1961), 662–64.

  IV. S. CHANDRASEKHAR and NORMAN R. LEBOVITZ, "On super-potentials in the theory of Newtonian gravitation," *Astrophys. J.*, **135** (1962), 238–47.

   V. S. CHANDRASEKHAR and NORMAN R. LEBOVITZ, "On the oscillations and the stability of rotating gaseous masses," *Astrophys. J.*, **135** (1962), 248–60.

  VI. S. CHANDRASEKHAR, "An approach to the theory of the equilibrium and the stability of rotating masses via the virial theorem and its extensions," *Proc. 4th U.S. Nat. Congress Appl. Mech.* (1962), pp. 9–14.

 VII. S. CHANDRASEKHAR and NORMAN R. LEBOVITZ, "On the super-potentials in the theory of Newtonian gravitation. II. Tensors of higher rank," *Astrophys. J.*, **136** (1962), 1032–36.

VIII. S. CHANDRASEKHAR and NORMAN R. LEBOVITZ, "The potentials and the superpotentials of homogeneous ellipsoids," *Astrophys. J.*, **136** (1962), 1037–47.

  IX. S. CHANDRASEKHAR, "On the point of bifurcation along the sequence of the Jacobi ellipsoids," *Astrophys. J.*, **136** (1962), 1048–68.

   X. S. CHANDRASEKHAR and NORMAN R. LEBOVITZ, "On the oscillations and the stability of rotating gaseous masses. II. The homogeneous, compressible model," *Astrophys. J.*, **136** (1962), 1069–81.

  XI. S. CHANDRASEKHAR and NORMAN R. LEBOVITZ, "On the oscillations and the stability of rotating gaseous masses. III. The distorted polytropes," *Astrophys. J.*, **136** (1962), 1082–104.

 XII. S. CHANDRASEKHAR and NORMAN R. LEBOVITZ, "On the occurrence of multiple frequencies and beats in the Beta Canis Majoris stars," *Astrophys. J.*, **136** (1962), 1105–07.

XIII. S. CHANDRASEKHAR and NORMAN R. LEBOVITZ, "On the stability of the Jacobi ellipsoids," *Astrophys. J.*, **137** (1963), 1142–61.

 XIV. S. CHANDRASEKHAR and NORMAN R. LEBOVITZ, "On the oscillations of the Maclaurin spheroid belonging to the third harmonics," *Astrophys. J.*, **137** (1963), 1162–71.

  XV. S. CHANDRASEKHAR and NORMAN R. LEBOVITZ, "The equilibrium and the stability of the Jeans spheroids," *Astrophys. J.*, **137** (1963), 1172–84.

XVI. S. Chandrasekhar, "The points of bifurcation along the Maclaurin, the Jacobi, and the Jeans sequences," *Astrophys. J.*, **137** (1963), 1185–202.

XVII. S. Chandrasekhar and Norman R. Lebovitz, "Non-radial oscillations and the convective instability of gaseous masses," *Astrophys. J.*, **138** (1963), 185–99.

XVIII. S. Chandrasekhar and Paul H. Roberts, "The ellipticity of a slowly rotating configuration," *Astrophys. J.*, **138** (1963), 801–8.

XIX. S. Chandrasekhar, "The equilibrium and the stability of the Roche ellipsoids," *Astrophys. J.*, **138** (1963), 1182–213.

XX. Norman R. Lebovitz, "On the principle of the exchange of stabilities. I. The Roche ellipsoids," *Astrophys. J.*, **138** (1963), 1214–17.

XXI. S. Chandrasekhar, "The equilibrium and the stability of the Darwin ellipsoids," *Astrophys. J.*, **140** (1964), 599–620.

XXII. S. Chandrasekhar, "The higher order virial equations and their applications to the equilibrium and stability of rotating configurations," *Lectures in Theoretical Physics, Vol. VI, Boulder, 1963* (Boulder, University of Colorado Press, 1964), pp. 1–72.

XXIII. S. Chandrasekhar and Norman R. Lebovitz, "On the ellipsoidal figures of equilibrium of homogeneous masses," *Astrophysica Norvegica*, **9** (1964), 323–32.

XXIV. S. Chandrasekhar, "The equilibrium and the stability of the Dedekind ellipsoids," *Astrophys. J.*, **141** (1965), 1043–55.

XXV. S. Chandrasekhar, "The equilibrium and the stability of the Riemann ellipsoids. I," *Astrophys. J.*, **142** (1965), 890–921.

XXVI. S. Chandrasekhar, "The post-Newtonian effects of general relativity on the equilibrium of uniformly rotating bodies. I. The Maclaurin spheroids and the virial theorem," *Astrophys. J.*, **142** (1965), 1513–18.

XXVII. Norman R. Lebovitz, "The Riemann ellipsoids," (lecture notes, Inst. Ap., Cointe-Sclessin, Belgium, 1965), p. 29.

XXVIII. S. Chandrasekhar, "The equilibrium and the stability of the Riemann ellipsoids. II," *Astrophys. J.*, **145** (1966), 842–77.

XXIX. Norman R. Lebovitz, "On Riemann's criterion for the stability of liquid ellipsoids," *Astrophys. J.*, **145** (1966), 878–85.

XXX. S. Chandrasekhar, "The post-Newtonian effects of general relativity on the equilibrium of uniformly rotating bodies. II. The deformed figures of the Maclaurin spheroids," *Astrophys. J.*, **147** (1967), 334–52.

XXXI. S. Chandrasekhar, "Virial relations for uniformly rotating fluid masses in general relativity," *Astrophys. J.*, **147** (1967), 383–84.

XXXII. S. Chandrasekhar, "The post-Newtonian effects of general relativity on the equilibrium of uniformly rotating bodies. III. The deformed figures of the Jacobi ellipsoids," *Astrophys. J.*, **148** (1967), 621–44.

XXXIII. S. Chandrasekhar, "Ellipsoidal figures of equilibrium—an historical account," *Comm. Pure & Appl. Math.*, **20** (1967), 251–65.

XXXIV. Norman R. Lebovitz, "Rotating fluid masses," *Annual Review of Astronomy and Astrophysics*, **5** (Palo Alto, California, 1967), 465–80.

XXXV. S. Chandrasekhar, "The virial equations of the fourth order," *Astrophys. J.*, **152** (1968), 293–304.

XXXVI. S. Chandrasekhar, "The effect of viscous dissipation on the stability of the Roche ellipsoids," *Publications of the Ramanujan Institute*, **1** (1969) (in press).

XXXVII. CARL E. ROSENKILDE, "The tensor virial-theorem including viscous stress and the oscillations of a Maclaurin spheroid," *Astrophys. J.*, **148** (1967), 825–32.

XXXVIII. LAWRENCE F. ROSSNER, "The finite-amplitude oscillations of the Maclaurin spheroids," *Astrophys. J.*, **149** (1967), 145–68.

XXXIX. MORRIS L. AIZENMAN, "The equilibrium and the stability of the Roche–Riemann ellipsoids," *Astrophys. J.*, **153** (1968), 511–44.

XL. S. CHANDRASEKHAR, "The stability of the congruent Darwin ellipsoids," *Astrophys. J.*, **157** (1969), 1419–34.

References in the text to the foregoing papers are by their roman numerals.

INDEX OF SYMBOLS

# A CATALOG OF SELECTED
DOVER BOOKS
IN ALL FIELDS OF INTEREST

A CATALOG OF SELECTED DOVER BOOKS IN ALL FIELDS OF INTEREST

LASERS AND HOLOGRAPHY, Winston E. Kock. Sound introduction to burgeoning field, expanded (1981) for second edition. 84 illustrations. 160pp. 5⅜ × 8¼. (EUK) 24041-X Pa. $3.50

FLORAL STAINED GLASS PATTERN BOOK, Ed Sibbett, Jr. 96 exquisite floral patterns—irises, poppie, lilies, tulips, geometrics, abstracts, etc.—adaptable to innumerable stained glass projects. 64pp. 8¼ × 11. 24259-5 Pa. $3.50

THE HISTORY OF THE LEWIS AND CLARK EXPEDITION, Meriwether Lewis and William Clark. Edited by Eliott Coues. Great classic edition of Lewis and Clark's day-by-day journals. Complete 1893 edition, edited by Eliott Coues from Biddle's authorized 1814 history. 1508pp. 5⅜ × 8½. 21268-8, 21269-6, 21270-X Pa. Three-vol. set $22.50

ORLEY FARM, Anthony Trollope. Three-dimensional tale of great criminal case. Original Millais illustrations illuminate marvelous panorama of Victorian society. Plot was author's favorite. 736pp. 5⅜ × 8½. 24181-5 Pa. $10.95

THE CLAVERINGS, Anthony Trollope. Major novel, chronicling aspects of British Victorian society, personalities. 16 plates by M. Edwards; first reprint of full text. 412pp. 5⅜ × 8½. 23464-9 Pa. $6.00

EINSTEIN'S THEORY OF RELATIVITY, Max Born. Finest semi-technical account; much explanation of ideas and math not readily available elsewhere on this level. 376pp. 5⅜ × 8½. 60769-0 Pa. $5.00

COMPUTABILITY AND UNSOLVABILITY, Martin Davis. Classic graduate-level introduction th theory of computability, usually referred to as theory of recurrent functions. New preface and appendix. 288pp. 5⅜ × 8½. 61471-9 Pa. $6.50

THE GODS OF THE EGYPTIANS, E.A. Wallis Budge. Never excelled for richness, fullness: all gods, goddesses, demons, mythical figures of Ancient Egypt; their legends, rites, incarnations, etc. Over 225 illustrations, plus 6 color plates. 988pp. 6⅛ × 9¼. (EBE) 22055-9, 22056-7 Pa., Two-vol. set $20.00

THE I CHING (THE BOOK OF CHANGES), translated by James Legge. Most penetrating divination manual ever prepared. Indispensable to study of early Oriental civilizations, to modern inquiring reader. 448pp. 5⅜ × 8½. 21062-6 Pa. $6.50

THE CRAFTSMAN'S HANDBOOK, Cennino Cennini. 15th-century handbook, school of Giotto, explains applying gold, silver leaf; gesso; fresco painting, grinding pigments, etc. 142pp. 6⅛ × 9¼. 20054-X Pa. $3.50

AN ATLAS OF ANATOMY FOR ARTISTS, Fritz Schider. Finest text, working book. Full text, plus anatomical illustrations; plates by great artists showing anatomy. 593 illustrations. 192pp. 7⅛ × 10¼. 20241-0 Pa. $6.50

EASY-TO-MAKE STAINED GLASS LIGHTCATCHERS, Ed Sibbett, Jr. 67 designs for most enjoyable ornaments: fruits, birds, teddy bears, trumpet, etc. Full size templates. 64pp. 8¼ × 11. 24081-9 Pa. $3.95

TRIAD OPTICAL ILLUSIONS AND HOW TO DESIGN THEM, Harry Turner. Triad explained in 32 pages of text, with 32 pages of Escher-like patterns on coloring stock. 92 figures. 32 plates. 64pp. 8¼ × 11. 23549-1 Pa. $2.95

SMOCKING: TECHNIQUE, PROJECTS, AND DESIGNS, Dianne Durand. Foremost smocking designer provides complete instructions on how to smock. Over 10 projects, over 100 illustrations. 56pp. 8¼ × 11. 23788-5 Pa. $2.00

AUDUBON'S BIRDS IN COLOR FOR DECOUPAGE, edited by Eleanor H. Rawlings. 24 sheets, 37 most decorative birds, full color, on one side of paper. Instructions, including work under glass. 56pp. 8¼ × 11. 23492-4 Pa. $3.95

THE COMPLETE BOOK OF SILK SCREEN PRINTING PRODUCTION, J.I. Biegeleisen. For commercial user, teacher in advanced classes, serious hobbyist. Most modern techniques, materials, equipment for optimal results. 124 illustrations. 253pp. 5⅝ × 8½. 21100-2 Pa. $4.50

A TREASURY OF ART NOUVEAU DESIGN AND ORNAMENT, edited by Carol Belanger Grafton. 577 designs for the practicing artist. Full-page, spots, borders, bookplates by Klimt, Bradley, others. 144pp. 8⅜ × 11¼. 24001-0 Pa. $5.95

ART NOUVEAU TYPOGRAPHIC ORNAMENTS, Dan X. Solo. Over 800 Art Nouveau florals, swirls, women, animals, borders, scrolls, wreaths, spots and dingbats, copyright-free. 100pp. 8⅜ × 11. 24366-4 Pa. $4.00

HAND SHADOWS TO BE THROWN UPON THE WALL, Henry Bursill. Wonderful Victorian novelty tells how to make flying birds, dog, goose, deer, and 14 others, each explained by a full-page illustration. 32pp. 6½ × 9¼. 21779-5 Pa. $1.50

AUDUBON'S BIRDS OF AMERICA COLORING BOOK, John James Audubon. Rendered for coloring by Paul Kennedy. 46 of Audubon's noted illustrations: red-winged black-bird, cardinal, etc. Original plates reproduced in full-color on the covers. Captions. 48pp. 8¼ × 11. 23049-X Pa. $2.25

SILK SCREEN TECHNIQUES, J.I. Biegeleisen, M.A. Cohn. Clear, practical, modern, economical. Minimal equipment (self-built), materials, easy methods. For amateur, hobbyist, 1st book. 141 illustrations. 185pp. 6⅛ × 9¼. 20433-2 Pa. $3.95

101 PATCHWORK PATTERNS, Ruby S. McKim. 101 beautiful, immediately useable patterns, full-size, modern and traditional. Also general information, estimating, quilt lore. 140 illustrations. 124pp. 7⅞ × 10¾. 20773-0 Pa. $3.50

READY-TO-USE FLORAL DESIGNS, Ed Sibbett, Jr. Over 100 floral designs (most in three sizes) of popular individual blossoms as well as bouquets, sprays, garlands. 64pp. 8¼ × 11. 23976-4 Pa. $2.95

AMERICAN WILD FLOWERS COLORING BOOK, Paul Kennedy. Planned coverage of 46 most important wildflowers, from Rickett's collection; instructive as well as entertaining. Color versions on covers. Captions. 48pp. 8¼ × 11.
20095-7 Pa. $2.50

CARVING DUCK DECOYS, Harry V. Shourds and Anthony Hillman. Detailed instructions and full-size templates for constructing 16 beautiful, marvelously practical decoys according to time-honored South Jersey method. 70pp. 9¼ × 12¼.
24083-5 Pa. $4.95

TRADITIONAL PATCHWORK PATTERNS, Carol Belanger Grafton. Cardboard cut-out pieces for use as templates to make 12 quilts: Buttercup, Ribbon Border, Tree of Paradise, nine more. Full instructions. 57pp. 8¼ × 11.
23015-5 Pa. $3.50

TOLL HOUSE TRIED AND TRUE RECIPES, Ruth Graves Wakefield. Popovers, veal and ham loaf, baked beans, much more from the famous Mass. restaurant. Nearly 700 recipes. 376pp. 5⅜ × 8½.   23560-2 Pa. $4.95

FAVORITE CHRISTMAS CAROLS, selected and arranged by Charles J.F. Cofone. Title, music, first verse and refrain of 34 traditional carols in handsome calligraphy; also subsequent verses and other information in type. 79pp. 8⅜ × 11.   20445-6 Pa. $3.50

CAMERA WORK: A PICTORIAL GUIDE, Alfred Stieglitz. All 559 illustrations from most important periodical in history of art photography. Reduced in size but still clear, in strict chronological order, with complete captions. 176pp. 8⅜ × 11¼.   23591-2 Pa. $6.95

FAVORITE SONGS OF THE NINETIES, edited by Robert Fremont. 88 favorites: "Ta-Ra-Ra-Boom-De-Aye," "The Band Played On," "Bird in a Gilded Cage," etc. 401pp. 9 × 12.   21536-9 Pa. $12.95

STRING FIGURES AND HOW TO MAKE THEM, Caroline F. Jayne. Fullest, clearest instructions on string figures from around world: Eskimo, Navajo, Lapp, Europe, more. Cat's cradle, moving spear, lightning, stars. 950 illustrations. 407pp. 5⅜ × 8½.   20152-X Pa. $5.95

LIFE IN ANCIENT EGYPT, Adolf Erman. Detailed older account, with much not in more recent books: domestic life, religion, magic, medicine, commerce, and whatever else needed for complete picture. Many illustrations. 597pp. 5⅜ × 8½.   22632-8 Pa. $7.95

ANCIENT EGYPT: ITS CULTURE AND HISTORY, J.E. Manchip White. From pre-dynastics through Ptolemies: scoiety, history, political structure, religion, daily life, literature, cultural heritage. 48 plates. 217pp. 5⅜ × 8½. (EBE)   22548-8 Pa. $4.95

KEPT IN THE DARK, Anthony Trollope. Unusual short novel about Victorian morality and abnormal psychology by the great English author. Probably the first American publication. Frontispiece by Sir John Millais. 92pp. 6½ × 9¼.   23609-9 Pa. $2.95

MAN AND WIFE, Wilkie Collins. Nineteenth-century master launches an attack on out-moded Scottish marital laws and Victorian cult of athleticism. Artfully plotted. 35 illustrations. 239pp. 6⅛ × 9¼.   24451-2 Pa. $5.95

RELATIVITY AND COMMON SENSE, Herman Bondi. Radically reoriented presentation of Einstein's Special Theory and one of most valuable popular accounts available. 60 illustrations. 177pp. 5⅜ × 8. (EUK)   24021-5 Pa. $3.95

THE EGYPTIAN BOOK OF THE DEAD, E.A. Wallis Budge. Complete reproduction of Ani's papyrus, finest ever found. Full hieroglyphic text, interlinear transliteration, word-for-word translation, smooth translation. 533pp. 6½ × 9¼.   (USO) 21866-X Pa. $8.95

COUNTRY AND SUBURBAN HOMES OF THE PRAIRIE SCHOOL PERIOD, H.V. von Holst. Over 400 photographs floor plans, elevations, detailed drawings (exteriors and interiors) for over 100 structures. Text. Important primary source. 128pp. 8⅜ × 11¼.   24373-7 Pa. $5.95

THE BOOK OF WOOD CARVING, Charles Marshall Sayers. Still finest book for beginning student. Fundamentals, technique; gives 34 designs, over 34 projects for panels, bookends, mirrors, etc. 33 photos. 118pp. 7¾ × 10⅝. 23654-4 Pa. $3.95

CARVING COUNTRY CHARACTERS, Bill Higginbotham. Expert advice for beginning, advanced carvers on materials, techniques for creating 18 projects—mirthful panorama of American characters. 105 illustrations. 80pp. 8⅜ × 11. 24135-1 Pa. $2.50

300 ART NOUVEAU DESIGNS AND MOTIFS IN FULL COLOR, C.B. Grafton. 44 full-page plates display swirling lines and muted colors typical of Art Nouveau. Borders, frames, panels, cartouches, dingbats, etc. 48pp. 9⅜ × 12¼. 24354-0 Pa. $6.95

SELF-WORKING CARD TRICKS, Karl Fulves. Editor of *Pallbearer* offers 72 tricks that work automatically through nature of card deck. No sleight of hand needed. Often spectacular. 42 illustrations. 113pp. 5⅜ × 8½. 23334-0 Pa. $3.50

CUT AND ASSEMBLE A WESTERN FRONTIER TOWN, Edmund V. Gillon, Jr. Ten authentic full-color buildings on heavy cardboard stock in H-O scale. Sheriff's Office and Jail, Saloon, Wells Fargo, Opera House, others. 48pp. 9¼ × 12¼. 23736-2 Pa. $3.95

CUT AND ASSEMBLE AN EARLY NEW ENGLAND VILLAGE, Edmund V. Gillon, Jr. Printed in full color on heavy cardboard stock. 12 authentic buildings in H-O scale: Adams home in Quincy, Mass., Oliver Wight house in Sturbridge, smithy, store, church, others. 48pp. 9¼ × 12¼. 23536-X Pa. $4.95

THE TALE OF TWO BAD MICE, Beatrix Potter. Tom Thumb and Hunca Munca squeeze out of their hole and go exploring. 27 full-color Potter illustrations. 59pp. 4¼ × 5½. (Available in U.S. only) 23065-1 Pa. $1.75

CARVING FIGURE CARICATURES IN THE OZARK STYLE, Harold L. Enlow. Instructions and illustrations for ten delightful projects, plus general carving instructions. 22 drawings and 47 photographs altogether. 39pp. 8⅜ × 11. 23151-8 Pa. $2.50

A TREASURY OF FLOWER DESIGNS FOR ARTISTS, EMBROIDERERS AND CRAFTSMEN, Susan Gaber. 100 garden favorites lushly rendered by artist for artists, craftsmen, needleworkers. Many form frames, borders. 80pp. 8¼ × 11. 24096-7 Pa. $3.50

CUT & ASSEMBLE A TOY THEATER/THE NUTCRACKER BALLET, Tom Tierney. Model of a complete, full-color production of Tchaikovsky's classic. 6 backdrops, dozens of characters, familiar dance sequences. 32pp. 9⅜ × 12¼. 24194-7 Pa. $4.50

ANIMALS: 1,419 COPYRIGHT-FREE ILLUSTRATIONS OF MAMMALS, BIRDS, FISH, INSECTS, ETC., edited by Jim Harter. Clear wood engravings present, in extremely lifelike poses, over 1,000 species of animals. 284pp. 9 × 12. 23766-4 Pa. $9.95

MORE HAND SHADOWS, Henry Bursill. For those at their 'finger ends," 16 more effects—Shakespeare, a hare, a squirrel, Mr. Punch, and twelve more—each explained by a full-page illustration. Considerable period charm. 30pp. 6½ × 9¼. 21384-6 Pa. $1.95

REASON IN ART, George Santayana. Renowned philosopher's provocative, seminal treatment of basis of art in instinct and experience. Volume Four of *The Life of Reason*. 230pp. 5⅜ × 8. 24358-3 Pa. $4.50

LANGUAGE, TRUTH AND LOGIC, Alfred J. Ayer. Famous, clear introduction to Vienna, Cambridge schools of Logical Positivism. Role of philosophy, elimination of metaphysics, nature of analysis, etc. 160pp. 5⅜ × 8½. (USCO) 20010-8 Pa. $2.75

BASIC ELECTRONICS, U.S. Bureau of Naval Personnel. Electron tubes, circuits, antennas, AM, FM, and CW transmission and receiving, etc. 560 illustrations. 567pp. 6½ × 9¼. 21076-6 Pa. $8.95

THE ART DECO STYLE, edited by Theodore Menten. Furniture, jewelry, metalwork, ceramics, fabrics, lighting fixtures, interior decors, exteriors, graphics from pure French sources. Over 400 photographs. 183pp. 8⅜ × 11¼. 22824-X Pa. $6.95

THE FOUR BOOKS OF ARCHITECTURE, Andrea Palladio. 16th-century classic covers classical architectural remains, Renaissance revivals, classical orders, etc. 1738 Ware English edition. 216 plates. 110pp. of text. 9½ × 12¾. 21308-0 Pa. $11.50

THE WIT AND HUMOR OF OSCAR WILDE, edited by Alvin Redman. More than 1000 ripostes, paradoxes, wisecracks: Work is the curse of the drinking classes, I can resist everything except temptations, etc. 258pp. 5⅜ × 8½. (USCO) 20602-5 Pa. $3.95

THE DEVIL'S DICTIONARY, Ambrose Bierce. Barbed, bitter, brilliant witticisms in the form of a dictionary. Best, most ferocious satire America has produced. 145pp. 5⅜ × 8½. 20487-1 Pa. $2.50

ERTÉ'S FASHION DESIGNS, Erté. 210 black-and-white inventions from *Harper's Bazar*, 1918-32, plus 8pp. full-color covers. Captions. 88pp. 9 × 12. 24203-X Pa. $6.50

ERTÉ GRAPHICS, Erté. Collection of striking color graphics: *Seasons, Alphabet, Numerals, Aces* and *Precious Stones*. 50 plates, including 4 on covers. 48pp. 9⅜ × 12¼. 23580-7 Pa. $6.95

PAPER FOLDING FOR BEGINNERS, William D. Murray and Francis J. Rigney. Clearest book for making origami sail boats, roosters, frogs that move legs, etc. 40 projects. More than 275 illustrations. 94pp. 5⅜ × 8½. 20713-7 Pa. $2.25

ORIGAMI FOR THE ENTHUSIAST, John Montroll. Fish, ostrich, peacock, squirrel, rhinoceros, Pegasus, 19 other intricate subjects. Instructions. Diagrams. 128pp. 9 × 12. 23799-0 Pa. $4.95

CROCHETING NOVELTY POT HOLDERS, edited by Linda Macho. 64 useful, whimsical pot holders feature kitchen themes, animals, flowers, other novelties. Surprisingly easy to crochet. Complete instructions. 48pp. 8¼ × 11. 24296-X Pa. $1.95

CROCHETING DOILIES, edited by Rita Weiss. Irish Crochet, Jewel, Star Wheel, Vanity Fair and more. Also luncheon and console sets, runners and centerpieces. 51 illustrations. 48pp. 8¼ × 11. 23424-X Pa. $2.50

DECORATIVE NAPKIN FOLDING FOR BEGINNERS, Lillian Oppenheimer and Natalie Epstein. 22 different napkin folds in the shape of a heart, clown's hat, love knot, etc. 63 drawings. 48pp. 8¼ × 11. 23797-4 Pa. $1.95

DECORATIVE LABELS FOR HOME CANNING, PRESERVING, AND OTHER HOUSEHOLD AND GIFT USES, Theodore Menten. 128 gummed, perforated labels, beautifully printed in 2 colors. 12 versions. Adhere to metal, glass, wood, ceramics. 24pp. 8¼ × 11. 23219-0 Pa. $2.95

EARLY AMERICAN STENCILS ON WALLS AND FURNITURE, Janet Waring. Thorough coverage of 19th-century folk art: techniques, artifacts, surviving specimens. 166 illustrations, 7 in color. 147pp. of text. 7⅞ × 10¾. 21906-2 Pa. $9.95

AMERICAN ANTIQUE WEATHERVANES, A.B. & W.T. Westervelt. Extensively illustrated 1883 catalog exhibiting over 550 copper weathervanes and finials. Excellent primary source by one of the principal manufacturers. 104pp. 6⅛ × 9¼. 24396-6 Pa. $3.95

ART STUDENTS' ANATOMY, Edmond J. Farris. Long favorite in art schools. Basic elements, common positions, actions. Full text, 158 illustrations. 159pp. 5⅜ × 8½. 20744-7 Pa. $3.95

BRIDGMAN'S LIFE DRAWING, George B. Bridgman. More than 500 drawings and text teach you to abstract the body into its major masses. Also specific areas of anatomy. 192pp. 6½ × 9¼. (EA) 22710-3 Pa. $4.50

COMPLETE PRELUDES AND ETUDES FOR SOLO PIANO, Frederic Chopin. All 26 Preludes, all 27 Etudes by greatest composer of piano music. Authoritative Paderewski edition. 224pp. 9 × 12. (Available in U.S. only) 24052-5 Pa. $7.50

PIANO MUSIC 1888-1905, Claude Debussy. Deux Arabesques, Suite Bergamesque, Masques, 1st series of Images, etc. 9 others, in corrected editions. 175pp. 9⅜ × 12¼. (ECE) 22771-5 Pa. $5.95

TEDDY BEAR IRON-ON TRANSFER PATTERNS, Ted Menten. 80 iron-on transfer patterns of male and female Teddys in a wide variety of activities, poses, sizes. 48pp. 8¼ × 11. 24596-9 Pa. $2.25

A PICTURE HISTORY OF THE BROOKLYN BRIDGE, M.J. Shapiro. Profusely illustrated account of greatest engineering achievement of 19th century. 167 rare photos & engravings recall construction, human drama. Extensive, detailed text. 122pp. 8¼ × 11. 24403-2 Pa. $7.95

NEW YORK IN THE THIRTIES, Berenice Abbott. Noted photographer's fascinating study shows new buildings that have become famous and old sights that have disappeared forever. 97 photographs. 97pp. 11⅜ × 10. 22967-X Pa. $7.50

MATHEMATICAL TABLES AND FORMULAS, Robert D. Carmichael and Edwin R. Smith. Logarithms, sines, tangents, trig functions, powers, roots, reciprocals, exponential and hyperbolic functions, formulas and theorems. 269pp. 5⅜ × 8½. 60111-0 Pa. $4.95

HANDBOOK OF MATHEMATICAL FUNCTIONS WITH FORMULAS, GRAPHS, AND MATHEMATICAL TABLES, edited by Milton Abramowitz and Irene A. Stegun. Vast compendium: 29 sets of tables, some to as high as 20 places. 1,046pp. 8 × 10½. 61272-4 Pa. $19.95

CHANCERY CURSIVE STROKE BY STROKE, Arthur Baker. Instructions and illustrations for each stroke of each letter (upper and lower case) and numerals. 54 full-page plates. 64pp. 8¼ × 11. 24278-1 Pa. $2.50

THE ENJOYMENT AND USE OF COLOR, Walter Sargent. Color relationships, values, intensities; complementary colors, illumination, similar topics. Color in nature and art. 7 color plates, 29 illustrations. 274pp. 5⅜ × 8½. 20944-X Pa. $4.95

SCULPTURE PRINCIPLES AND PRACTICE, Louis Slobodkin. Step-by-step approach to clay, plaster, metals, stone; classical and modern. 253 drawings, photos. 255pp. 8⅛ × 11. 22960-2 Pa. $7.50

VICTORIAN FASHION PAPER DOLLS FROM HARPER'S BAZAR, 1867-1898, Theodore Menten. Four female dolls with 28 elegant high fashion costumes, printed in full color. 32pp. 9¼ × 12¼. 23453-3 Pa. $3.50

FLOPSY, MOPSY AND COTTONTAIL: A Little Book of Paper Dolls in Full Color, Susan LaBelle. Three dolls and 21 costumes (7 for each doll) show Peter Rabbit's siblings dressed for holidays, gardening, hiking, etc. Charming borders, captions. 48pp. 4¼ × 5½. 24376-1 Pa. $2.25

NATIONAL LEAGUE BASEBALL CARD CLASSICS, Bert Randolph Sugar. 83 big-leaguers from 1909-69 on facsimile cards. Hubbell, Dean, Spahn, Brock plus advertising, info, no duplications. Perforated, detachable. 16pp. 8¼ × 11.
24308-7 Pa. $2.95

THE LOGICAL APPROACH TO CHESS, Dr. Max Euwe, et al. First-rate text of comprehensive strategy, tactics, theory for the amateur. No gambits to memorize, just a clear, logical approach. 224pp. 5⅜ × 8½. 24353-2 Pa. $4.50

MAGICK IN THEORY AND PRACTICE, Aleister Crowley. The summation of the thought and practice of the century's most famous necromancer, long hard to find. Crowley's best book. 436pp. 5⅜ × 8½. (Available in U.S. only)
23295-6 Pa. $6.50

THE HAUNTED HOTEL, Wilkie Collins. Collins' last great tale; doom and destiny in a Venetian palace. Praised by T.S. Eliot. 127pp. 5⅜ × 8½.
24333-8 Pa. $3.00

ART DECO DISPLAY ALPHABETS, Dan X. Solo. Wide variety of bold yet elegant lettering in handsome Art Deco styles. 100 complete fonts, with numerals, punctuation, more. 104pp. 8⅛ × 11. 24372-9 Pa. $4.50

CALLIGRAPHIC ALPHABETS, Arthur Baker. Nearly 150 complete alphabets by outstanding contemporary. Stimulating ideas; useful source for unique effects. 154 plates. 157pp. 8⅜ × 11¼. 21045-6 Pa. $5.95

ARTHUR BAKER'S HISTORIC CALLIGRAPHIC ALPHABETS, Arthur Baker. From monumental capitals of first-century Rome to humanistic cursive of 16th century, 33 alphabets in fresh interpretations. 88 plates. 96pp. 9 × 12.
24054-1 Pa. $4.50

LETTIE LANE PAPER DOLLS, Sheila Young. Genteel turn-of-the-century family very popular then and now. 24 paper dolls. 16 plates in full color. 32pp. 9¼ × 12¼. 24089-4 Pa. $3.50

THE RIME OF THE ANCIENT MARINER, Gustave Doré, S.T. Coleridge. Doré's finest work, 34 plates capture moods, subtleties of poem. Full text. 77pp. 9¼ × 12. 22305-1 Pa. $4.95

SONGS OF INNOCENCE, William Blake. The first and most popular of Blake's famous "Illuminated Books," in a facsimile edition reproducing all 31 brightly colored plates. Additional printed text of each poem. 64pp. 5¼ × 7. 22764-2 Pa. $3.50

AN INTRODUCTION TO INFORMATION THEORY, J.R. Pierce. Second (1980) edition of most impressive non-technical account available. Encoding, entropy, noisy channel, related areas, etc. 320pp. 5⅜ × 8½. 24061-4 Pa. $4.95

THE DIVINE PROPORTION: A STUDY IN MATHEMATICAL BEAUTY, H.E. Huntley. "Divine proportion" or "golden ratio" in poetry, Pascal's triangle, philosophy, psychology, music, mathematical figures, etc. Excellent bridge between science and art. 58 figures. 185pp. 5⅜ × 8½. 22254-3 Pa. $3.95

THE DOVER NEW YORK WALKING GUIDE: From the Battery to Wall Street, Mary J. Shapiro. Superb inexpensive guide to historic buildings and locales in lower Manhattan: Trinity Church, Bowling Green, more. Complete Text; maps. 36 illustrations. 48pp. 3⅞ × 9¼. 24225-0 Pa. $2.50

NEW YORK THEN AND NOW, Edward B. Watson, Edmund V. Gillon, Jr. 83 important Manhattan sites: on facing pages early photographs (1875-1925) and 1976 photos by Gillon. 172 illustrations. 171pp. 9¼ × 10. 23361-8 Pa. $7.95

HISTORIC COSTUME IN PICTURES, Braun & Schneider. Over 1450 costumed figures from dawn of civilization to end of 19th century. English captions. 125 plates. 256pp. 8⅜ × 11¼. 23150-X Pa. $7.50

VICTORIAN AND EDWARDIAN FASHION: A Photographic Survey, Alison Gernsheim. First fashion history completely illustrated by contemporary photographs. Full text plus 235 photos, 1840-1914, in which many celebrities appear. 240pp. 6½ × 9¼. 24205-6 Pa. $6.00

CHARTED CHRISTMAS DESIGNS FOR COUNTED CROSS-STITCH AND OTHER NEEDLECRAFTS, Lindberg Press. Charted designs for 45 beautiful needlecraft projects with many yuletide and wintertime motifs. 48pp. 8¼ × 11. 24356-7 Pa. $2.50

101 FOLK DESIGNS FOR COUNTED CROSS-STITCH AND OTHER NEEDLE-CRAFTS, Carter Houck. 101 authentic charted folk designs in a wide array of lovely representations with many suggestions for effective use. 48pp. 8¼ × 11. 24369-9 Pa. $2.25

FIVE ACRES AND INDEPENDENCE, Maurice G. Kains. Great back-to-the-land classic explains basics of self-sufficient farming. The one book to get. 95 illustrations. 397pp. 5⅜ × 8½. 20974-1 Pa. $4.95

A MODERN HERBAL, Margaret Grieve. Much the fullest, most exact, most useful compilation of herbal material. Gigantic alphabetical encyclopedia, from aconite to zedoary, gives botanical information, medical properties, folklore, economic uses, and much else. Indispensable to serious reader. 161 illustrations. 888pp. 6½ × 9¼. (Available in U.S. only) 22798-7, 22799-5 Pa., Two-vol. set $16.45

SURREAL STICKERS AND UNREAL STAMPS, William Rowe. 224 haunting, hilarious stamps on gummed, perforated stock, with images of elephants, geisha girls, George Washington, etc. 16pp. one side. 8¼ × 11.   24371-0 Pa. $3.50

GOURMET KITCHEN LABELS, Ed Sibbett, Jr. 112 full-color labels (4 copies each of 28 designs). Fruit, bread, other culinary motifs. Gummed and perforated. 16pp. 8¼ × 11.   24087-8 Pa. $2.95

PATTERNS AND INSTRUCTIONS FOR CARVING AUTHENTIC BIRDS, H.D. Green. Detailed instructions, 27 diagrams, 85 photographs for carving 15 species of birds so life-like, they'll seem ready to fly! 8¼ × 11.   24222-6 Pa. $2.75

FLATLAND, E.A. Abbott. Science-fiction classic explores life of 2-D being in 3-D world. 16 illustrations. 103pp. 5⅜ × 8.   20001-9 Pa. $2.00

DRIED FLOWERS, Sarah Whitlock and Martha Rankin. Concise, clear, practical guide to dehydration, glycerinizing, pressing plant material, and more. Covers use of silica gel. 12 drawings. 32pp. 5⅜ × 8½.   21802-3 Pa. $1.00

EASY-TO-MAKE CANDLES, Gary V. Guy. Learn how easy it is to make all kinds of decorative candles. Step-by-step instructions. 82 illustrations. 48pp. 8¼ × 11.
                      23881-4 Pa. $2.50

SUPER STICKERS FOR KIDS, Carolyn Bracken. 128 gummed and perforated full-color stickers: GIRL WANTED, KEEP OUT, BORED OF EDUCATION, X-RATED, COMBAT ZONE, many others. 16pp. 8¼ × 11.   24092-4 Pa. $2.50

CUT AND COLOR PAPER MASKS, Michael Grater. Clowns, animals, funny faces...simply color them in, cut them out, and put them together, and you have 9 paper masks to play with and enjoy. 32pp. 8¼ × 11.   23171-2 Pa. $2.25

A CHRISTMAS CAROL: THE ORIGINAL MANUSCRIPT, Charles Dickens. Clear facsimile of Dickens manuscript, on facing pages with final printed text. 8 illustrations by John Leech, 4 in color on covers. 144pp. 8⅜ × 11¼.
                      20980-6 Pa. $5.95

CARVING SHOREBIRDS, Harry V. Shourds & Anthony Hillman. 16 full-size patterns (all double-page spreads) for 19 North American shorebirds with step-by-step instructions. 72pp. 9¼ × 12¼.   24287-0 Pa. $4.95

THE GENTLE ART OF MATHEMATICS, Dan Pedoe. Mathematical games, probability, the question of infinity, topology, how the laws of algebra work, problems of irrational numbers, and more. 42 figures. 143pp. 5⅜ × 8½. (EBE)
                      22949-1 Pa. $3.50

READY-TO-USE DOLLHOUSE WALLPAPER, Katzenbach & Warren, Inc. Stripe, 2 floral stripes, 2 allover florals, polka dot; all in full color. 4 sheets (350 sq. in.) of each, enough for average room. 48pp. 8¼ × 11.   23495-9 Pa. $2.95

MINIATURE IRON-ON TRANSFER PATTERNS FOR DOLLHOUSES, DOLLS, AND SMALL PROJECTS, Rita Weiss and Frank Fontana. Over 100 miniature patterns: rugs, bedspreads, quilts, chair seats, etc. In standard dollhouse size. 48pp. 8¼ × 11.   23741-9 Pa. $1.95

THE DINOSAUR COLORING BOOK, Anthony Rao. 45 renderings of dinosaurs, fossil birds, turtles, other creatures of Mesozoic Era. Scientifically accurate. Captions. 48pp. 8¼ × 11.   24022-3 Pa. $2.50

25 KITES THAT FLY, Leslie Hunt. Full, easy-to-follow instructions for kites made from inexpensive materials. Many novelties. 70 illustrations. 110pp. 5⅜ × 8½.
22550-X Pa. $2.25

PIANO TUNING, J. Cree Fischer. Clearest, best book for beginner, amateur. Simple repairs, raising dropped notes, tuning by easy method of flattened fifths. No previous skills needed. 4 illustrations. 201pp. 5⅜ × 8½. 23267-0 Pa. $3.50

EARLY AMERICAN IRON-ON TRANSFER PATTERNS, edited by Rita Weiss. 75 designs, borders, alphabets, from traditional American sources. 48pp. 8¼ × 11.
23162-3 Pa. $1.95

CROCHETING EDGINGS, edited by Rita Weiss. Over 100 of the best designs for these lovely trims for a host of household items. Complete instructions, illustrations. 48pp. 8¼ × 11. 24031-2 Pa. $2.25

FINGER PLAYS FOR NURSERY AND KINDERGARTEN, Emilie Poulsson. 18 finger plays with music (voice and piano); entertaining, instructive. Counting, nature lore, etc. Victorian classic. 53 illustrations. 80pp. 6½ × 9¼. 22588-7 Pa. $1.95

BOSTON THEN AND NOW, Peter Vanderwarker. Here in 59 side-by-side views are photographic documentations of the city's past and present. 119 photographs. Full captions. 122pp. 8¼ × 11. 24312-5 Pa. $6.95

CROCHETING BEDSPREADS, edited by Rita Weiss. 22 patterns, originally published in three instruction books 1939-41. 39 photos, 8 charts. Instructions. 48pp. 8¼ × 11. 23610-2 Pa. $2.00

HAWTHORNE ON PAINTING, Charles W. Hawthorne. Collected from notes taken by students at famous Cape Cod School; hundreds of direct, personal *apercus*, ideas, suggestions. 91pp. 5⅜ × 8½. 20653-X Pa. $2.50

THERMODYNAMICS, Enrico Fermi. A classic of modern science. Clear, organized treatment of systems, first and second laws, entropy, thermodynamic potentials, etc. Calculus required. 160pp. 5⅜ × 8½. 60361-X Pa. $4.00

TEN BOOKS ON ARCHITECTURE, Vitruvius. The most important book ever written on architecture. Early Roman aesthetics, technology, classical orders, site selection, all other aspects. Morgan translation. 331pp. 5⅜ × 8½. 20645-9 Pa. $5.50

THE CORNELL BREAD BOOK, Clive M. McCay and Jeanette B. McCay. Famed high-protein recipe incorporated into breads, rolls, buns, coffee cakes, pizza, pie crusts, more. Nearly 50 illustrations. 48pp. 8¼ × 11. 23995-0 Pa. $2.00

THE CRAFTSMAN'S HANDBOOK, Cennino Cennini. 15th-century handbook, school of Giotto, explains applying gold, silver leaf; gesso; fresco painting, grinding pigments, etc. 142pp. 6⅛ × 9¼. 20054-X Pa. $3.50

FRANK LLOYD WRIGHT'S FALLINGWATER, Donald Hoffmann. Full story of Wright's masterwork at Bear Run, Pa. 100 photographs of site, construction, and details of completed structure. 112pp. 9¼ × 10. 23671-4 Pa. $6.95

OVAL STAINED GLASS PATTERN BOOK, C. Eaton. 60 new designs framed in shape of an oval. Greater complexity, challenge with sinuous cats, birds, mandalas framed in antique shape. 64pp. 8¼ × 11. 24519-5 Pa. $3.50

HOW THE OTHER HALF LIVES, Jacob A. Riis. Journalistic record of filth, degradation, upward drive in New York immigrant slums, shops, around 1900. New edition includes 100 original Riis photos, monuments of early photography. 233pp. 10 × 7⅞. 22012-5 Pa. $7.95

CHINA AND ITS PEOPLE IN EARLY PHOTOGRAPHS, John Thomson. In 200 black-and-white photographs of exceptional quality photographic pioneer Thomson captures the mountains, dwellings, monuments and people of 19th-century China. 272pp. 9⅜ × 12¼. 24393-1 Pa. $12.95

GODEY COSTUME PLATES IN COLOR FOR DECOUPAGE AND FRAM-ING, edited by Eleanor Hasbrouk Rawlings. 24 full-color engravings depicting 19th-century Parisian haute couture. Printed on one side only. 56pp. 8¼ × 11. 23879-2 Pa. $3.95

ART NOUVEAU STAINED GLASS PATTERN BOOK, Ed Sibbett, Jr. 104 projects using well-known themes of Art Nouveau: swirling forms, florals, peacocks, and sensuous women. 60pp. 8¼ × 11. 23577-7 Pa. $3.50

QUICK AND EASY PATCHWORK ON THE SEWING MACHINE: Susan Aylsworth Murwin and Suzzy Payne. Instructions, diagrams show exactly how to machine sew 12 quilts. 48pp. of templates. 50 figures. 80pp. 8¼ × 11. 23770-2 Pa. $3.50

THE STANDARD BOOK OF QUILT MAKING AND COLLECTING, Marguerite Ickis. Full information, full-sized patterns for making 46 traditional quilts, also 150 other patterns. 483 illustrations. 273pp. 6⅞ × 9⅞. 20582-7 Pa. $5.95

LETTERING AND ALPHABETS, J. Albert Cavanagh. 85 complete alphabets lettered in various styles; instructions for spacing, roughs, brushwork. 121pp. 8¾ × 8. 20053-1 Pa. $3.95

LETTER FORMS: 110 COMPLETE ALPHABETS, Frederick Lambert. 110 sets of capital letters; 16 lower case alphabets; 70 sets of numbers and other symbols. 110pp. 8⅛ × 11. 22872-X Pa. $4.50

ORCHIDS AS HOUSE PLANTS, Rebecca Tyson Northen. Grow cattleyas and many other kinds of orchids—in a window, in a case, or under artificial light. 63 illustrations. 148pp. 5⅜ × 8½. 23261-1 Pa. $2.95

THE MUSHROOM HANDBOOK, Louis C.C. Krieger. Still the best popular handbook. Full descriptions of 259 species, extremely thorough text, poisons, folklore, etc. 32 color plates; 126 other illustrations. 560pp. 5⅜ × 8½. 21861-9 Pa. $8.50

THE DORE BIBLE ILLUSTRATIONS, Gustave Doré. All wonderful, detailed plates: Adam and Eve, Flood, Babylon, life of Jesus, etc. Brief King James text with each plate. 241 plates. 241pp. 9 × 12. 23004-X Pa. $8.95

THE BOOK OF KELLS: Selected Plates in Full Color, edited by Blanche Cirker. 32 full-page plates from greatest manuscript-icon of early Middle Ages. Fantastic, mysterious. Publisher's Note. Captions. 32pp. 9¾ × 12¼. 24345-1 Pa. $4.50

THE PERFECT WAGNERITE, George Bernard Shaw. Brilliant criticism of the Ring Cycle, with provocative interpretation of politics, economic theories behind the Ring. 136pp. 5⅜ × 8½. (Available in U.S. only) 21707-8 Pa. $3.00

JAPANESE DESIGN MOTIFS, Matsuya Co. Mon, or heraldic designs. Over 4000 typical, beautiful designs: birds, animals, flowers, swords, fans, geometrics; all beautifully stylized. 213pp. 11⅛ × 8¼.   22874-6 Pa. $7.95

THE TALE OF BENJAMIN BUNNY, Beatrix Potter. Peter Rabbit's cousin coaxes him back into Mr. McGregor's garden for a whole new set of adventures. All 27 full-color illustrations. 59pp. 4¼ × 5½. (Available in U.S. only)   21102-9 Pa. $1.75

THE TALE OF PETER RABBIT AND OTHER FAVORITE STORIES BOXED SET, Beatrix Potter. Seven of Beatrix Potter's best-loved tales including Peter Rabbit in a specially designed, durable boxed set. 4¼ × 5½. Total of 447pp. 158 color illustrations. (Available in U.S. only)   23903-9 Pa. $10.80

PRACTICAL MENTAL MAGIC, Theodore Annemann. Nearly 200 astonishing feats of mental magic revealed in step-by-step detail. Complete advice on staging, patter, etc. Illustrated. 320pp. 5⅜ × 8½.   24426-1 Pa. $5.95

CELEBRATED CASES OF JUDGE DEE (DEE GOONG AN), translated by Robert Van Gulik. Authentic 18th-century Chinese detective novel; Dee and associates solve three interlocked cases. Led to van Gulik's own stories with same characters. Extensive introduction. 9 illustrations. 237pp. 5⅜ × 8½.

23337-5 Pa. $4.50

CUT & FOLD EXTRATERRESTRIAL INVADERS THAT FLY, M. Grater. Stage your own lilliputian space battles. By following the step-by-step instructions and explanatory diagrams you can launch 22 full-color fliers into space. 36pp. 8¼ × 11.   24478-4 Pa. $2.95

CUT & ASSEMBLE VICTORIAN HOUSES, Edmund V. Gillon, Jr. Printed in full color on heavy cardboard stock, 4 authentic Victorian houses in H-O scale: Italian-style Villa, Octagon, Second Empire, Stick Style. 48pp. 9¼ × 12¼.

23849-0 Pa. $3.95

BEST SCIENCE FICTION STORIES OF H.G. WELLS, H.G. Wells. Full novel *The Invisible Man*, plus 17 short stories: "The Crystal Egg," "Aepyornis Island," "The Strange Orchid," etc. 303pp. 5⅜ × 8½. (Available in U.S. only)

21531-8 Pa. $4.95

TRADEMARK DESIGNS OF THE WORLD, Yusaku Kamekura. A lavish collection of nearly 700 trademarks, the work of Wright, Loewy, Klee, Binder, hundreds of others. 160pp. 8¾ × 8. (Available in U.S. only)   24191-2 Pa. $5.95

THE ARTIST'S AND CRAFTSMAN'S GUIDE TO REDUCING, ENLARGING AND TRANSFERRING DESIGNS, Rita Weiss. Discover, reduce, enlarge, transfer designs from any objects to any craft project. 12pp. plus 16 sheets special graph paper. 8¼ × 11.   24142-4 Pa. $3.50

TREASURY OF JAPANESE DESIGNS AND MOTIFS FOR ARTISTS AND CRAFTSMEN, edited by Carol Belanger Grafton. Indispensable collection of 360 traditional Japanese designs and motifs redrawn in clean, crisp black-and-white, copyright-free illustrations. 96pp. 8¼ × 11.   24435-0 Pa. $3.95

TWENTY-FOUR ART NOUVEAU POSTCARDS IN FULL COLOR FROM CLASSIC POSTERS, Hayward and Blanche Cirker. Ready-to-mail postcards reproduced from rare set of poster art. Works by Toulouse-Lautrec, Parrish, Steinlen, Mucha, Cheret, others. 12pp. 8¼× 11.   24389-3 Pa. $2.95

READY-TO-USE ART NOUVEAU BOOKMARKS IN FULL COLOR, Carol Belanger Grafton. 30 elegant bookmarks featuring graceful, flowing lines, foliate motifs, sensuous women characteristic of Art Nouveau. Perforated for easy detaching. 16pp. 8¼ × 11.   24305-2 Pa. $2.95

FRUIT KEY AND TWIG KEY TO TREES AND SHRUBS, William M. Harlow. Fruit key covers 120 deciduous and evergreen species; twig key covers 160 deciduous species. Easily used. Over 300 photographs. 126pp. 5⅜ × 8½.   20511-8 Pa. $2.25

LEONARDO DRAWINGS, Leonardo da Vinci. Plants, landscapes, human face and figure, etc., plus studies for Sforza monument, *Last Supper*, more. 60 illustrations. 64pp. 8¼ × 11⅛.   23951-9 Pa. $2.75

CLASSIC BASEBALL CARDS, edited by Bert R. Sugar. 98 classic cards on heavy stock, full color, perforated for detaching. Ruth, Cobb, Durocher, DiMaggio, H. Wagner, 99 others. Rare originals cost hundreds. 16pp. 8¼ × 11.   23498-3 Pa. $3.25

TREES OF THE EASTERN AND CENTRAL UNITED STATES AND CANADA, William M. Harlow. Best one-volume guide to 140 trees. Full descriptions, woodlore, range, etc. Over 600 illustrations. Handy size. 288pp. 4½ × 6⅜.   20395-6 Pa. $3.95

JUDY GARLAND PAPER DOLLS IN FULL COLOR, Tom Tierney. 3 Judy Garland paper dolls (teenager, grown-up, and mature woman) and 30 gorgeous costumes highlighting memorable career. Captions. 32pp. 9¼ × 12¼.

      24404-0 Pa. $3.50

GREAT FASHION DESIGNS OF THE BELLE EPOQUE PAPER DOLLS IN FULL COLOR, Tom Tierney. Two dolls and 30 costumes meticulously rendered. Haute couture by Worth, Lanvin, Paquin, other greats late Victorian to WWI. 32pp. 9¼ × 12¼.   24425-3 Pa. $3.50

FASHION PAPER DOLLS FROM GODEY'S LADY'S BOOK, 1840-1854, Susan Johnston. In full color: 7 female fashion dolls with 50 costumes. Little girl's, bridal, riding, bathing, wedding, evening, everyday, etc. 32pp. 9¼ × 12¼.

      23511-4 Pa. $3.95

THE BOOK OF THE SACRED MAGIC OF ABRAMELIN THE MAGE, translated by S. MacGregor Mathers. Medieval manuscript of ceremonial magic. Basic document in Aleister Crowley, Golden Dawn groups. 268pp. 5⅜ × 8½.

      23211-5 Pa. $5.00

PETER RABBIT POSTCARDS IN FULL COLOR: 24 Ready-to-Mail Cards, Susan Whited LaBelle. Bunnies ice-skating, coloring Easter eggs, making valentines, many other charming scenes. 24 perforated full-color postcards, each measuring 4¼ × 6, on coated stock. 12pp. 9 × 12.   24617-5 Pa. $2.95

CELTIC HAND STROKE BY STROKE, A. Baker. Complete guide creating each letter of the alphabet in distinctive Celtic manner. Covers hand position, strokes, pens, inks, paper, more. Illustrated. 48pp. 8¼ × 11.   24336-2 Pa. $2.50

READY-TO-USE BORDERS, Ted Menten. Both traditional and unusual interchangeable borders in a tremendous array of sizes, shapes, and styles. 32 plates. 64pp. 8¼ × 11. 23782-6 Pa. $3.50

THE WHOLE CRAFT OF SPINNING, Carol Kroll. Preparing fiber, drop spindle, treadle wheel, other fibers, more. Highly creative, yet simple. 43 illustrations. 48pp. 8¼ × 11. 23968-3 Pa. $2.50

HIDDEN PICTURE PUZZLE COLORING BOOK, Anna Pomaska. 31 delightful pictures to color with dozens of objects, people and animals hidden away to find. Captions. Solutions. 48pp. 8¼ × 11. 23909-8 Pa. $2.25

QUILTING WITH STRIPS AND STRINGS, H.W. Rose. Quickest, easiest way to turn left-over fabric into handsome quilt. 46 patchwork quilts; 31 full-size templates. 48pp. 8¼ × 11. 24357-5 Pa. $3.25

NATURAL DYES AND HOME DYEING, Rita J. Adrosko. Over 135 specific recipes from historical sources for cotton, wool, other fabrics. Genuine premodern handicrafts. 12 illustrations. 160pp. 5⅜ × 8½. 22688-3 Pa. $2.95

CARVING REALISTIC BIRDS, H.D. Green. Full-sized patterns, step-by-step instructions for robins, jays, cardinals, finches, etc. 97 illustrations. 80pp. 8¼ × 11. 23484-3 Pa. $3.00

GEOMETRY, RELATIVITY AND THE FOURTH DIMENSION, Rudolf Rucker. Exposition of fourth dimension, concepts of relativity as Flatland characters continue adventures. Popular, easily followed yet accurate, profound. 141 illustrations. 133pp. 5⅜ × 8½. 23400-2 Pa. $3.00

READY-TO-USE SMALL FRAMES AND BORDERS, Carol B. Grafton. Graphic message? Frame it graphically with 373 new frames and borders in many styles: Art Nouveau, Art Deco, Op Art. 64pp. 8¼ × 11. 24375-3 Pa. $3.50

CELTIC ART: THE METHODS OF CONSTRUCTION, George Bain. Simple geometric techniques for making Celtic interlacements, spirals, Kellstype initials, animals, humans, etc. Over 500 illustrations. 160pp. 9 × 12. (Available in U.S. only) 22923-8 Pa. $6.00

THE TALE OF TOM KITTEN, Beatrix Potter. Exciting text and all 27 vivid, full-color illustrations to charming tale of naughty little Tom getting into mischief again. 58pp. 4¼ × 5½. (USO) 24502-0 Pa. $1.75

WOODEN PUZZLE TOYS, Ed Sibbett, Jr. Transfer patterns and instructions for 24 easy-to-do projects: fish, butterflies, cats, acrobats, Humpty Dumpty, 19 others. 48pp. 8¼ × 11. 23713-3 Pa. $2.50

MY FAMILY TREE WORKBOOK, Rosemary A. Chorzempa. Enjoyable, easy-to-use introduction to genealogy designed specially for children. Data pages plus text. Instructive, educational, valuable. 64pp. 8¼ × 11. 24229-3 Pa. $2.50